Since its refurbishment in December 1993, the Hubble Space Telescope has revealed spectacular and intriguing details in every object it has turned its acute gaze upon. What discoveries has the HST made so far? How does this telescope actually work? And what is the scientific mission of the HST? This lavishly illustrated volume is the first to answer these questions in a complete review of the most exciting science to come from the Hubble Space Telescope. A superb collection of dramatic images taken by the HST is supported by a lively and informative, but non-technical, guide to what these images tell us about our colorful and intricate universe.

Beginning with an outline of the scientific goal of the HST, we are led through the excitement of its launch, the dismay on deployment, and the relief on refitting. Then, following a clear explanation of just how the HST makes its observations, our exploration of the universe begins. We are presented with fairly familiar objects – such as planets, star clusters and supernovae – but in detail never previously seen, as well as exotic objects such as black holes, active galactic nuclei, peculiar stars, optical jets and gravitational lenses that have eluded astronomers until quite recently.

From the local Solar System and nearby stars, to the most distant quasars and early universe, this highly illustrated volume presents a penetrating and colorful view of the universe as never seen before. *Hubble Vision* will capture the imagination of all those interested in the astronomical quest of understanding our universe – from the general reader and amateur astronomer through to the professional scientist.

# HUBBLE VISION

Astronomy with the Hubble Space Telescope

# HUBBLE VISION

## Astronomy with the Hubble Space Telescope

CAROLYN COLLINS PETERSEN
*University of Colorado*

AND

JOHN C. BRANDT
*University of Colorado*

*Published by the Press Syndicate of the University of Cambridge*
*The Pitt Building, Trumpington Street, Cambridge CB2 1RP*
*40 West 20th Street, New York, NY 10011-4211, USA*
*10 Stamford Road, Oakleigh, Melbourne 3166, Australia*



*First published 1995*

*Printed in Great Britain at the University Press, Cambridge*

*A catalogue record for this book is available from the British Library*

*Library of Congress cataloguing in publication data*

Petersen, Carolyn Collins.
Hubble vision: astronomy with the Hubble Space Telescope /
Carolyn Collins Petersen and John C. Brandt.
p. cm.
Includes bibliographical references and index.

ISBN 0 521 49643 8

1. Hubble Space Telescope. 2. Space astronomy. I. Brandt, John
C. II. Title.
QB500.268.P38 1995
522'.2919–dc20 95-32568 CIP

ISBN 0 521 49643 8 hardback

WV

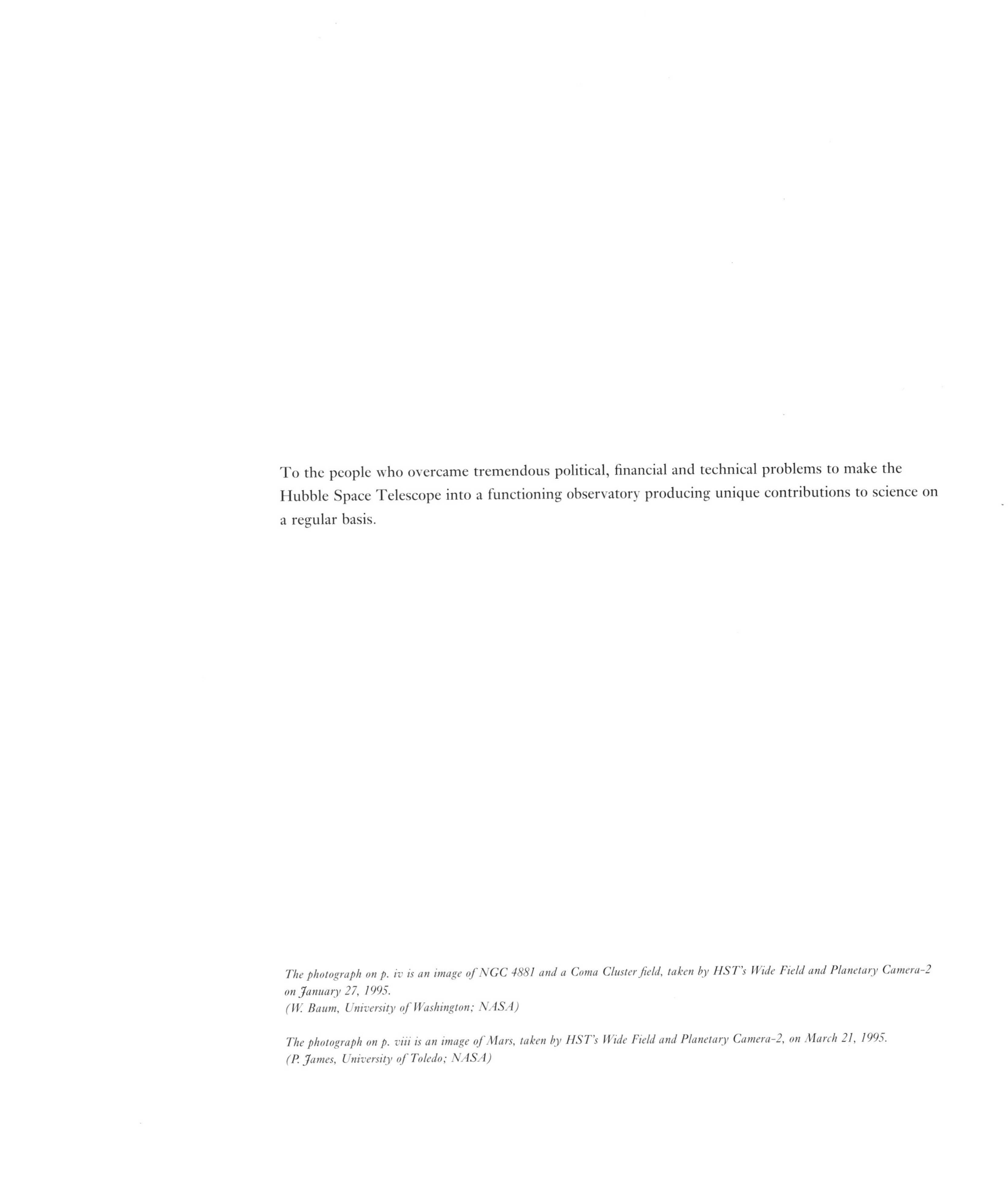

To the people who overcame tremendous political, financial and technical problems to make the Hubble Space Telescope into a functioning observatory producing unique contributions to science on a regular basis.

*The photograph on p. iv is an image of NGC 4881 and a Coma Cluster field, taken by HST's Wide Field and Planetary Camera-2 on January 27, 1995.*
*(W. Baum, University of Washington; NASA)*

*The photograph on p. viii is an image of Mars, taken by HST's Wide Field and Planetary Camera-2, on March 21, 1995.*
*(P. James, University of Toledo; NASA)*

# Contents

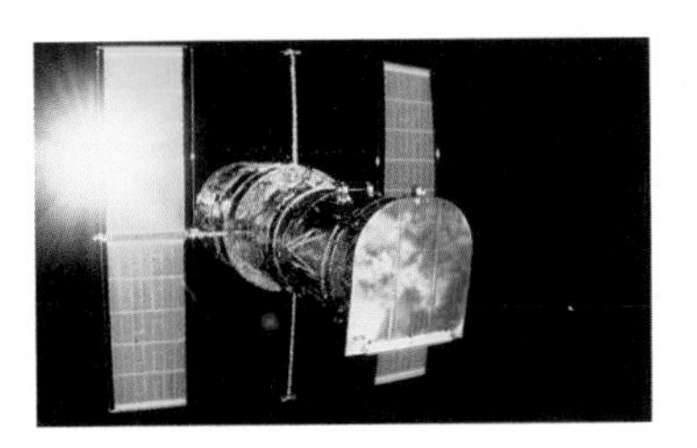

# Preface

We are privileged to live and work in a time of great astronomical discoveries. There are weeks when one can hardly turn the pages of a major newspaper without seeing a story about some intriguing object in the heavens. Since 1990, many of those discoveries have been made with an observatory that began its life under a cloud of technical problems and a storm of controversy – the Hubble Space Telescope.

Our goal in writing *Hubble Vision* was to present a straightforward description of the telescope and its scientific contributions in a way that makes the story accessible to the nonscientist. Because of the extensive publicity associated with the telescope, much of it unfavorable, we have not undertaken this task lightly. To avoid recounting the inevitable politics that surround projects of this magnitude, we have focused on the *science observations* made with the Hubble Space Telescope – as described to us by people from the Space Telescope Science Institute, the NASA Goddard Space Flight Center, the Marshall Space Flight Center, researchers at a range of universities and research contractors in the USA, Canada and Europe.

Both of us had very different reasons for wanting to write *Hubble Vision*. Carolyn Collins Petersen's initial motivations for this project came from seeing daily electronic mail updates of observations made with the Hubble Space Telescope (often referred to as 'HST'). These updates conflicted with concurrent media reports that implied the project was a failure, and with a view held by some in the science community that the telescope was consuming scarce resources without a similar scientific gain. In short, it seemed that (in the beginning of the mission, at least) HST was getting lots of attention for its problems but not getting much recognition for its science contributions.

To counter what seemed to be widespread ignorance of HST's successes, Petersen wrote a planetarium program in 1991 called 'Hubble: Report From Orbit' that detailed some of the most important discoveries made with HST, along with the operational problems experienced by the telescope and its users. That program was the basis for the production of an award-winning educational video, which – like the planetarium show – answered a good many questions about Hubble from members of the 'general public' who were not privy to the behind-the-scenes communications regarding observations. The success of these education efforts – including a newly updated planetarium program released in late 1994, also called 'Hubble Vision' – underscored one fact: despite official and media skepticism about the telescope, there has always been a significant public interest in astronomy and space science. This has often been demonstrated by our experiences in places such as food markets and bank lobbies, where chance conversations about what we do with HST often turn into spirited question-and-answer sessions about things astronomical.

In addition, John Brandt was initially intrigued by conversations with seat mates during numerous airline trips to Baltimore-Washington International – the most convenient airport for travel to and from the Goddard Space Flight Center and the Space Telescope Science Institute. During these flights, the subject of HST would invariably come up. People were amazed that someone would be traveling to HST meetings and even more amazed that people were doing good science with the telescope. After all, everyone knew that the Hubble Space Telescope was a 'piece of expensive, worthless junk'.

Confusion about Hubble's status during its first years was the result of highly variable press coverage over the years. Descriptions of the mission have ranged from inflammatory critiques to joking monologues. More recently, however, laudatory stories have focused on HST's achievements. In the July 19, 1993, issue of *TIME Magazine*, writer Dennis Overbye referred to the Hubble Space Telescope as one of the 'genuine marvels' returning 'spectacular results'. Since the December 1993 First Servicing Mission of HST, the media seem to have abandoned HST as the butt of journalistic jokes and have begun to take reports of the telescope's achievements more seriously than in the past.

It is easy to conclude that much of the public discussion about HST mirrors the current national tendency in the United States to reach sometimes erroneous conclusions about so-called 'Big Science' projects based on incomplete, misleading or blatantly false information.

It might help to look at how others see the project, particularly in light of other space mission setbacks: the problems with the Galileo High-Gain antenna, the loss of the Mars Observer spacecraft and the myriad of technical snags that occasionally ground the Space Shuttle. Reflecting on America's inclination to 'overblame' itself when something goes wrong, British author James Burke once said,

> You Americans really amaze me for two reasons. You seem to value education so little in your country. You do not invest in books and classroom materials, and your teachers are paid less than convenience store clerks. And you do such wonderful things, technologically – but you are so hard on yourselves. The Hubble Space Telescope – so there is a tiny flaw in the mirror. Your attitude is, 'there it is – we failed!' No one in the world possesses the skills and know-

> how to have accomplished what you did! America represents the greatest collection of minds that has ever existed in the history of this planet!

So, here we are, two people on the inside of the Hubble Space Telescope project, who have taught and studied science for many years: one a science writer, and the other an astronomer who has been intimately involved with the HST since 1977. Our goal is to bring the other side of HST to everyone: the readers, the taxpayers of several countries who support the organizations that support HST – the people who represent a great collection of minds around the world. What do we tell you?

We did not want to tell a political story; that has already been done. Today, the core of the HST story is not in its construction, or the political processes that brought it into being, but in the science it is doing. The universe is a constantly changing place, and HST, along with its sister orbiting observatories – including International Ultraviolet Explorer, Hipparcos, Yohkoh, Compton Gamma Ray Observatory, Cosmic Background Explorer – in addition to the excellent collection of ground-based facilities around the world, are studying and measuring the growth and evolution of the cosmos. Of course, we cannot and do not ignore the problems that the telescope has faced. As we have heard from a number of scientists who use HST, it is a working observatory – and it faces challenges that no ground-based facility will ever experience.

It is easy to be judgmental when discussing Hubble – after all, the telescope still operates in the full glare of publicity. In the early days, every media story about HST took pains to remind us about the mirror problems and the tremendous cost of the program. No other facility faced this sort of scrutiny, and no other observatory started its public life with such a spectacularly disastrous birth.

Now, having just said that we want to tell a science story, it would be shortsighted of us to ignore the political considerations completely. So, we confine many of the judgmental (read: political) aspects of the HST problems to this Preface. That leaves the rest of the book for the 'science goodies'. After all, that is why HST is up there – to get the goodies. It is not always easy. This work is not the complete history of the Space Telescope project, although for the sake of our story, we present the 'executive summary' in Chapter 1. There is just not enough room here for a complete blow-by-blow timeline, and anyway it has been done better in other places. For those readers who delight in reading about the politics and management of large programs, we highly recommend Robert Smith's detailed book, *The Space Telescope* (Cambridge University Press, 1993).

Understanding the nature of the beast is essential to overall understanding of just how Hubble does its work. HST has had technical and managerial problems from the beginning. Of course, every complex human institution has problems, and HST is certainly not unique there. In any large project, mistakes are always made. The procedures to catch mistakes, including sound engineering practice and testing, were not always available during planning and construction, and the potential impact of including them on schedule and budgets was disastrously unthinkable. We should emphasize that the problem lies in the system, and is not the fault of the individuals trying to make things work: the Hubble Space Telescope project is blessed with a collection of the most talented and gifted scientific and engineering minds available, working within an organizational system that is not perfect.

As everyone knows by now, the major technical problems on HST were the spherical aberration of the primary mirror and the spacecraft jitter produced by the solar panels. After the First Servicing Mission, the focus was improved to meet the original design expectations and the new solar panels solved the jitter problems. A regimen of other repairs and replacements made the telescope 'as good as new' – indeed, better than the original – and ready to meet the observation challenges ahead.

So where are we now? Some regard the functioning of HST as a miracle, and others regard it as the result of lots of good, hard work. The truth is, the Hubble Space Telescope, before and after the First Servicing Mission, has been scientifically productive – this book documents that quite well. To quote Maryland senator Barbara Mikulski, 'The trouble with Hubble is over!'

Despite major progress in many areas, there is still a long way to go. Adversarial relationships between people and organizations still exist. The overall system is highly bureaucratic; just requesting observation time can be a nightmare even for the scientists connected to HST, and is even more traumatic for the uninitiated.

Before we leave this very brief discussion of HST politics, we should talk about two other subjects. The first is cost, which brings to mind a statement attributed to the late United States Senator Everett Dirksen:

> A billion here, a billion there and pretty soon you're talking about real money.

With HST, we *are* talking real money but it is not always clear how much. Assigning a precise 'cost' is impossible because NASA keeps books in a complex way. For example, launch costs are not considered part of project budgets. Inflation, which affects everyone, changes the buying power of our money – a 1987 dollar was worth a lot more than a 1995 dollar. So, simply adding up the yearly budgets will not do. Yet, we have tried to come up with a general statement about costs: to date (exclusive of launches), the cost of HST is about $2.6 billion, and the current rate for the use of HST adds up to about $8 per second.

To put this in perspective, the total NASA budget consistently has been about 1% of the US government budget. HST's budget has dropped to about 2% of the NASA budget. So, if you want to judge HST in context, it boils down to whether or not the project is worth 0.02% of the US budget. It may look a small amount, but it is a lot of money.

There is no doubt about it – large space-science projects are expensive and they need to be considered carefully. In the case of Hubble Space Telescope, the original expectations have been met, and, in some cases, superseded, and it is doing 'frontier science' that cannot be achieved any other way.

The second subject is the topic of the Hubble Space Telescope and the future of NASA. Many people in the space-science community feel that the successful servicing mission was necessary for NASA's future. What is next? The budget for Hubble will almost certainly decline, and properly maintaining the telescope through its 15-year lifetime (until 2005) will be a challenge. Unfortunately, the same bureaucratic constraints and attitudes that have contributed to the difficulties with HST are still in place. In addition, things may have returned to 'business as usual' at NASA because recent budget mandates are gutting many aspects of the US space program – including HST. The long-term outlook for NASA is uncertain, and US national priorities are shifting. The first golden age of space research that many scientists have known is over. This is not to paint a total 'gloom and doom' picture, however. For the short term at least, we can rejoice in the discoveries of a very robust HST and hope that the future of space exploration improves.

The full story of the Hubble Space Telescope is a long, sometimes tortuous, often interesting, and incredibly productive scientific journey. Our story may *begin* here on Earth, with a capsule history of the telescope's conception and birth, and a brief look at the tools and concepts of astronomy behind the telescope, but it does not *end* here. HST's scientific odyssey is really the prime focus of our work here, and this takes us well beyond our home neighborhood of planets to the stars and galaxies. HST's intriguing accomplishments – ranging from the aftermath of Comet Shoemaker–Levy 9's date with Jovian destiny, to the discovery of the most distant galaxies and a new calculation for the Hubble Constant – are the real story here. The good news about HST's journey is that there is a great deal of new science to discuss. The bad news is – like travelers who return from faraway places with lots of pictures and postcards – we cannot possibly talk about all of it in one sitting. As much as we would like to include all noteworthy results and present a complete history, there is not room here. So, we have chosen to present a simplified survey of HST's observations to illustrate the scientific side of the Hubble Space Telescope's odyssey.

We conclude our tale with a look at future plans for HST. To keep from turning this into a textbook, we have included an extended glossary at the end of the book that explains many of the concepts used in our discussions. When we talk about actual HST discoveries, we let some of the researchers who are the prime recipients of HST's data largesse talk about their work. Theirs are the voices not always heard in the media, yet they are the people who are using Hubble's data to 'push the envelope' of astronomical discovery.

Carolyn Collins Petersen
John C. Brandt

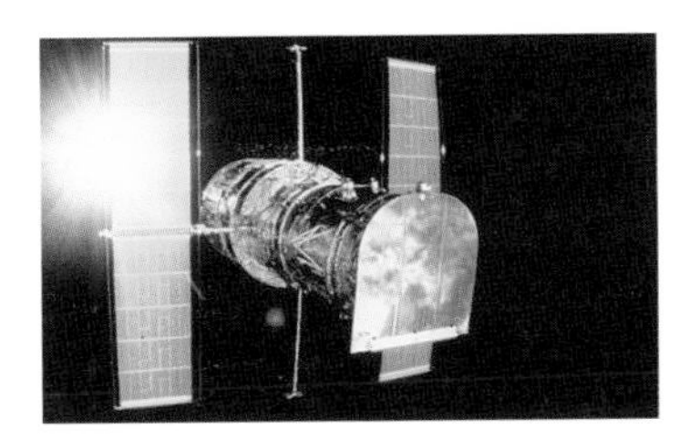

# Acknowledgments

As usual, when authors take on a task like this one, a large number of individuals and institutions are called on to contribute their viewpoints, tolerate questions, and put up with the stresses that accompany any creative effort. Of course, our spouses come to mind first. Mark Collins Petersen and Dorothy Bell Brandt endured inevitable hassles and supported us both during this project. To them we owe our love and thanks. Larry, Calicat and Rusty provided valuable 'fur fixes' during the times when typing and editing this manuscript took a toll.

Dozens of scientists, engineers, program managers, observers and research personnel involved with HST took time to talk with us about their work and their experience using this most complex telescope. Our 'conceptual gestalt' of HST was built up through the conversations we had with these and many, many others both inside and outside the project.

Mike A'Hearn, Ron Allen, Claude Arpigny, Reta Beebe, Fritz Benedict, Bob Bless, Al Boggess, Alex Brown, Bob Brown, Margaret Burbidge, Jason Cardelli, Ken Carpenter, Bob Chapman, Art Code, Peter Conti, David Crisp, Cindy Cunningham, Roger Doxsey, Alan Dressler, Dennis Ebbets, Steve Edberg, Sandra Faber, Michel Festou, Otto Franz, Riccardo Giacconi, Heidi Hammel, Rich Harms, William Hathaway, Suzanne Hawley, Sara R. Heap, Jeff Hoffman, Bill Jeffreys, Mark Johnston, James Kaler, Peter Kandefer, Bill Keathley, Rob Landis, Tod Lauer, David Leckrone, Steve Lee, David Levy, Jeffrey Linsky, Duccio Macchetto, Richard McCray, Steve Maran, Bruce Margon, Georges Meylan, Heather Morrison, Claude Nicollier, Robert O'Dell, Jean Olivier, Cora Randall, Nancy Roman, Blair Savage, Jim Secosky, Carolyn Shoemaker, Gene Shoemaker, Sue Simken, Martin Snow, Ted Snow, Lyman Spitzer, Peter Stockman, Alex Storrs, Rodger Thompson, Larry Trafton, Bill van Altena, Hal Weaver, Ed Weiler, Jim Westphal, Ray Weymann, Robert Williams and Ben Zellner took time out from busy schedules to share their vision of what it means to work with HST and to explain HST's place in astronomy history. We appreciate their gracious cooperation, and we apologize to anyone we interviewed who has been left off the list.

Many other scientists shared their science results and comments in the form of papers, press releases and hallway conversations, and we found these materials to be invaluable. We also thank the scientists who supplied us with illustrations, some custom-made just for this book. Special thanks go to Ray Villard and his staff at Space Telescope Science Institute's Office of Public Outreach for their extraordinary and enthusiastic cooperation.

Finally, we thank Mark C. Petersen for extensive manuscript review and suggestions, Tim Kuzniar for custom artwork, and Darien Gould for interview transcriptions. We are grateful to the Cambridge University Press staff, in particular Adam Black, Irene Pizzie and Simon Mitton. Astronomers Tom Ake, Ken Carpenter, Sandra Faber, Carl Hansen, James Kaler, David Leckrone, Steve Shore and Robert Smith all offered superb assistance in reading parts of the manuscript and in bringing mistakes and potential problems to our attention. Still, in a project this complex, a few errors are bound to creep in, and any remaining are, of course, our responsibility.

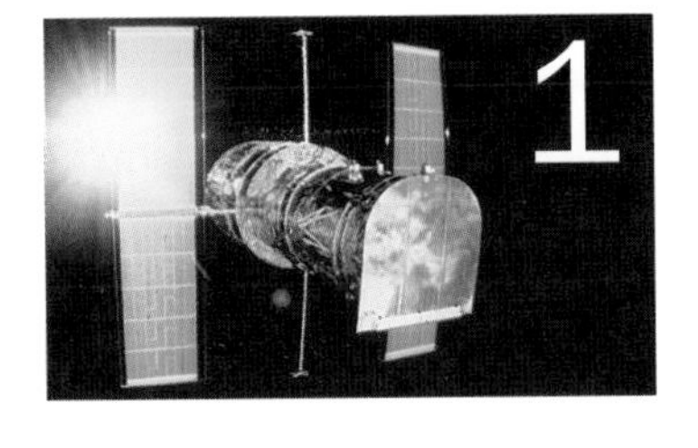

# 1 Space Telescope: the dream and the reality

Stars scribble in our eyes the frosty sagas, the gleaming cantos of unvanquished space.

*Hart Crane*

We are all in the gutter, but some of us are looking at the stars.

*Oscar Wilde*

## First light, first frustrations

Stargazing is a ritual as old as human culture. On a clear, dark night we lift our gaze to the skies, and suddenly our souls are touched by the splendor of the stars. We are the descendants of countless generations of astronomers, and, as it did for our ancestors, the universe dangles tantalizing mysteries before our eyes. It dares us to come out and explore. Then, preposterously, it challenges us with seemingly impassable distances between Earth and anything but the closest planets. Yet, distances and light continue to fascinate us. We see something in the sky and ask, What is it? How far away is it? How fast is it traveling through space? How old is the universe? How big is it? Will it ever end?

Astronomy concerns itself with answering these questions and many others. It would be exciting if astronomers could explore the cosmos from the bridge of the some space-faring vessel like Star Trek's *Enterprise* – whipping between planets and star systems at warp speed and examining interstellar mysteries with long-range scanners and advanced probes. Unfortunately, experiencing the evolution of stars and galaxies on any short-term scale is impossible for us because we are a short-lived species. However, that has not discouraged stargazers from trying to make sense of the cosmos and exploring it with some of the most fantastic tools ever invented.

Imagine that somewhere in the world, probably atop a mountain, well away from light pollution, a group of astronomers, engineers, night assistants and electronics experts angles a telescope toward the sky for the first time. If everything has been built to specification – the mirror is ground correctly, the instruments are aligned precisely, and the pointing programs execute accurately – the telescope will open its eye to the universe. The coordinates of some distant planet, or star, or galaxy have been programmed into the telescope's drive computer.

When all is ready, the observation begins. Light from that distant object enters the telescope tube, bounces off the mirrors and into the apertures of complex instruments designed to analyze and transform it into data. Years of planning, building, tweaking, experimenting with new technology, all boils down to the moment of truth, the event called 'first light'.

Generally, first light for any kind of telescope is a relatively private moment. Only the astronomers, night assistants and designers are there to see the first images from a newly commissioned facility. Actually, first light is most important to the people who will spend their time running the telescope, learning its nuances and correcting its glitches. No members of the press are waiting outside the door, breathless for pictures. There are no government dignitaries, no administrators, just the people who designed the telescope and its instruments. First light for a telescope is like first flight for a new aircraft, or a ship's shakedown cruise. For the telescope's users, first light is an astronomer's version of the big dress rehearsal for all the nights to come.

If the first-light images from a ground-based telescope are not the crisp, crystal-clear pictures everyone is used to seeing in the pages of astronomy magazines, it is not exactly a surprise. First-light images rarely look spectacular, but they are important because they are really test images that tell astronomers how the telescope is performing.

May 20, 1990, was the Hubble Space Telescope's 'first light'. The event was like nothing any other telescope team had ever experienced. It was an exhilarating and confusing time, and the culmination of years of planning, development and training by thousands of people.

At the appointed time, HST – from its orbital vantage point some 600 kilometers above the Earth's surface – turned its flywheels and gracefully moved into position to catch the light from a star cluster over 1300 light years away. Deep inside the telescope, relays clicked, motors whirred, electronic circuits came to life and HST signaled its readiness. In control rooms at the Goddard Space Flight Center in Greenbelt, Maryland, the Marshall Space Flight Center in Huntsville, Alabama, and the Space Telescope Science Institute in Baltimore, Maryland, everyone watched computer screens as the pre-programmed test observation sequence executed. At Goddard, members of the press watched the scientists, ready to translate the experience for the waiting public.

It was a moment scientists and technical teams had been anticipating for years. To be sure, no one expected the first-light images to be very exciting or spectacular. They were

**Figure 1.1.** The drama of first light for HST plays out on the faces of those awaiting it. On the left, a small group of scientists, including John Bahcall of The Institute for Advanced Study, Princeton, are a study in nervous tension just before the image of NGC 3532 appeared on monitor screens. On the right, a happy and relieved Jim Westphal (Principal Investigator for the Wide Field and Planetary Camera team), flashes his trademark grin after the image appears. (Courtesy Richard Tresch Fienberg, *Sky and Telescope Magazine*)

more for the benefit of the engineers and controllers to evaluate the telescope's performance. Representatives of the HST teams, such as program scientist Edward Weiler, tried to prepare everyone for disappointing images by commenting to reporters that scientists would probably learn nothing of any scientific value from the images. 'We looked at that image as an engineering test,' he said later. Another astronomer, Holland Ford, tried to explain that operating a complex instrument like HST was not going to be easy. 'It's a safe prediction that not everything will work,' he told reporters after launch. Despite these warnings, there was a great sense of excitement about the event and a feeling in the HST community that any complications would be mere 'shakedown cruise' problems. Any hurdles would be overcome as the network of spacecraft teams and scientists learned to work together to manage the telescope. Everybody was optimistic about Hubble's fine future.

The HST teams were not the only ones excited about the event. Fueled by a never-ending stream of 'Hubble hype' in the press, the general public was ready to be amazed by glorious astronomical images. For months, HST was subjected to the most impressive amount of publicity since the first Moon landings. It was going to be the telescope to reveal the secrets of the stars, probe the innards of galaxies, and give scientists a wide-ranging eye on the universe.

Far above Earth, light from the star cluster NGC 3532, located in the southern hemisphere constellation Carina, streamed into the telescope. It hit the primary mirror, and bounced onto a smaller secondary mirror. From there, the light traveled back through an opening in the primary and bounced off another mirror and into the Wide Field and Planetary Camera, which encoded the light into digital data and stored it on a recorder. When the telescope was in range of a Tracking and Data Relay Satellite, it 'downloaded' the data to the satellite, which sent the information on its way to Earth. Moments later, the image appeared on the monitors at the Space Telescope Science Institute. It looked like a hash of pixels on a gray screen. Members of the camera team processed the image

**Figure 1.2.** The first-light image of a region in the open cluster NGC 3532 from WF/PC-1. (NASA; ESA)

through a set of computer programs, and an electric sense of excitement spread through the room like wildfire. Finally, an image of a starfield emerged, and applause broke out amid shouts of 'Look at that!' and 'It works! Hubble works!' Amid a growing cacophony of other comments, everyone tried to figure out just where in the target starfield HST was looking.

Over at Goddard, Jim Westphal watched as the image shimmered onto his screen. Cameras whirred and microphones waited to pick up his words. For a brief moment, no one breathed, and everyone strained to make some sense of the image while waiting for Jim's interpretation. He peered at the monitor and studied the image. As soon as he could make out the stars in the field, he broke into a big grin and announced: 'I'm pleased as Punch!'

That image of NGC 3532 was hailed as a great success by the press. The image was not the prettiest sight in the history of astronomy, and everyone knew that the telescope needed some focusing adjustments, but the first step had been taken. The Hubble Space Telescope worked. At that point, first light was a full-blown media event and everyone got into the act. Astronomers proclaimed HST as the instrument that would solve the great and hitherto unanswered questions in astronomy. NASA administrators were cautiously optimistic but expressed similar sentiments. It was a unique time in the history of publicly built telescopes because few other observatories had ever operated in the full glare of press lights and public attention. By some insiders' accounts, no other telescope operators were ever as pressured to deliver first-light photos so quickly.

However, behind closed doors and far away from the hoopla, people puzzled over the look of the images. Members of the Wide Field and Planetary Camera Team (dubbed 'Wiff-Pickers' because of their instrument's acronym – WF/PC) were not quite so excited about the image. Things did not look right. People were feeling downright uneasy about what they had.

Star images normally look like a bright spot of light surrounded by a little scattered light. This is because most of the star's light is focused by the telescope's mirrors into a little 'core' or bright spot. HST's first-light image showed a sharp spike of light surrounded by a huge halo and strange tendrils of light extending out from the central core – *not* the way a star should look through a correctly focused telescope. As the telescope relayed more images to the ground for analysis, the uneasiness increased.

It was all very puzzling, and the Wiff-Pickers felt frustrated because they did not understand what they were seeing. To make matters worse, the WF/PC team members were trying to figure out what they were seeing at the same time that controllers were still becoming acquainted with HST's in-orbit behavior.

Other difficulties cropped up with the telescope. For one thing, HST's Fine Guidance Sensors did not always lock onto their guide stars properly. Sometimes they would scan a field with multiple stars and select the wrong star. That was a software problem, but like other software 'bugs' it took precious programming time to search out the offending code and rewrite it.

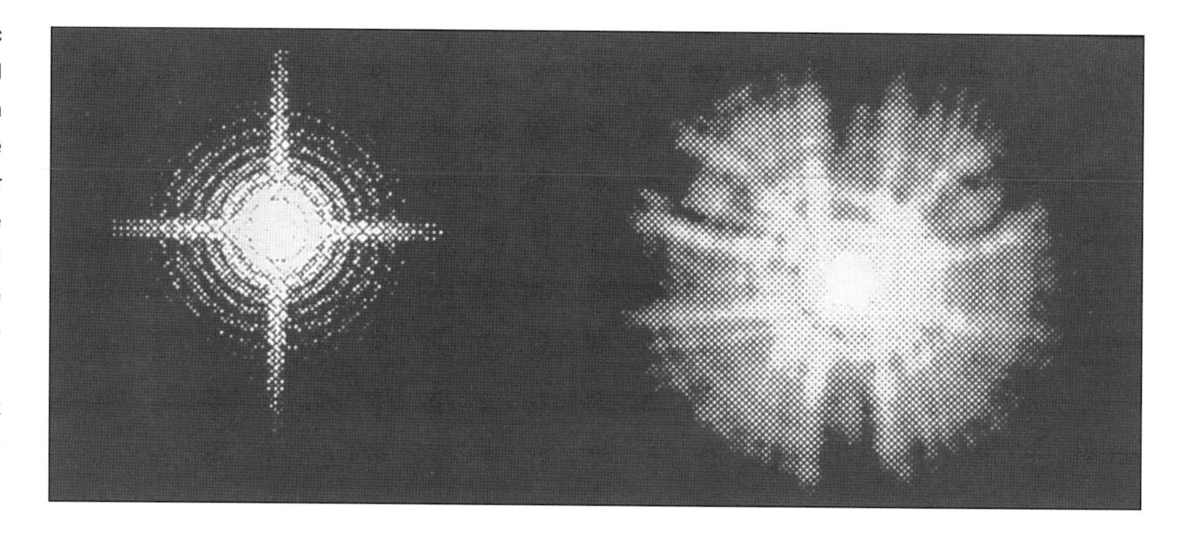

**Figure 1.3.** A dramatic demonstration of what spherical aberration looks like when a bright star provides the light. The image on the left is what the star should have looked like. The image on the right is an actual HST star image. Notice how the light is spread out over a wide area in the aberrated image. (NASA–Goddard Space Flight Center)

The other major problem was a flapping solar panel. Shortly after deployment, controllers noticed that the telescope jittered and moved around more than it should. Every spacecraft has jitter, especially during its early weeks in orbit, but HST's shaking was worse than expected, and it affected the telescope's ability to stay locked on target.

As frustrating as these two problems were, they were not catastrophic and they *were* fixable. It was a different situation, however, with the telescope's focusing. From the day of first light in May until the last week of June, the world at large heard about some minor 'shakedown cruise-type' problems with the telescope. Few people on the outside heard even a hint of the scientists' misgivings about the images. To be fair, there were still many tests to be done and adjustments to be made, and it was hard to tell if an image looked strange because the focusing just needed adjusting, or if some other, larger problem existed. Still, concerns that something was seriously wrong with the mirror were spreading among insiders during whispered hallway conferences and subdued lunchtime discussions.

This crescendo of concern first began to build among HST team members after a project meeting held at Goddard Space Flight Center on the day after first light. During that meeting, uncertainties about the mirror were aired in a discussion about the way the images looked. Team member Roger Lynds – an astronomer and optical expert – stood up and said he thought HST had a serious spherical aberration problem. It was the first time anyone mentioned those words in connection with HST's main mirror, and Lynds's suggestion was dismissed abruptly. Yet, Space Telescope Science Institute scientist Chris Burrows came to the same conclusion almost immediately when he started to analyze the first-light image and realized that something was curiously wrong with it. He wrote a software program that calculated the kinds of problems that would create such an image. The report based on his analysis independently confirmed Lynds's diagnosis of spherical aberration.

Of course, no one wanted to hear such a depressing diagnosis because it implied that a major mistake had been made. A spherically aberrated mirror is one that has been ground incorrectly. The curve at the center of the mirror does not match the curve at the edge. Essentially, the flawed mirror does not reflect light into a tiny point of encircled energy called the 'core'. Instead, it spreads the light out over a larger area. On HST, the center focused well, but the edge was out of focus, making a big halo.

It was simply unthinkable that HST's main mirror, the 'heart' of the telescope, could have been ground incorrectly without someone noticing before launch. No one – least of all Hughes-Danbury Optical Systems (formerly Perkin-Elmer), the contractors responsible for the mirror, wanted to admit to that kind of mistake. So, engineers from NASA and Hughes-Danbury started to look for other causes. For example, there was a suspicion that the secondary mirror might have been off-center or tipped in some way. At the same time, controllers and engineers worked on a series of focusing tests for the mirrors and the cameras, and Hughes-Danbury engineers looked into ways of using mechanisms called 'primary-mirror actuators' to change the focus of the main mirror.

While the tests continued, HST executed a series of observations of the star Iota Carinae using the WF/PC. At the same time, it also ran a science observation of the star cluster NGC 188. Everyone turned to the Wiff-Pickers to explain the images, and the team obliged by analyzing the images and discussing interpretations at the daily meetings. Sandra Faber, an astronomer and WF/PC team member, remembers those days fondly. 'Every picture that came in was interesting and always promoted lots of discussion. We were kind of basking in the glory of the moment because we were the only people who were returning visible data from the telescope. There was a little period in there when every 11 a.m. meeting at Goddard, we looked forward to the latest picture or analysis of the picture from the WF/PC team. We liked that, being only human.'

Due to the team's careful analysis of each image's characteristics during these meetings, engineers took great interest in what the implications were for the telescope. The images continued to show the strange streamers of light and defocused cores, and when the Faint Object Camera took its first-light image on June 17, the same problems appeared. That ruled out any suggestions that the problem lay with any other part of the telescope – the WF/PC, for example – and set the stage for more primary-mirror tests. Around the same time, Hughes-Danbury engineers announced that the primary-mirror actuators they were depending on to fix the mirror could only partially correct the focus problem.

The Wiff-Pickers, along with researchers from the Space Telescope Science Institute, argued that if this was a ground-based telescope with a problem such as the one HST was exhibiting, and people did not know what was going on, there was one definitive test that could answer a lot of questions: the running of the telescope through its full range of focus. This is a familiar process to anyone who has used a camera lens or owned a telescope or pair of binoculars. First, run the optics completely out of focus so that everything looks blurry. Then, slowly move the optics all the way out of focus the other direction, taking test images at every step. The procedure shows where focusing problems occur with the

**Figure 1.4.** The first-light image of stars in the cluster NGC 188 from the FOC, taken June 17, 1990. (STScI)

optics, and often reveals other flaws as well. In particular, it would reveal the characteristic fingerprints of spherical aberration.

Running the telescope through focus, however, made people feel uncomfortable. It is all very well and good to apply the procedure to a telescope on Earth, where someone can go in and fix things if something goes wrong. However, testing a telescope on orbit in this way was considered a risky move. 'There was very big resistance to this by the project managers,' Faber said later. 'They're very conservative people, and they didn't like the idea of running the telescope way out of focus, and then maybe the focus mechanism would fail, and we wouldn't be able to get it back. Then we'd have a *really* useless telescope.'

Despite these concerns, Faber and Burrows pushed for the test, and eventually the program managers agreed to do it. While the focus images were being taken, WF/PC team member John Holtzman created a series of images using a computer program to simulate spherical aberration. He and Faber analyzed the simulations and created a model of the focus run.

The focus test itself produced a series of star images marred by tendrils. When these images were compared to the simulations, the similarities were astounding. With the earlier conclusion that the actuators could not be used to fix the focus, and the new evidence from the computer simulations and the focus test, it was becoming very clear very quickly that spherical aberration was the answer – however dreaded – to the puzzling questions surrounding HST's strange-looking images. On June 19, 1990, the day Hughes-Danbury reported the inability of the primary-mirror actuators to fix the problem, Faber wrote in her diary, 'This is when we realized we are well and truly dead.' Six days later, when the computer simulations and focus test images were presented to a meeting of project scientists and managers, she wrote that the evidence left the group 'stunned'.

At that point, it was all over except the shouting. The word spread through the teams like wildfire, but, to everyone's great amazement, the news had not reached the press. Certainly there had been comments circulating about 'tendrils' in the images – hints that some science writers have said in retrospect should have tipped them off to a serious focusing problem. Aside from the scientists on the scene who were working with the engineers from Hughes-Danbury to understand just what they were seeing, very few people in the larger astronomical community understood just how serious the problem was. The actual announcement was left to NASA management, and a press conference was arranged for June 27. It was scheduled after a meeting of the HST's Science Working Group, during which the top project scientists would be officially informed of the problem.

The press conference was a funereal and depressing event, with NASA scientists almost at a loss for words to describe their feelings. Anyone watching the conference was left with the immediate feeling that HST was dead in the water. The press picked up on that mood immediately, setting the stage for highly critical hyperbole. The story led the news for days, with the media screaming out the problems in a variety of innovative headlines:

**The Hubble: one sick puppy!**

**Pix nixed as Hubble sees double**

**Star crossed: NASA's $1.5 billion blunder**

Taking advantage of the media spotlight, politicians and pundits rushed to denounce the project as a waste of money, an outrage and a 'techno-turkey'. Almost overnight, political cartoonists and late-night talk show hosts were making jokes at the telescope's expense. The long-awaited telescope that would see to the distant reaches of the universe became linked in the public's mind with bungling, government-sponsored science, a sort of orbital Mr Magoo.

The aberration was indeed an embarrassment of the worst sort. It was the equivalent of a rookie mistake, caused by a mix of technical problems, politics and just plain human stubbornness in the face of error. It should not have happened, but it did.

The announcement concerning the spherical aberration caused a firestorm of public criticism of NASA, the telescope, and – in a case of 'guilt-by-association' – of all 'Big Science' projects. Some of that criticism, which continues to this day, came from members of the astronomical community who had worked hard – some for more than fifteen years – to get the telescope funded and built in the first place. Like Sandra Faber, they dreamed of using Hubble to unravel cosmic mysteries billions of light years from Earth. Every scientist with time on the telescope had questions about how the aberration would affect their observations, the cameras and the spectrographs. Everyone expected the worst, and it looked like astronomers were left standing at the door to the universe, peering at the stars with myopic eyes while waiting for beleaguered optical experts and engineers to solve a completely unexpected mystery – 'The Case of the Misground Mirror'.

Those were depressing days for the people connected with Hubble Space Telescope. Everyone involved with the project was in a state of grief and disappointment over what looked like a colossal missed opportunity Ironically, there occurred a loss of focus, not just for the mirror, but in the observing programs of scientists who had awaited HST for years. There was lost science and a flagging confidence in that 'can-do' attitude that had been the hallmark of the American space effort.

Most embarrassing of all was the inexorable erosion of public interest in what astronomers do best – unlocking the secrets of the universe and showing them off to an interested community. A general public that had grown used to applauding every step into space was weary of shocks. First there was the loss of the space shuttle *Challenger* and its crew, the subsequent grounding of the shuttle fleet, and years of problems with other missions. Then, with the staggering blow of a spherically aberrated, jittery, cranky telescope, dreams of exploring the ends of the universe seemed to be turning into just so many ashes on the dustheap of failed science. The promised wonders of the universe that HST's designers and promoters had predicted were starting to look like pipe dreams.

A few days before the public announcement of spherical aberration, a group of scientists met for dinner at a restaurant near Goddard. It was a disconsolate and worried group of people who sat speculating about what would happen. About that dinner, Faber wrote in her diary:

> The younger guys are hit especially hard. They are wondering about their careers. They are wondering how long their jobs will last. There's a lot of talk about the 'clone' [a second-generation Wide Field and Planetary Camera that was already under construction]. Would NASA try to kill the program or have the patience to fix up the clone [to compensate for the effects of spherical aberration] and replace it? This would be costly, but it could be a big, American-style rescue effort and it might be appealing.

Talk turned to the future of HST, and to the uncertain future of 'Big Science'. The team indulged in a lot of speculation. Some of their pronouncements about investigations and the backlash against scientific research were almost prophetic. Someone suggested that there would be a big investigation – indeed, the Allen Committee was formed to evaluate the problem. Someone else expressed the fear that backlash against HST as a 'Big Science' project might affect the funding for other government science projects, such as the Super Conducting SuperCollider – the first of a new generation of particle accelerators. The group members speculated on whether NASA would sue Perkin-Elmer. It did, but, with the passing of time, events of a decade earlier were hard to prove. (In 1993, NASA came to an unexciting settlement with the company.)

These were all long-term predictions. The big question was: what do we do now? Incredibly, despite the spherical aberration and other problems with the telescope, scientists found that the telescope was not a total loss. It could still do observations, but it focused only 10–15% of the light into a central core, instead of the tightly focused 75 or 80% expected by astronomers. The rest was scattered into tendrils and spikes. If something could be salvaged from the data that HST could provide, how would astronomers go about digging it out?

Science and the human spirit have always flourished through adversity, and – as we all know – HST was not a lost cause. Quietly, behind the scenes and largely out of the media spotlight, some of the brightest people in the science business turned their attention to figuring out ways to compensate for the telescope's flaws, even before the big news of HST's problems was released. One possibility, suggested by University of Arizona scientist Tod Lauer, was to use 'deconvolution', a computerized image processing technique that essentially puts the light back where it belongs. The drawback was that if the data were deconvolved too much, useful information might be lost in the process. However, deconvolution could turn lousy images into acceptable ones. If examples of deconvolved images had been available on the day spherical aberration was announced, it is possible that some of the publicity HST endured would not have been quite so unsavory. Another particularly useful approach was to increase exposure times, gathering more light from an object to increase the data rate.

These processes later became important parts of the astronomers' toolbox during HST's first years on orbit, but, in the dark days of summer 1990, few people outside of the HST community knew about them. Instead, everyone wanted to know, How did this happen? How could the main mirror of a such an expensive telescope be ground wrong? Was somebody not paying attention?

## How the mistake was made

Grinding a telescope mirror is an arcane art, as many amateur astronomers find out when they get the urge to make their own telescopes. The optical engineer starts out with a 'blank' – a piece of optical-quality glass. That glass has to be ground and polished repeatedly until it is the proper shape to focus light. When all the grinding and polishing is complete, a reflective coating is applied. If this is done correctly, the mirrors will focus most incoming light into a tiny point. This well-focused core of light is the 'Holy Grail' for the scientists using the telescope. The better the focus, the better the information.

HST's primary mirror was ground by a team of optical engineers at Perkin-Elmer Corporation of Danbury, Connecticut. Late in 1978, the team took delivery of the 1-ton, 2.4-meter-diameter, $US1 million mirror blank that would become the heart of the Hubble Space Telescope. Their job: to grind and polish it until the mirror's shape focused incoming light properly. To make sure that the mirror was ground and polished accurately, Perkin-Elmer built a special device called a 'reflective null corrector'. This consisted of two small mirrors and a tiny lens hung above HST's main mirror. A light beam was flashed onto the primary mirror through the tiny lens in the corrector. This test created a pattern of light called an interferogram which looked like a big fingerprint. If the 'fingerprint' did not look right, opticians would continue to grind the mirror until it looked as expected.

This process should have guaranteed a perfect mirror, but, in installing the lens in the null corrector, the technicians made a mistake. The lens had to be set at a precise distance from the null corrector's mirrors. To determine that distance, the technicians used a rod with a special cap that had a tiny hole in it. An alignment light beam was sent through the hole in the cap down to the rod. To make sure that the laser beam did not bounce off the cap, special paint was applied to it. The light was supposed to go through the hole and bounce off the polished tip of the rod, which would then tell the technicians where to set the lens. If it failed to do this, they would know right away to re-aim the light.

It was an intricate measuring process, and it failed because a tiny spot of paint had worn off the cap. So, when the light went down to the cap, it bounced off the worn spot instead of going through the hole and down to the tip of the rod. To the technicians, it looked like the light was bouncing off the rod, and so they attempted to hand-correct the placement

**Figure 1.5.** The HST primary mirror at Perkin-Elmer (now Hughes-Danbury). (STScI)

of the lens. When the lens would not go where the corrector said it should, the technicians inserted three little washers into the setup to make it work. The difference in distance between where the lens was set and where it should have been set was only 1.3 millimeters, but it was enough to throw off the whole process. The result was that the opticians thought they needed to grind off more glass. It was a near-fatal mistake. HST's mirror was ground too flat, and nobody found out until after the telescope was deployed in 1990. It was a perfectly ground mirror – but it was ground perfectly wrong. As one scientist put it, 'Hubble Space Telescope had the best spherical aberration money could buy'.

That is, then, the technical reason why HST has an aberrated mirror. However, people, not computers, did that technical work, and understanding the human factor in Hubble's troubles is important. Simply put, HST faced technical hurdles and entrenched managerial problems even before it was designed. Every complex human institution has such problems – but they are particularly hard to accept with a telescope that started out with such high expectations. HST did not start out as a jittery, spherically aberrated mess. In fact, it started out as a solution to one of the problems astronomers deal with: the atmosphere.

## HST's place in history

Until the space age, astronomers plied their trade from the surface of the Earth. They worked behind the large, unblinking eyes of US-based facilities like Arizona's Kitt Peak National Observatory, the observatories atop Mauna Kea in Hawaii, California's Lick Observatory, the observatories on Palomar Mountain and Mount Wilson, Wisconsin's Yerkes Observatory, and a variety of excellent institutions around the world: the Anglo-Australian Observatory, the Royal Observatory at Edinburgh, European Southern Observatory in Chile, Armagh Observatory in Northern Ireland, and the international facilities at Cerro Tololo Inter-American Observatory.

**Figure 1.6.** Kitt Peak National Observatory near Tucson, Arizona, USA. (National Optical Astronomy Observatories)

**Figure 1.7.** Aerial view of the Cerro Tololo Inter-American Observatory in Chile. (National Optical Astronomy Observatories)

Magnificent science has been and continues to be done at these facilities, which are the backbone of astronomy research. In fact, the astronomers who use HST also use ground-based facilities, and are well-versed in achieving 'good science' at those observatories.

Even the best ground-based facility, however, is hampered by the blanket of gases that we breathe to keep alive – the Earth's atmosphere. To put it simply, for some types of astronomy, the atmosphere is a big problem. It interferes with the light that astronomers want to see in three distinct ways: it bends it, absorbs certain wavelengths, and it also gives off radiation that interferes with incoming light.

Scientists have solved some of the atmospheric problems by locating observatories on high mountains, so placing them in thinner regions of the atmosphere. One interesting fringe benefit of this is that astronomers get to work in some of the most beautiful and desolate places on Earth. Even at high altitudes, however, atmospheric interference remains

**Figure 1.8.** The International Ultraviolet Explorer (artist's conception). (NASA)

a problem. Most astronomers interested in ultraviolet light, gamma or X-ray wavelengths do their research from orbiting spacecraft, such as the International Ultraviolet Explorer, ROSAT, the Cosmic Background Explorer and now HST.

The idea of a large space telescope in orbit around the Earth is not a new one. In 1923, German rocket scientist Hermann Oberth outlined ideas for a space station and an orbital space telescope in his book *Die Rakete zu den Planetraümen* (*The Rocket Into Planetary Space*). He imagined a telescope attached to a station in geosynchronous orbit (where today's telecommunications and weather satellites are located), and even discussed the effect of station jitter on observations of dim, distant objects. To solve that problem, he came up with some rather ingenious ideas for attaching a space-based telescope to a small asteroid, and he even suggested that future scientists tow the asteroid to Earth orbit so that the telescope 'night crew' would not be bored while living on a station orbiting between Mars and Jupiter!

Oberth's writings attracted a great deal of attention and criticism at a time when any idea of traveling beyond the atmosphere was best left to people who were more familiar with such crazy ideas – science-fiction writers. However, he persisted in his thinking, and revisited the idea of a large space telescope in 1957, in a book called *Menschen Im Weltraüm* (*Man In Space*), writing

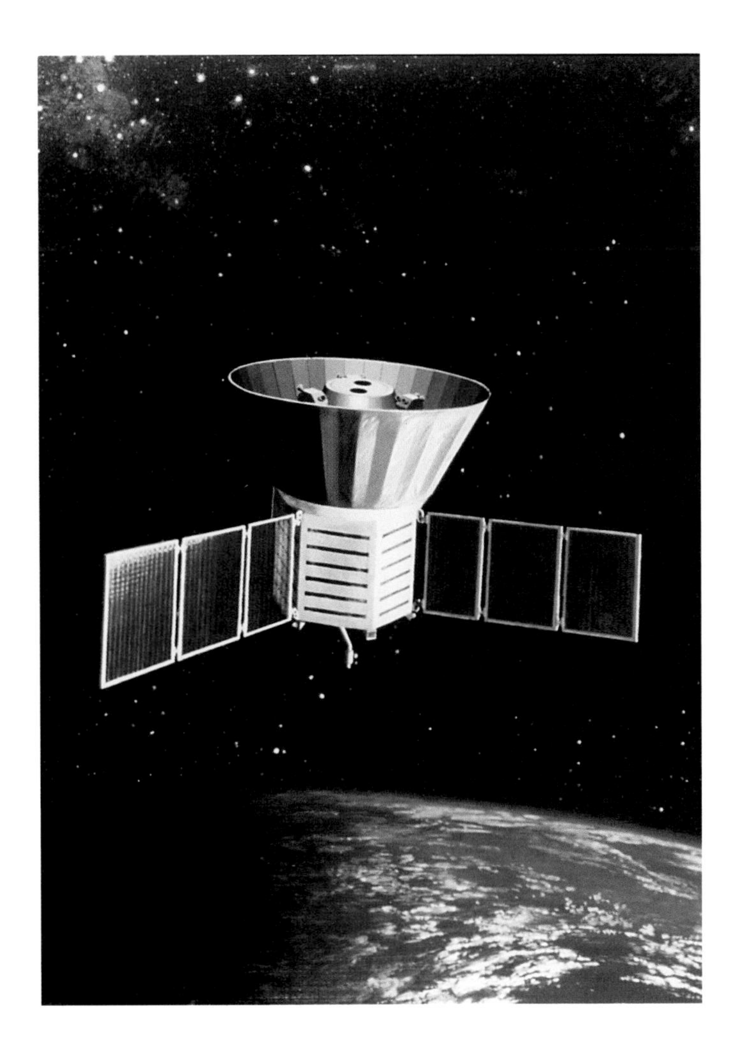

**Figure 1.9.** The Cosmic Background Explorer (artist's conception). (NASA)

> It is only too obvious that astronomical research would gain remarkable benefits from a space station. The atmosphere, gravitation, and the movement and rotation of the Earth affect observation of the sky. The immense difficulties involved in producing even larger telescopes, and the fact that their use may be interfered with by the atmosphere have been shown quite recently in the case of the 200-inch Mount Palomar telescope.

The advent of World War II turned any thoughts of peaceful astronomical use of rockets toward military applications. Other possible uses for rocketry remained in the realm of science fiction. Robert Smith, in his book *The Space Telescope*, points out that at least one scientist published an article detailing the possibility of putting a 300-inch telescope on the Moon. Smith writes,

> It was perhaps fitting that the idea was published in an issue of *Astounding Science Fiction*.

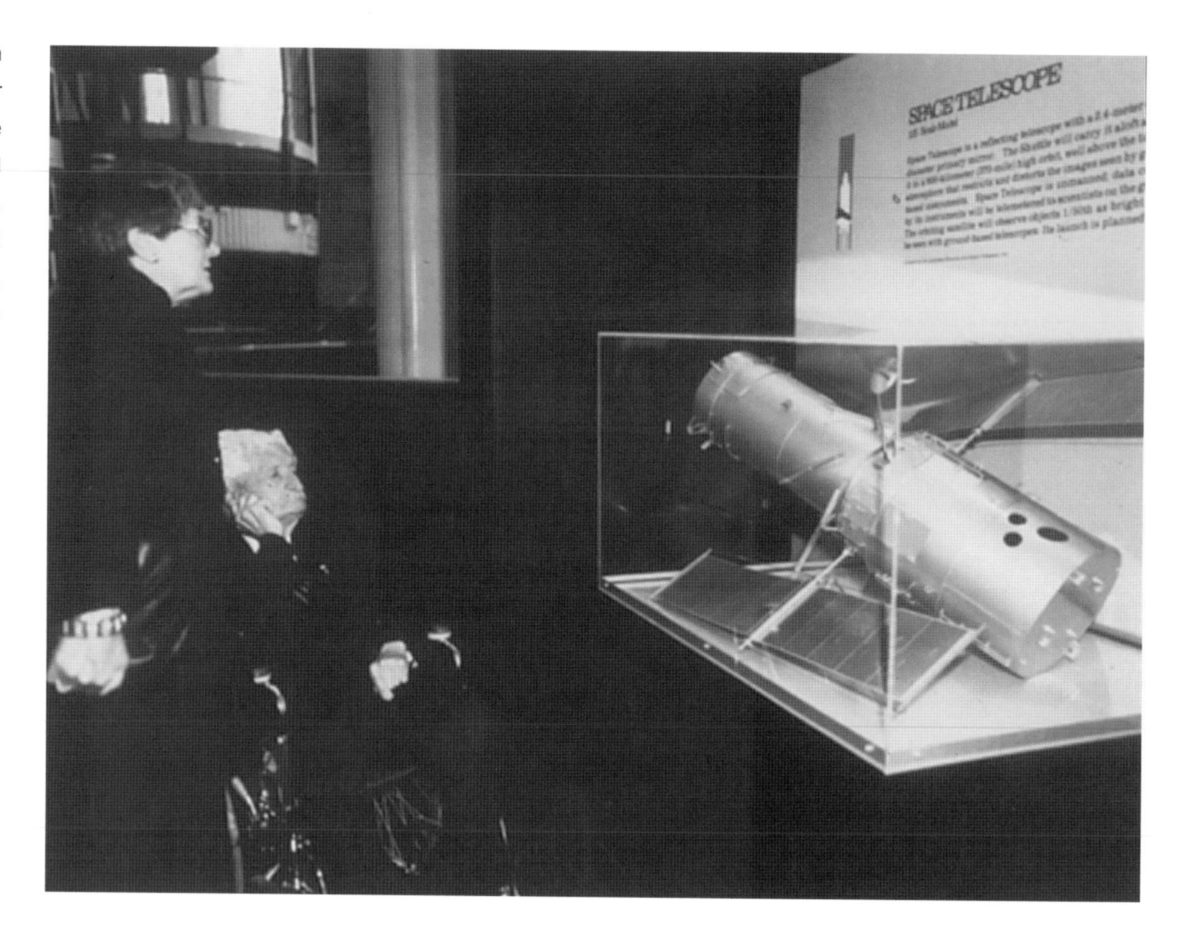


**Figure 1.10.** Hermann Oberth (1894–1989) and his daughter viewing an exhibit on the Space Telescope at the National Air and Space Museum (Washington, D.C.). (Courtesy of the National Air and Space Museum, Smithsonian Institution)

Another scientist (and science-fiction fan) interested in the idea of a space telescope was astronomer Lyman Spitzer. He spent the war years researching underwater weapons, but kept up an interest in rocket advances and science fiction. In 1946, Spitzer wrote a paper suggesting a space telescope for RAND, a 'think tank' operated by the Douglas Aircraft Company. The paper, which was quickly classified, was called 'Astronomical Advantages of an Extra-Terrestrial Observatory', and it probably served as the 'birth announcement' for the idea of what became the Hubble Space Telescope.

Spitzer's paper was prophetic, because it suggested nearly every field of study that HST is doing – from ultraviolet studies of supernovae and analysis of eclipsing binary stars, to measurements of the structure of distant galaxies, globular clusters and quasars. Spitzer does not find his own scientific prescience surprising, even now. As he put it recently, 'Anybody who was foolish enough to think that rockets might someday be useful for sending things into space would naturally end up with a list very similar to what I wrote.'

In his RAND report, Spitzer recognized that such a telescope would have to be a large and radically different instrument from anything previously used. He wrote:

> Most astronomical problems could be investigated more rapidly and effectively with such a hypothetical instrument than with present equipment. However, there are many problems which could be investigated only with such a large telescope of very high resolving power. It should be emphasized, however, that the chief contribution of such a radically new and more powerful instrument would be, not to supplement our present ideas of the universe we live in, but rather to uncover new phenomena not yet imagined, and perhaps to modify profoundly our basic concepts of space and time.

In May 1990, Appendix V of Spitzer's larger 1946 paper was reprinted in *The Astronomical Quarterly*, and he was invited to contribute his thoughts more than four decades after his original publication. He wrote:

> The chief effect [of this paper] was on me. My studies convinced me that a large space telescope would revolutionize astronomy and might well be launched in my lifetime.

Spitzer spent the 1960s pushing for a large orbital telescope in front of NASA and Congress. His lobbying efforts went on during a period of NASA research and development for the first generation of orbiting telescopes, called the Orbiting Astronomical Observatories (OAOs).

Studies of these orbiting observatories went on against the backdrop of NASA's highly publicized manned space exploration efforts and the headlong rush to put a man on the Moon. It was a heady and interesting time for advocates of a space telescope. In the early part of the 1960s, few things seemed impossible, no matter how technologically complex they might appear. During this time, the telescope – then called the Large Space Telescope (LST) – was by far the most technically complex unmanned project under consideration by the agency. Its crowning glory would be a 3-meter (120-inch) main mirror.

Nancy Roman, who was NASA's Chief of Astronomy at the time, later becoming Program Scientist, with responsibility over a broad range of projects, said of the first official proposal for a space telescope, 'Basically, Spitzer proposed what really was LST back in 1946, but I don't think anybody took it seriously until 1962.'

Roman admitted to being openly skeptical of such a project because she felt that promoters of an orbital telescope were underestimating the engineering complexity of such a mission. There was also the matter of determining just what kind of science could be done with a space telescope. In 1966, Lyman Spitzer chaired the first meeting of a committee from the prestigious National Academy of Sciences looking into the possibility of such a telescope. Called the 'Ad Hoc Committee on the Large Space Telescope', the group studied the range of possible uses for LST, and met with other astronomers to discuss ways to implement the project. In 1969, the committee published a report called 'Scientific Uses of the Large Space Telescope', that gathered all the ideas and plans for LST into several major proposals.

Of prime importance to astronomers was the search for an accurate measurement of astronomical distances. This is one of the guiding principles in astronomy. On Earth, we have all sorts of distance units – centimeters, meters, kilometers, inches, feet, miles, etc. – allowing us to figure out how far away something is. In space, it is more difficult to measure distances between objects. We can use light-travel time between the Earth and other objects in the Solar System by measuring the time it takes for a radar signal to travel from Earth to an object, bounce off the object and return to Earth. For more distant objects,

**Figure 1.11.** Lyman Spitzer, photographed in 1985. (Lyman Spitzer, Princeton University)

we can measure their parallax, which is the apparent shift in their position in the sky as the Earth goes around in its orbit around the Sun. Parallax measurements work well for the region of the galaxy immediately around us, but to determine the distance to more distant stars and other galaxies, we must turn to the so-called 'standard candles' such as Cepheid variable stars. Astronomers derive the distances to these stars – and hence their galaxies – by comparing the difference between their intrinsic brightness (absolute magnitude) and their observed brightness (observed magnitude).

For very distant galaxies and quasars, there is no convenient yardstick, so astronomers use something called the 'Hubble Constant'. It works like this: the most distant objects in the universe seem to be receding from us at high speed. The distance a galaxy is from us is directly related to the speed at which it is moving away from us, and this relationship is scaled by the Hubble Constant. However, it is not quite a constant because astronomers have yet to agree on one value for this all-important relationship. The drive to define this constant was part of the original motivation for the Space Telescope: a powerful orbiting telescope would be able to see the standard candles in more distant galaxies and would eventually allow astronomers to pin down a more precise value for the Hubble Constant; then it could be used to determine accurate distances to galaxies. As we discuss later in Chapter 6, HST has taken some important first steps in defining the Hubble Constant.

Astronomers also wanted their powerful Space Telescope to study the structure, scale and evolution of the universe. Because the light from the most distant objects takes so long to reach us, these objects appear today as they actually existed millions and billions of years ago. The way they were grouped together *then* – their structure – is what we see today. Essentially, astronomers are able to look back in time.

Ultimately, accurate distance determinations, further understanding of distant objects and other measurements that we shall discuss later, are vital keys to determining the age and fate of the universe, and to answering important astronomical questions such as, Will the distances between galaxies expand forever? Or will the galaxies spread out only to a certain point and then start a headlong contraction back to a point called the 'Big Crunch'? These are powerful questions, and, in the days of HST's early development, they were compelling guiding principles.

The Ad Hoc Committee was not looking to boost a space telescope at the cost of the ground-based installations. Indeed, two other goals were to build a series of intermediate-sized orbiting telescopes, and to increase the number of ground-based observatories to work in tandem with the space-based instruments.

As a member of the Ad Hoc Committee, NASA's Nancy Roman eventually decided that a large space telescope could be built, and she took up the battle for the project. 'I didn't really become a fighter for space telescope until the early 70s,' she said. 'Before that I didn't think that we were ready for a telescope of the size and complexity that everybody wanted, until one of the Orbiting Astronomical Observatories had flown successfully.'

Roman had good reason to be cautious, since she had responsibility for scientific oversight of the Orbiting Astronomical Observatory program. Two successful missions – OAO-II and Copernicus – flew in the 1960s and early 1970s, and made up the first generation of space telescopes. They were not exactly the intermediate-sized projects called for by Spitzer's committee, but they did provide valuable experience in building orbital telescopes.

By the 1970s, support for a Large Space Telescope was concentrated largely in the astrophysics community, with little invited input from the planetary science community. There was a general feeling that planetary scientists had gobbled up a big share of science money with their probes to the Moon, Mars, Mercury and Venus. Astronomers who studied stars and galaxies wanted to build *their* big science project. So, although Roman and others saw them as an important constituency for a large telescope, planetary scientists were not too welcome at the planning table at the beginning. 'Certainly the people who were pushing a space telescope were not interested in planets,' explained Roman. 'From where I sat, planets seemed to be a useful component from the beginning. My memory is that I was interested in using it [Space Telescope] for planets long before it became a political necessity.'

Once it became obvious that the project would not fly without a broad base of support, the working groups rallied support from the planetary scientists. However, by that time political support was beginning to fade. The somewhat receptive political climate for projects like LST began to change to one of outright skepticism. The financial and spiritual costs of the Vietnam War affected Congressional willingness to fund 'Big Science' projects, and the staggering costs of the Moon missions did not help matters. Against this backdrop of shrinking budgets and what seemed like the beginnings of national apathy toward big-time space missions, NASA and the working groups of scientists studying the telescope were asking Congress to spend somewhere between $US300 and $US700 million for a 3-meter telescope. To a member of Congress at that time, it must have seemed like the astronomical community was everywhere, asking for money: ground-based facilities, such as the Very Large Array radio telescope in New Mexico, were under construction, and there were constant appeals for funds to upgrade existing observatories. This made the request for money to build LST a rather difficult lump to swallow, and the automatic Congressional response was 'Find a way to make it cheaper'. To do that meant making the telescope smaller, which meant compromising the science that could be done with it. In many scientific circles the Congressional response, although predictable, was greeted with skepticism and an attitude that if the telescope could not be built as it was proposed, then no good science would come from it.

Still, even with a smaller space telescope, there was the tantalizing possibility that astronomers could get *something* out of the project, so the community looked for ways to keep the project going. To do this, astronomers had to become politically aware and organize grass-roots support for the Large Space Telescope. This they did, largely through the leadership and lobbying efforts of scientists like Lyman Spitzer and the Institute for Advanced Study astronomer John Bahcall.

In 1974, a group of scientists called the Science Working Group for LST agreed to look at the development of a 2.4-meter (94.5-inch) telescope. Downsizing the mirror was painful to consider because it meant a loss of scientific capability, but the result was a space telescope project that engineers and scientists predicted would cost no more than $US300 million through the end of its first year of operation. That was a more realistic amount of money for Congress to consider.

The downsizing required a number of design compromises and scientific 'scale-backs' to keep the program on the funding track, and $US300 million remained the magic number for a long time. After several more years of studies, fine-tuning, Congressional lobbying by members of the science community, and repeated attempts to delete the telescope from the NASA budget, Congress approved an updated budget of $US425–475 million to build the project. By the summer of 1977, the way was finally cleared for the construction of the telescope.

There *was* one catch: in return for the slightly higher price tag, Congress wanted NASA to seek a foreign partner in the project. In 1977, NASA and the European Space Agency signed an agreement setting up a collaboration between the two agencies to run Space Telescope as a joint NASA/ESA program. In essence, ESA agreed to supply 15% of the development cost for the program, making this contribution in the form of the Faint Object Camera and the solar arrays. In return, ESA astronomers would be guaranteed 15% of available observing time. ESA also agreed to provide some staff members to the yet-to-be formed Space Telescope Science Institute as its part of the bargain. The stage was set, and it was time to make the telescope a reality. Launch was scheduled for 1983; it was going to be a busy six years until then.

(As a short historical aside, it is interesting to trace the line of name changes that the project went through in its developmental years. The space telescope first emerged as 'Large Orbital Telescope', then became the 'Large Space Telescope' (LST). Eventually that was shortened to 'Space Telescope' (ST). In 1983 it was officially named the Edwin P. Hubble Space Telescope, after the US astronomer who first discovered the relationship between the velocity and distance of receding galaxies that describes the expansion of the universe.)

## Building HST

How do you go about building a space telescope? The most important thing you need (besides abundant funding) is a set of mirrors. It cannot be just any set; it must be a specially made optical system built to withstand the rigors of launch and an orbital deployment in a microgravity environment. It needs to be mounted in a spacecraft assembly along with an array of scientific instruments, guidance sensors, communications antennae, solar panels to supply power, an on-board computer and a variety of other equipment. When that has been accomplished, you test it and then launch it.

**Figure 1.12**. Edwin Hubble (1889–1953) at the 48-inch Schmidt telescope on Palomar Mountain. (The Observatories of the Carnegie Institution of Washington)

This was the process initiated by NASA in early 1977. As soon as the project was approved, the agency faced the gargantuan task of designing the telescope, and selecting contractors to build the spacecraft and the instruments that would fly aboard it. The agency also had to determine which NASA center would supervise the contractors, and had to find a home for an institute to run the telescope after it was built and launched. NASA virtually needed to amass a 'standing army' of people necessary to make HST a reality.

NASA had a complex engineering and scientific undertaking on its hands. Fortunately, there were capable people within the agency to work with contractors and oversee the construction of the telescope. For example, as Program Scientist, Nancy Roman was there to help expedite the process and to make sure the contractors kept the science objectives in mind. 'I felt my job, not only with Space Telescope, but with other projects as well, was to keep the two sides talking to each other,' she explained. 'I used to say when I first joined NASA that I felt I was acting as an interpreter between the engineers and the scientists because, while they both wanted the same thing in the end, they didn't speak the same language.'

Early on in the process of getting the engineers to design and build what the scientists wanted, the NASA Marshall Space Flight Center at Huntsville, Alabama, was selected to be the lead management center for the project. The NASA Goddard Space Flight Center in Greenbelt, Maryland, was chosen to be the operational 'nerve center' for the telescope, and was given responsibility to develop the scientific instruments.

The scientific expectations for the telescope were impressive, and so was the preliminary 'hype'. During the first days of development, Marshall public relations people handed out folders boasting that Space Telescope's 2.4-meter mirror would enable scientists to see 15 billion light years into space. Objects 50 times fainter than anything seen by the most powerful ground-based telescopes would be within reach of the telescope's powerful eye. HST would supply images ten times sharper than the best Earth-based observatories. Unlike facilities on Earth, which typically observe around 2000 hours per year and lose large blocks of observing time to cloud cover and instrument problems, planners claimed that the 2.4-meter observatory would be able to rack up an impressive 4500 hours of observation time per year.

With such optimistic claims being made about it, Space Telescope shaped up as a difficult project right from the start. As complex as the engineering process would be, an equally intricate web of contractors and university/private contractor groups vied for the privilege of designing and building the scientific instruments. Scientists competed for membership on the Investigation Definition Teams that would define the scientific goals and oversee construction of the instruments. Those instruments were:

- the Wide Field and Planetary Camera (WF/PC-1);
- the Faint Object Camera (FOC) (to be built by the European Space Agency);
- the Faint Object Spectrograph (FOS);
- the High Resolution Spectrograph (now called the Goddard High Resolution Spectrograph or GHRS);
- the High Speed Photometer (HSP).

Construction of the Space Telescope began in 1979, with the creation of the primary mirror and the building of the optical telescope assembly. To form an idea of the extensive network of companies and individuals working on the telescope, consider the following numbers: 21 major subcontractors, one university and three NASA centers spanning 21 states and 12 other countries were responsible for some aspect of the telescope's components, software design and other elements – from the mirror down to the batteries, star-trackers, data gathering systems, latches, solar arrays and a myriad of other electronic devices essential to the smooth operation of the telescope. A detailed schematic of HST as it was proposed is shown in Figures 1.13 (a) and (b).

Almost from the start, the Space Telescope program ran into financial and political difficulties. Remember that the project was sold to Congress in 1976 at a funding level of $US475 million to pay for its development and a 1983 launch. However, in 1983, the telescope was not ready to go, and the launch had to be delayed again and again. By 1986 the budget had grown to nearly twice the original estimate. What happened?

Many complications stemmed from managerial entanglements, which in turn arose from the expensive nature of the project and the technical complexity of the telescope. Naturally,

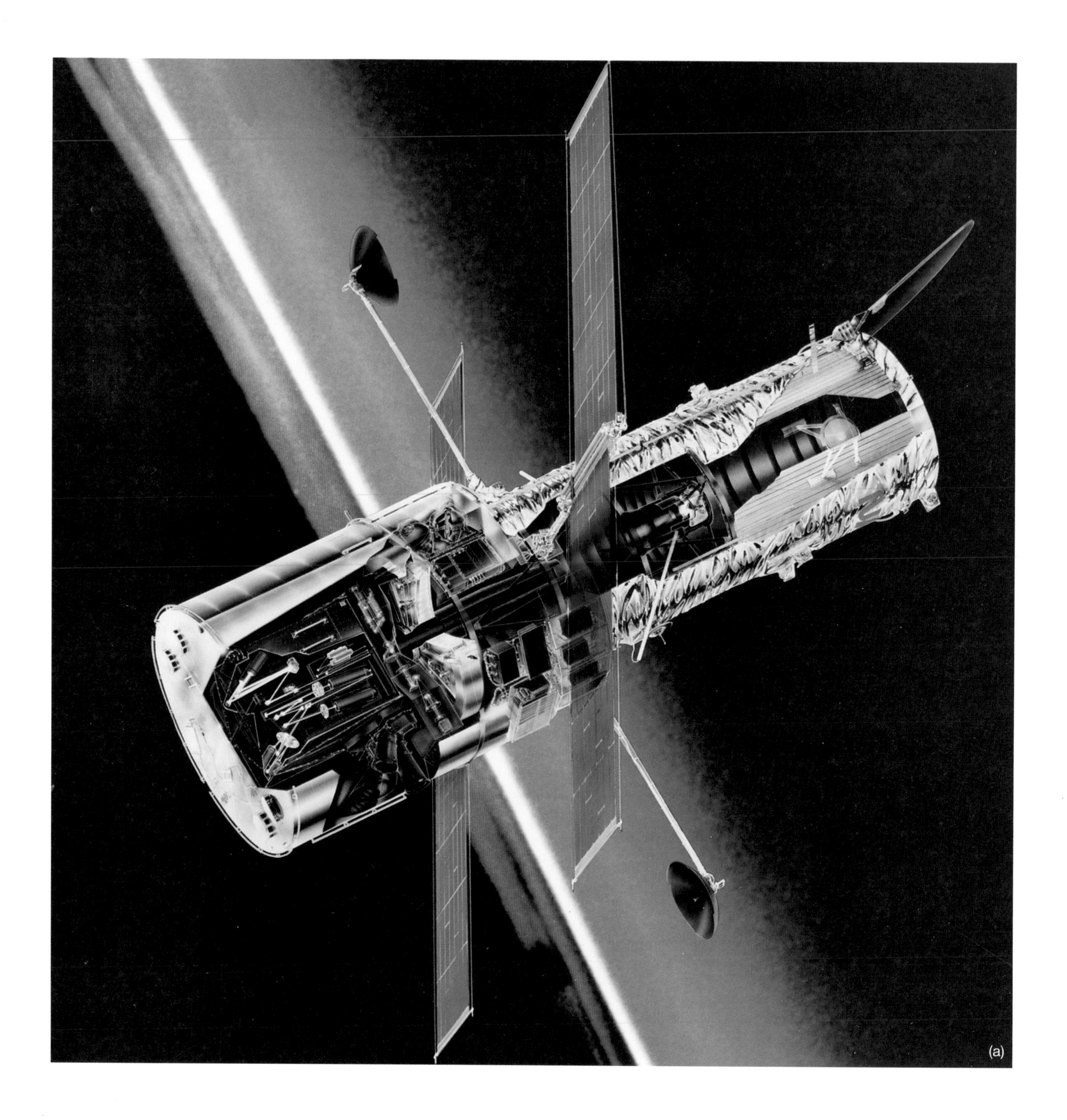
(a)

**Figure 1.13.** (a, facing page) Cutaway schematic of the Hubble Space Telescope. (b, this page) Key to the cutaway schematic illustration. (NASA)

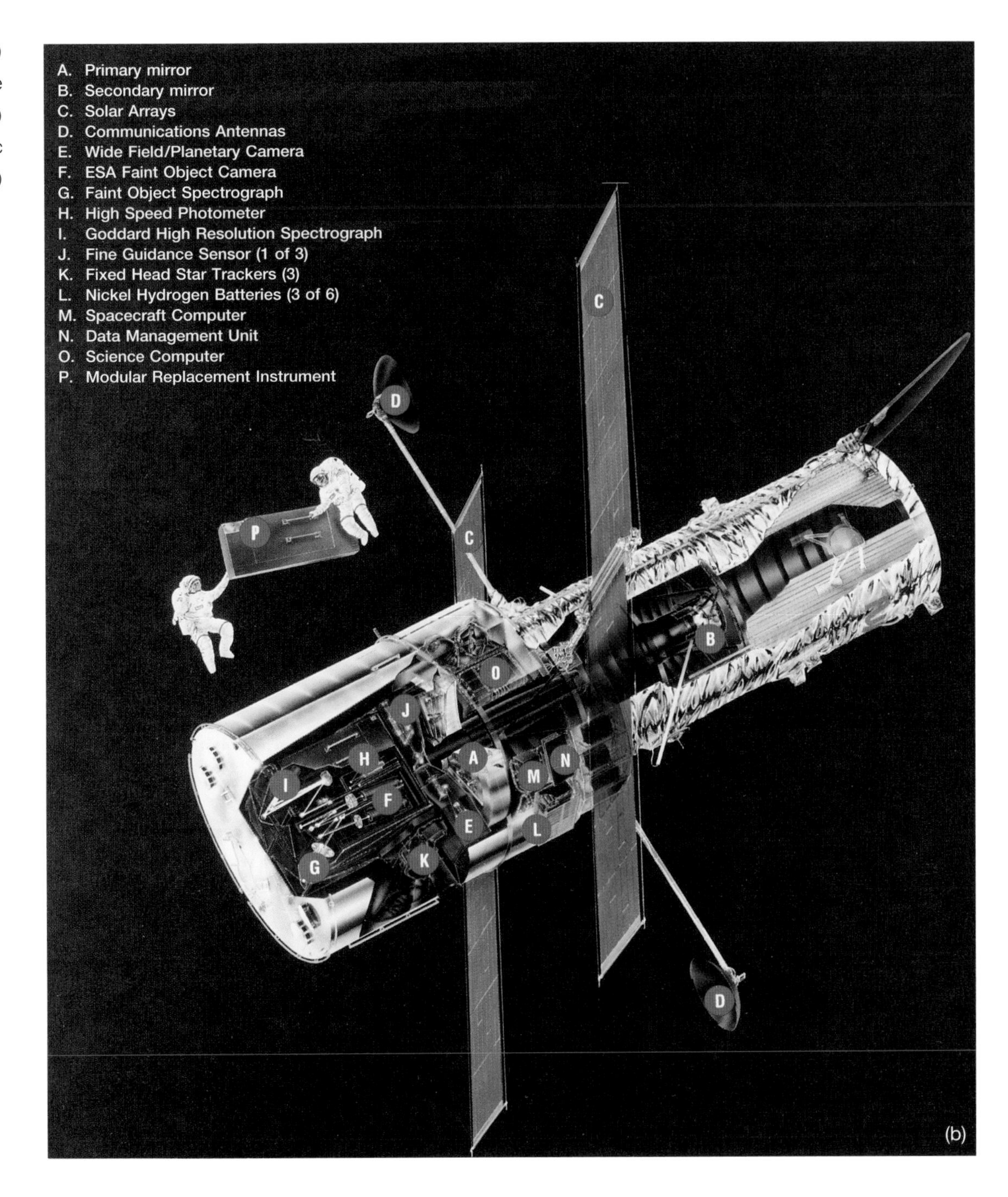

**Figure 1.14.** Scenes from the construction of the telescope at Lockheed Missiles and Space Company, now Lockheed Martin. (STScI)

when large sums of money are involved in a government project, a cumbersome administration takes charge of spending that money and directing the actual work. It takes time for commands to travel from the top of the hierarchy to the bottom in such a bureaucracy, and, furthermore, HST was built under tremendous deadline pressure.

Time pressures do not exist gracefully alongside technological development. Nearly every group involved in designing and building the telescope ran into difficulties in design and construction. These stretched out the time it took to build each of the telescope's intricate mechanisms. Space Telescope was a first-generation design, and, as such, it was expected that scientists and engineers would spend time ironing bugs out of its systems. But solving problems took longer than expected. New technologies had to be developed for instruments such as the Fine Guidance Sensors, which would allow the telescope to lock on to guide stars. The design of the entire optical telescope assembly was part of what NASA managers called 'advanced development'.

Designing and building new technology costs more than using tested designs, and many items – computer memories, for example – change rapidly. Consequently, what worked for previous space vehicles became out of date and inappropriate for HST, leading to the necessary development of new technology. So it went with nearly every part of the telescope, and the budget continued to expand. By 1985, the cost of the telescope was estimated to be $US1.175 billion, which was just enough to cover assembly and the on-orbit verification program. By 1986, that figure had grown to $US1.6 billion just to design and develop the telescope. That was the year the *Challenger* exploded, postponing HST's launch until 1990.

The shuttle tragedy may have led to unexpected benefits for the telescope, since it still was not ready for launch in 1986. Engineers took advantage of the enforced delays to make improvements to systems on board the spacecraft, and to test others to make sure they were still in working order. Still, the delays and cost overruns affected the telescope in many ways. For the Faint Object Spectrograph, for example, the lengthy launch postponement meant that a critical mirror oxidized while in storage. This reduced the reflectivity, a problem which was not discovered until after the telescope was launched.

Storage expenses and costs of testing and retesting all had to be added into the budget. By the time the telescope was finally launched into space, had operated for three and a half years and was refurbished, it cost $US2.3 billion. At the time of writing (1995), the total yearly budget for HST is about $US235 million dollars. Looking at it another way, the total telescope costs nearly $US8 per second to use!

Politically, the Space Telescope program was a maze of turf battles. Contributing to the general difficulty of building the telescope were the adversarial relationships between NASA agencies and contractors. Marshall Space Flight Center, which had overall responsibility for the contracts to design and build the telescope, fought with Goddard Space Flight Center, which bridled at being a subcontractor to a sister NASA center. NASA Headquarters viewpoints conflicted with the European Space Agency's goals. The Investigation Definition Teams – groups of scientists charged with the responsibility of designing the on-board instruments – fought with the newborn Space Telescope Science Institute. The Space Telescope community itself received criticism from ground-based astronomers and others who felt that precious science money was going into one of the largest boondoggles ever funded by NASA. Even other space observatory teams felt left out.

To defend HST from continual assaults and budget criticism, NASA put pressure on contractors to cut costs and boost efficiency Designs changed and costs were cut, but the telescope suffered each time this happened. In the case of Perkin-Elmer and the mirror grinding, pressure made a bad situation worse. The company, which had underbid the project at $US69.5 million so it would be chosen to do the job, was running way behind schedule in its attempt to provide a finished mirror. Costs were mounting, and the company managers told NASA that the project would cost more than they had bid. NASA agreed to pay more after a review of the process, but urged Perkin-Elmer to finish the job quickly. The pressure was on the company to perform – so, in the interests of saving time and money, crucial tests of the mirror were omitted. The tests performed by Perkin-Elmer were inadequate, and, as we saw earlier, incorrect. The company finished polishing the mirror in 1981, claiming it exceeded NASA's specifications. Three years later, it delivered the completed Optical Telescope Assembly to NASA and presented a final bill for more than $US300 million, around five times the original bid.

## The Space Telescope Science Institute

From the start, it was clear that the astronomy community did not want NASA to coordinate the use of the telescope. Historically, the relationship between scientists and NASA has been a wary one, and nowhere was it more contentious than during planning for the Space Telescope Science Institute. In the early years of the space agency, the National Academy of Sciences had formed a Space Science Board to recommend future policy. NASA managers were not happy taking orders from an outside group and wanted to use their own scientists to make decisions. This did not sit well with those outside the agency, and, in any case, it just was not good politics for NASA to ignore the viewpoints of those working elsewhere.

With the continuing deployment of the OAO series, and the growth of ground-based observatories receiving federal money through the National Science Foundation, the formation of 'national laboratory'-style institutes to oversee those observatories was an accepted practice. Several groups formed consortia to manage the facilities, such as Associated Universities Incorporated and the Association of Universities for Research in Astronomy (AURA).

The concept of an institute to implement the science program of the Space Telescope first came to light in a 1972 report issued by HST Project Scientist Bob O'Dell. It was assumed that Goddard Space Flight Center would be the science operations center as well as the control center for the telescope, but that assumption was challenged by various members of the science community. At stake were large amounts of telescope time on what was anticipated to be the premier observatory in astronomy, as well as large sums of money to be spent on the observatory operations. To give just a couple of examples of the scientists' concerns: some felt that NASA scientists would 'skim the cream' of observing time on a space telescope, depriving outsiders of access; there was also a feeling that was stated generally as 'NASA is so big, it'll just mess this up'. Obviously, some measure of independent input and oversight was needed, so another working group was formed to look into ways in which an institute could be formed. That committee looked at every possibility, ranging from giving NASA complete responsibility, to finding an outside consortium to run the telescope.

Discussions and studies continued over a period of years while various factions fought for control over the institute. Despite the science community's unease with NASA, the Goddard Space Flight Center wanted to house the Institute. Other spacecraft were being controlled from the Center, and managers pointed out that Goddard people had plenty of experience. In 1976, yet another committee was formed to give NASA advice on the formation of an institute. This group, led by Donald Hornig, was formed by the National Academy of Sciences. The Hornig Committee heard testimony from all sides of the institute debate before retreating to Woods Hole, Massachusetts, to hammer out a final recommendation.

Called the Hornig Report (insiders dubbed it 'the purple peril' because of the color of the cover and the threat it represented to Goddard's hopes for housing the institute) it recommended an institute independent of Goddard Space Flight Center, but with close ties to the NASA center. This was not what some within NASA wanted, but the agency accepted the recommendations while keeping its right to control telescope operations on orbit.

The stage was set for the development of the Space Telescope Science Institute. After a competition among several consortia and agencies, the AURA group was selected in 1981 to run the Institute at Johns Hopkins University in Baltimore, Maryland. Computer Sciences Corporation (CSC) would support the controllers, managers, programmers, and other professionals needed to administer the telescope operations.

Now all that was left to do was to launch the telescope and start observing with it.

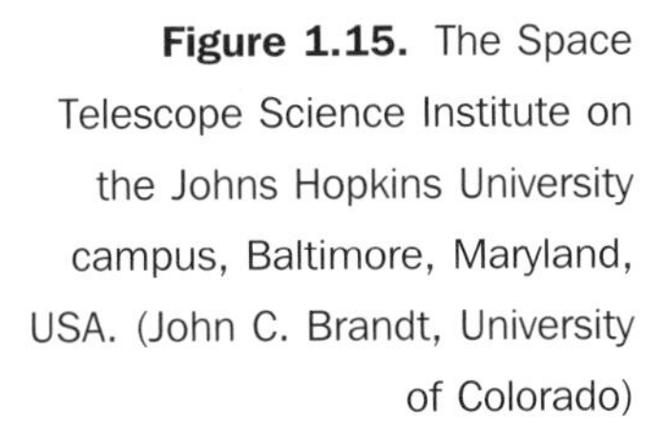

**Figure 1.15.** The Space Telescope Science Institute on the Johns Hopkins University campus, Baltimore, Maryland, USA. (John C. Brandt, University of Colorado)

**Figure 1.16.** Launch of HST on 24 April, 1990. (NASA–Kennedy Space Center)

## Launch, deployment and 'attitude shift'

No one who has ever seen a space shuttle launch can adequately describe the intense sights and sounds they experience. Large-screen IMAX movies come the closest to recording the event, but there comes a point when no movie can duplicate the earthshaking power of a shuttle leaving the ground and moving to orbit in less than 10 minutes. People attending a launch watch it in very different ways. Some cluster in groups at VIP sites around the Kennedy Space Center. Others watch from the press stands, and thousands gather along causeways and highways leading into the Center to experience a shuttle blasting off. Some choose to watch the launch alone, away from the crowds. Others hold hands and hug as the shuttle thunders its way into the sky.

In late April, 1990, HST was launched as thousands watched. NASA's announcer saluted the mission: '...and liftoff of the space shuttle *Discovery* with the Hubble Space Telescope – our window on the universe.' A day later, the telescope was deployed in orbit over South America, and people all over the world watched as HST's solar arrays were unrolled and the telescope was set free. For the scientists and the teams watching the event, it was the realization of a dream. It was an emotional moment. Some laughed; others cried. All were acutely aware that it was the beginning of a scientific adventure they would never forget. It just was not the adventure they originally planned on having.

In a roundabout way, this brings us back to the mirror problems, the jitter and the legion of other difficulties that people had with HST. Where in that history of technical oversight, painstaking planning and innovative design did HST go wrong? There are some who say that the mirror was ground incorrectly because there was not enough money to do it right. Others point out that the opticians at Perkin-Elmer were under a lot of pressure to do the job right, but to do it fast. Recall that the Perkin-Elmer people improvised during the measurements of the mirror's surface 'fingerprint'. If they had not been under the gun to produce a working mirror for NASA, perhaps today's HST would not have spent three years as 'the hobbled Space Telescope'.

Just as HST's reputation has changed since launch, so too have the human members of HST's team. For most of the scientists who lived through the dark days of summer 1990, it was a roller coaster of emotions. They went from the highs of launch to the lows of tragedy and grief when the spherical aberration was discovered. Looking back over that time, Sandra Faber wrote,

> It was the most frantic, most exhilarating, yet most frustrating time of my life. We were working 100 hour weeks, and when I fell into bed at one a.m., I could not sleep. I was turning over a million theories in my mind, plus a million strategies for getting new data and analyzing it. I was on a continuous adrenalin high. This lasted the whole summer, well into fall.

**Figure 1.17.** Deployment of HST on 25 April, 1990. (NASA–Johnson Space Center)

If not for the efforts of hard-working and creative people, HST might well have continued to be thought of as a complete failure. When some within NASA and the science community sought to write it off, team members pulled together and came up with ways to salvage the impaired telescope. What changed? Who can say? Perhaps it was that all-too-human propensity to unite in the face of adversity.

Faber, who still works with HST data as a Wiff-Picker, remembers watching attitudes shift between factions within the telescope community. 'The Space Telescope was a troubled project from the start,' she noted. 'It was full of little groups that had nothing to do for years except fight with one another while waiting for the telescope to go up.'

Somewhere between the tense turf battles during the telescope's formative years and the ugly finger-pointing that went on after Hubble's mirror problem was discovered, Faber believes there was '...an about-face'. She continues, 'People really began to pull together. They forgot their animosities and got on with the job of understanding a very complex and difficult project that was in deep trouble. There's the nice human side of the story in all this.'

Scientists' perceptions about HST have changed, too. Despite the difficulty of receiving data from the telescope during the first three years, and the hassles inherent in 'deconvolving' the data, astronomers have lined up three deep to use the observatory. HST, in fact, was hugely oversubscribed, even in its impaired state.

Part of HST's human story is the rapid and united effort that scientists and engineers made to design repairs for HST's faulty mirror. In particular, the Space Telescope Science Institute played a vital role in pushing for a refurbishment mission. The day after spherical aberration was announced to the world, Space Telescope Science Institute astronomer Holland Ford began pressing for ways to repair the optics. After some months, an HST strategy panel was formed. Ford and another Institute astronomer, Bob Brown, chaired the panel, which included Lyman Spitzer and Jim Crocker. The panel's charge: to find a 'fix' for HST's mirror.

According to Spitzer, all possible ways to deal with the mirror were up for consideration. There were even proposals to bring the telescope back to Earth, a dangerous technical undertaking: 'We knew that some highly placed people were pushing for bringing the telescope back down,' he said. 'One of the main arguments that we had against it was a political one – that once we got it down, could you ever get it back up again?'

The Corrective Optics Space Telescope Axial Replacement (COSTAR) was the most palatable idea proposed to the panel. Basically, COSTAR was a device that would take the light coming from the aberrated mirror and 're-focus' it for use by the spectrographs and the Faint Object Camera. However, the incorporation of COSTAR required removing the High Speed Photometer, which meant expanding an already crowded servicing mission schedule. Nevertheless, the panel presented the COSTAR proposal to NASA, which then evaluated the idea and approved it.

The 'fix' for the Wide Field and Planetary Camera was deceptively simple. It involved a 'clone' called WF/PC-2, which was intended as an emergency replacement should the original fail. Building the clone was underway, although it was not a high-priority development at the time. After a meeting of the repair strategy panel, members of the 'Clone Team' pointed out that WF/PC-2 could be retrofitted to account for the effects of spherical aberration. To do this, one mirror in each optical quadrant of the replacement WF/PC could be refigured. At that point, interest in the WF/PC development mushroomed, and the project was accepted for the first servicing mission.

## Memories of a mission

Everyone has memories of 'defining moments' in the history of the Hubble Space Telescope. For some scientists, those moments will take place during hallway conversations and paper sessions at professional meetings for years to come. Others will experience their moments when they glimpse a distant quasar or galaxy frozen in an HST image for an instant in time. For others, it will be the first time they take spectra of a galactic jet and realize that they have indisputable evidence of another black hole.

**Figure 1.18.** The First Servicing Mission. (NASA–Johnson and Kennedy Space Centers) (a) Shuttle launch on 2 December, 1993.

(b) The Moon and HST from *Endeavor*. (c) The astronaut crew (front row, from left to right) Mission Specialists Claude N. Nicollier of Switzerland and Kathryn D. Thornton (red shirts) and Jeffrey A. Hoffman (white shirt); (back row, from left to right) Payload Commander F. Story Musgrave, Commander Richard O. Covey, Pilot Kenneth D. Bowersox and Mission Specialist Thomas D. Akers. (d) Photograph of HST showing faulty solar cell wing.

**Figure 1.18.** (e) HST in the shuttle bay for servicing.

For many members of the public, those defining moments came at the end of 1993, whilst watching the First Hubble Servicing Mission unfold on television. HST team members converged on the Kennedy Space Center in Florida to watch one of the most anticipated missions in the history of manned space exploration. What they got was one of the best-planned and beautifully executed shuttle missions ever flown.

The atmosphere was restrained but upbeat, and for many the prospect of seeing a 'night-time launch' outweighed the security procedures and regulations surrounding any visit to the space center. The launch, on December 2, 1993 went like clockwork and space shuttle *Endeavor* lifted off a fraction of a second after the opening of the launch window at 4:47 a.m.

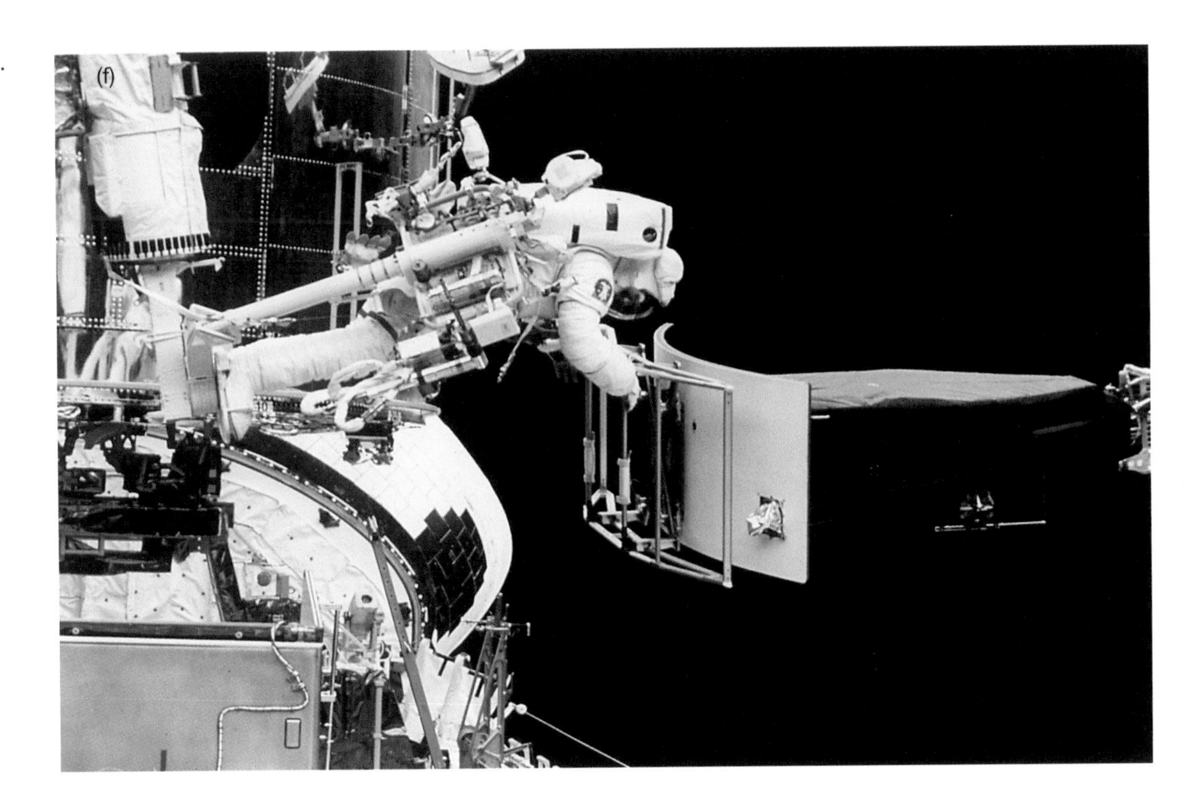


(f) The installation of WF/PC-2.

The sounds and images of astronauts working in space made the First Servicing Mission a very exciting and dramatic television experience. For 11 days, Story Musgrave, Jeff Hoffman, Kathryn (K.T.) Thornton and Tom Akers entranced the world with their extra-vehicular repair expertise while Dick Covey, Ken Bowersox and Claude Nicollier worked the mission from inside the shuttle. They had 11 tasks to accomplish, and they were so well-trained that they made a tough job look easy. In the process of repairing HST, they also went a long way toward repairing NASA's tarnished public relations image.

## Anatomy of a servicing mission

Refurbishing HST was more than an 11-day space mission. It had its roots in HST's earliest planning days, when engineers came up with an on-orbit telescope design that would allow for several such servicing missions. At first launch in 1990, plans called for at least three such missions during HST's 15-year lifetime.

A servicing mission is not unlike having a servicing plan for a car. Owners who want to maintain their cars in good working order usually plan on replacing critical components, such as the tires and brakes, at regular intervals. The owner might also upgrade the sound system, or put in new upholstery. Personal computer owners go through this sort of exercise too. In the early 1980s, the first computers for home use were either Apple or CP/M-based systems. Then DOS-based systems came on the market, and since that time all

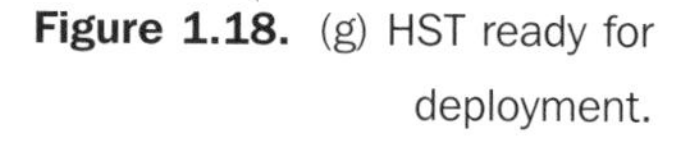

**Figure 1.18.** (g) HST ready for deployment.

computers feature faster operating systems, color monitors, bigger storage drives and more efficient keyboards. Today's computer user can buy CD-ROM drives, sound cards, speakers, screen-saver software and an amazing array of other applications to optimize computer use. Of course, maintenance of these complex computers is important, and, over time, upgrading these systems becomes equally important.

It was the same way with HST – parts wore out, became older and needed replacing. Mission planners knew there would be new advances in computer technology and camera designs. At least two servicing missions were planned to 'swap out' instruments: HST's 'modular' design allows for upgrades as new technologies become available. HST's

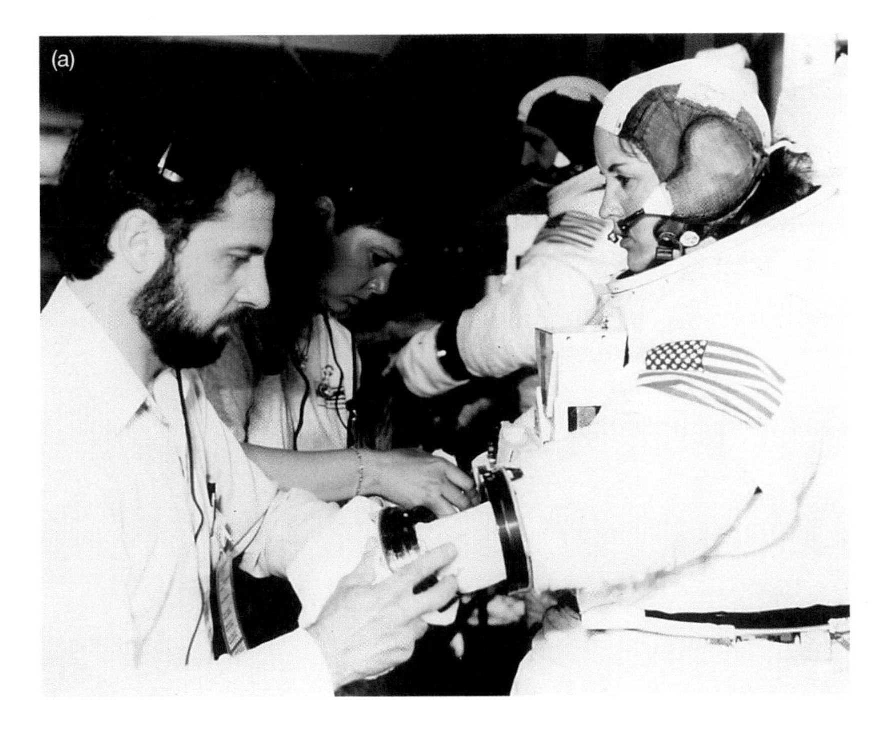

**Figure 1.19.** Training for the First Servicing Mission. (NASA-Johnson Space Center) (a) Suiting up for practice. Astronauts Kathy Thornton and Tom Akers get ready for an underwater rehearsal session. (b) Wearing training versions of the shuttle's Extravehicular Mobility Units, Thornton and Akers practice on a full-scale mockup of the Hubble Space Telescope in a neutral buoyancy simulator at the Johnson Space Center.

designers also made it possible to improve the computer and the solar arrays. What they did not foresee, of course, was the spherical aberration. That problem, along with a flapping solar array, made it absolutely necessary to get the First Servicing Mission (FSM) off the ground as soon as possible.

Detailed planning for the FSM began in the summer of 1990. As problems developed, they were added to the growing 'laundry list' of things the astronauts would repair or replace. The astronauts had 11 tasks to complete, and they trained for a year to learn how to accomplish them. In neutral buoyancy tanks in Alabama and Texas (Figures 1.19b and c) and on frictionless floors at the Johnson Space Center in Houston, the astronauts learned how to work in simulated reduced gravity with the various components they would use during the mission. At the Goddard Space Flight Center, they inserted the *actual* flight hardware into an accurate model of HST.

The astronauts even took advantage of virtual-reality computer simulations to achieve the most realistic feel of what it would be like to work with an orbiting telescope in a

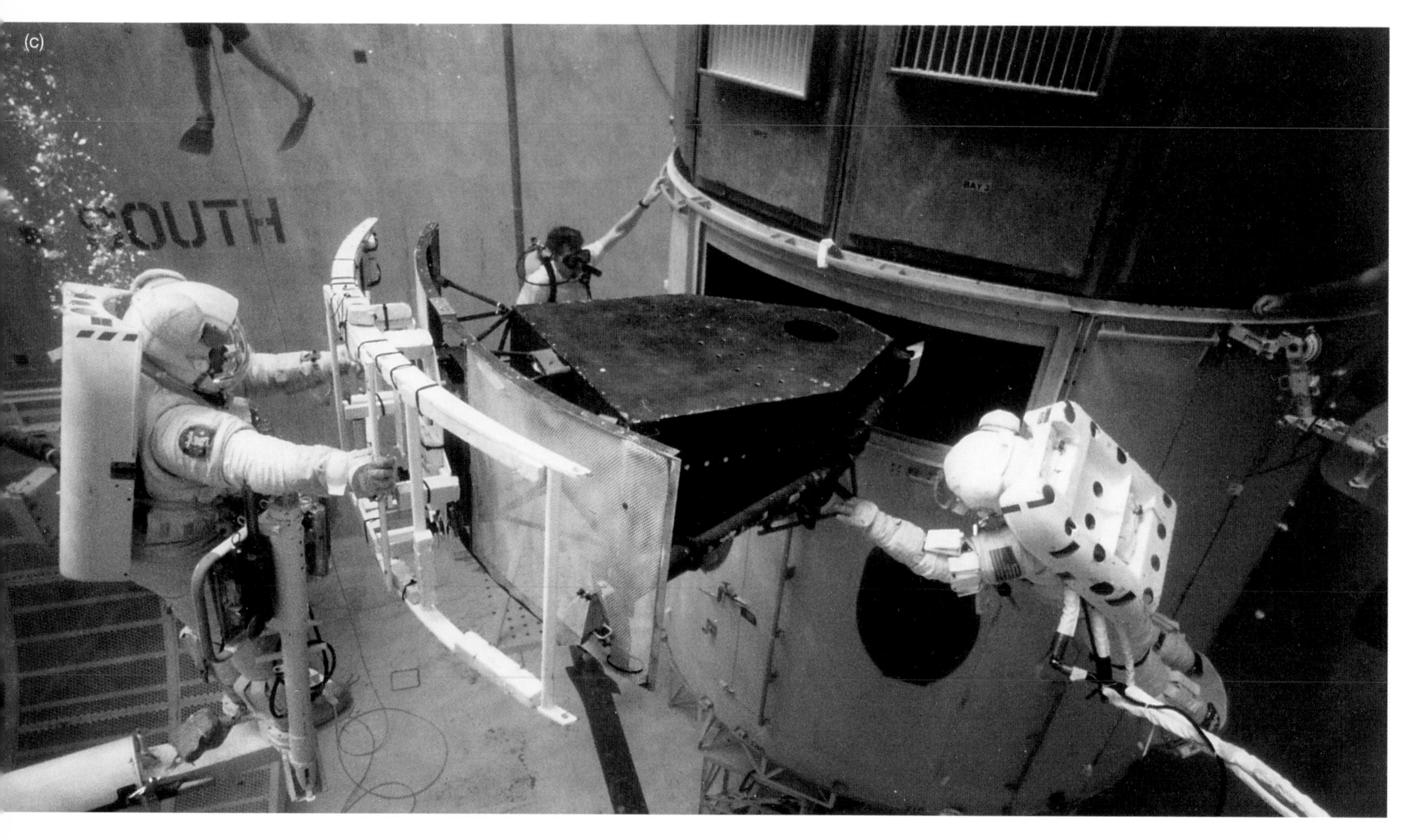


**Figure 1.19.** (c) Astronauts Story Musgrave and Jeffrey Hoffman in the Neutral Buoyancy Simulator at the Marshall Space Flight Center, Huntsville, Alabama, USA. They are preparing to remove a mockup of a WF/PC from a full-scale model of HST.

microgravity environment. The hundreds of hours of practice and rehearsal contributed to the very smooth and spectacularly beautiful mission. The astronauts accomplished their tasks, often ahead of schedule and with good humor.

What were the main elements of the First Servicing Mission? The three main goals of the mission were to restore the planned scientific capabilities of the telescope, to restore the reliability of all HST systems, and to confirm that HST could be serviced while in orbit. Specifically, the tasks for the First Servicing Mission were to replace the Wide Field and Planetary Camera with the 'clone' instrument; to replace the High Speed Photometer with COSTAR, which would correct the aberration for the Faint Object Camera and the spectrographs; to replace faulty gyroscope packages; to replace the solar panels; to repair the GHRS; and to perform a host of other minor repairs.

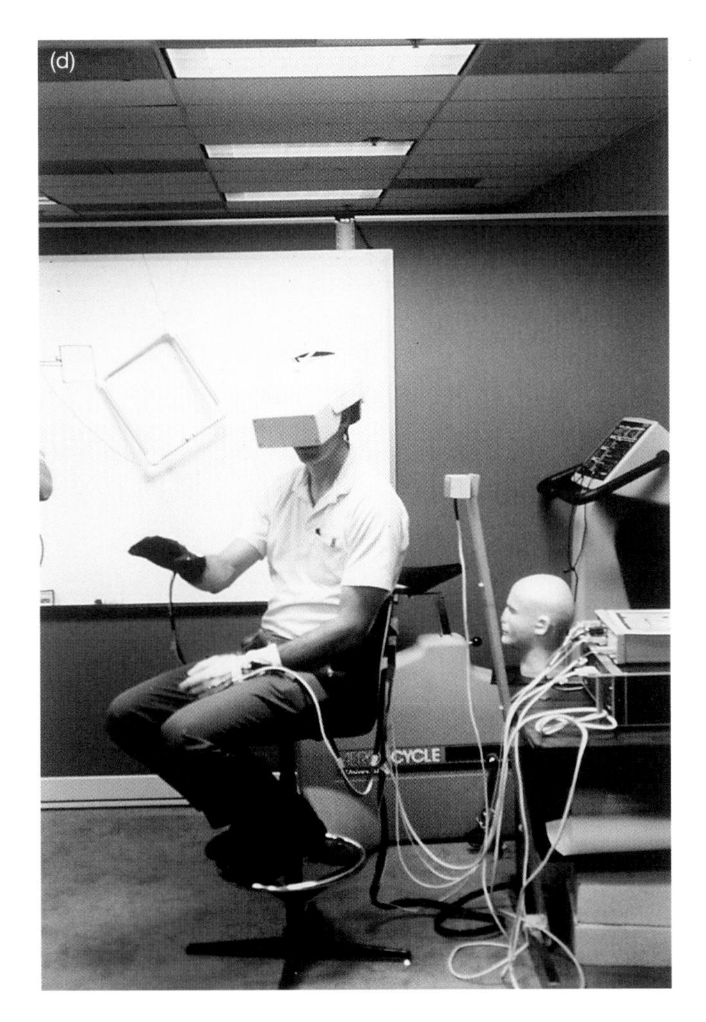

(d) Crew members prepared for the servicing mission using a virtual reality system that assisted in refining positioning patterns for the shuttle's Remote Manipulator System (the robot arm). This was the first time that virtual reality systems were used in training for a shuttle mission. Here, astronaut Jeffrey Hoffman uses a special helmet and gloves to practice positioning near the telescope while on the end of the robot arm. (e) Swiss scientist and mission specialist Claude Nicollier works with mission pilot Ken Bowersox at the controls of the Remote Manipulator Facility in the Shuttle Mockup and Integration Laboratory at the NASA–Johnson Space Center.

## First light again

After the exhilaration of launch and the servicing mission came the hardest wait of all – waiting to see if HST had truly been repaired. HST Project Scientist Dave Leckrone, in a press conference shortly after the mission, stated that the repair teams had 'finished eye surgery on the patient'. He expected that there would be a six- to eight-week wait until controllers could take off the bandages and test the patient's eyesight. Everyone waited while controllers learned to use the newly refurbished instrument and began the ticklish process of deploying the corrective optics.

For a small group of scientists who were lucky enough to be at the Space Telescope Science Institute very early one mid-December morning in 1993, their memories are forever linked to the moment when the 'first-light' image from the newly installed WF/PC-2 appeared on a computer screen. To the uninitiated eye, it looked like a hash of pixels,

**Figure 1.20.** The scene at the Space Telescope Science Institute when the first images were received after refurbishment. (Ray Villard, STScI)

but to the jubilant scientists crowded around the monitor it was a *focused* hash of pixels. After three and a half years of frustration, hard work, training and planning, HST was capable of doing all the science it was sent to do, and more.

Following the servicing mission euphoria, telescope controllers started performing basic tests of HST's new instruments and capabilities. Operations specialist Bill Hathaway judged the post-servicing mission orbital verification period as much more relaxed than the first harried moments of HST's original first light: 'A lot of the how-to and who-does-what-when procedures were quite routine by then,' he wrote. 'This time there was a lot more confidence that we could handle the verification.'

Like others at the Institute, Hathaway had fears that the mission would not work. These concerns were laid completely to rest by the success of the mission, but, until that time, he was prepared for the worst. 'The real uncertainty of November 1993 [before the mission] was greatly dispelled by the incredibly smooth job the astronauts pulled off,' he said. 'Myself, I had a 23-foot van in my back woods so I could, if needed, pack all my stuff and move back to the wilds.'

Like many people, Program Scientist Ed Weiler remained skeptical that HST was completely repaired, even after the successful mission. Two weeks after the astronaut crew returned to Earth, an extremely detailed raw image of a distant galaxy arrived late on New Year's Eve. At that point, even Ed had to admit that the repair was a success. 'When that image came up, even us skeptics were convinced that we had fixed the telescope,' he said. 'That was one heck of a way to start a new year and a new era for HST.'

Bill Hathaway's impression was less restrained. 'You've probably seen the [video] tape of the sharp, electrifying reactions of the scientist to the first focused star image,' he wrote. 'Me, I was more impressed by the set of images of familiar galaxies and clusters and nebulae. It felt almost too easy... [but] the tingles and wonder were well deserved.'

Most of the world received its first look at the spectacular images from the newly refurbished Hubble Space Telescope during a happy media event on January 13, 1994 – nearly three and a half years after the depressing spherical aberration press conference of June 27, 1990. NASA dignitaries proudly commended the teams of scientists and astronauts who made the repair mission a success. Maryland Democratic Senator Barbara Mikulski, a strong supporter of the Space Telescope Science Institute and the Goddard Space Flight Center, while at the same time one of the agency's toughest critics, proudly presented the first images to the world. She was exuberant as she announced the end of Hubble's troubles. 'The pictures are remarkable,' she said. 'The science that will come from these pictures is of historical significance. We will now be able to do twenty-first-century science.'

At the same press conference, Dave Leckrone pronounced HST fixed. 'Back in December, I told you that we had finished eye surgery on the patient and that it would be six to eight weeks before we could lift the bandages,' he said. 'The bandages came off ahead of schedule, and our patient has a new vision of incredible clarity.'

There is no doubt about it – the early results from HST were sensational and demonstrated just how clear Hubble's vision has become. The suite of images released in early 1994 shows just exactly how far HST has come from the dark days of 1990.

Because it was important to show just how much the telescope's vision had improved, one of the first images released was of a single star. The Faint Object Camera showed a

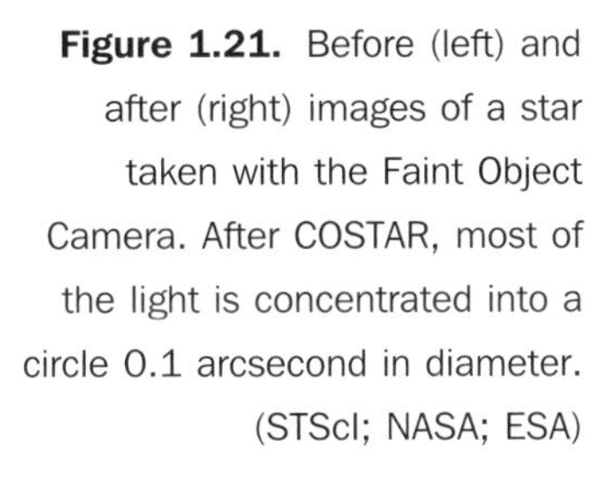
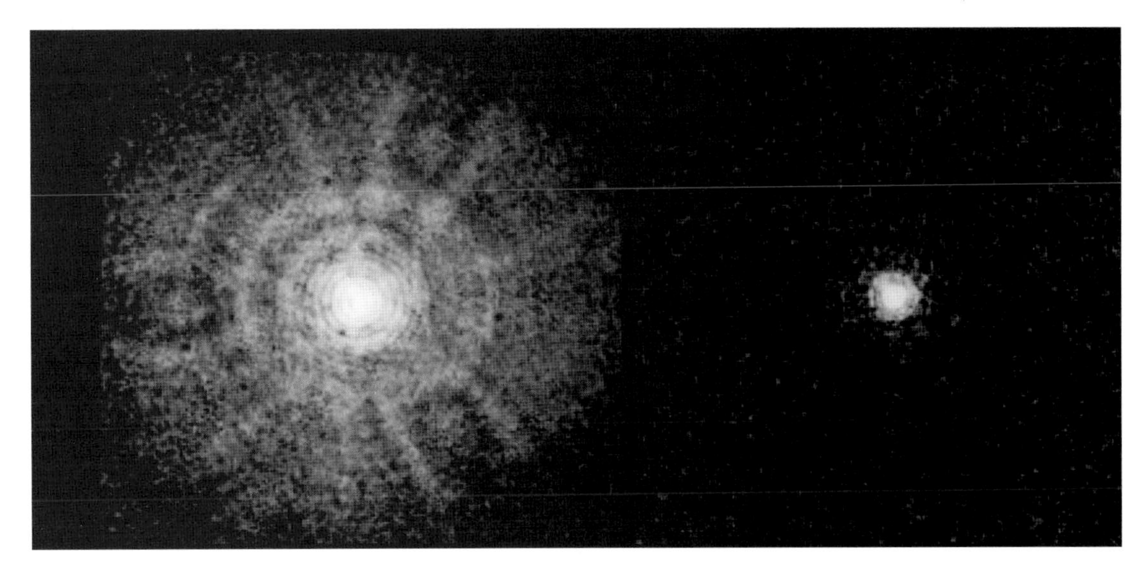

**Figure 1.21.** Before (left) and after (right) images of a star taken with the Faint Object Camera. After COSTAR, most of the light is concentrated into a circle 0.1 arcsecond in diameter. (STScI; NASA; ESA)

point-like light source. This in itself would not be much cause for excitement, but right next to it was an image of the same star, taken before the corrective optics (COSTAR) had been installed. The difference was dramatic, as can be seen in Figure 1.21. Where light from the star had once been spread out over a circle 1 to 2 arcseconds in diameter, the telescope's corrective optics concentrated the light into a circle 0.1 arcsecond across. At a Ball Aerospace press conference in early January, 1994, scientist Charles Pellerin summed up the improvement: 'When we put the servicing mission together, we had a goal of 60% encircled energy,' he said. 'The original HST was designed for 70%, and if we had gotten that 70% we would have been enormously happy. The theoretical maximum is about 85 to 87%. The last word that I got is that we're estimating at 84% performance, which is incredible.'

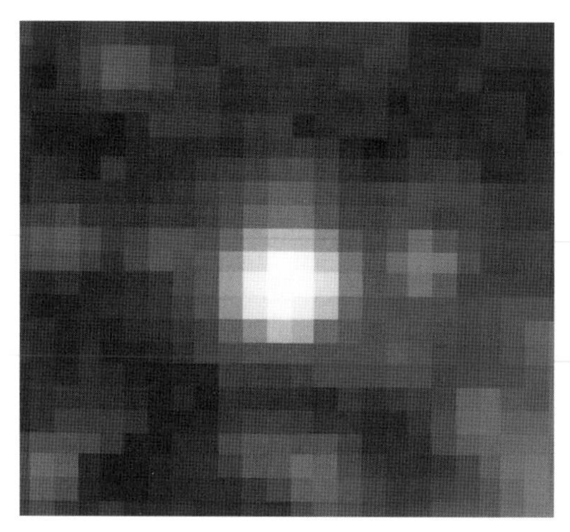

Ground image at 0.6 arcsecond resolution

WF/PC-1 image (before FSM)

WF/PC-2 image (after FSM)

**Figure 1.22.** The star Melnick 34 in 30 Doradus as seen from the ground, by WF/PC-1 and by WF/PC-2. The improvement in image quality is clearly demonstrated. (STScI; NASA; ESA)

Figure 1.22 shows the first results from WF/PC-2; the 84% encircled energy is clear evidence of the improvement in HST's vision. However, it is one thing to look at a star and see the difference between the ground-based views and the 'old' and 'new' images from HST, but how about some other things, such as galaxies and star clusters and stellar explosions – what some astronomers have dubbed the 'hot stuff' of the universe?

Tens of millions of light years away from Earth lies a group of galaxies called the Virgo Cluster. In Burnham's *Celestial Handbook*, this area of the sky is described as a place where 'one may gaze upon the radiance of a hundred vastly remote star cities, twinkling across the millions of light years'. Stargazers with telescopes can browse at will through the Cluster, exploring a wealth of galactic treasures. But what if you had HST at your disposal and could target one of these galaxies for serious study?

On November 30, 1993, just a few days before the First Servicing Mission, HST looked at the spiral galaxy M100, located in the upper portion of the Virgo Cluster. Figure 1.23

**Figure 1.23.** A WF/PC-2 image of nearly all of M100, a spiral galaxy in the Virgo Cluster. The image is about 2.5 arcminutes across. (STScI; NASA; ESA)

shows the entire galaxy, taken after the mission. The photograph on the left in Figure 1.24 shows a blurry, somewhat indistinct, image of the central portion of M100, taken with WF/PC-1, whereas on the right is the WF/PC-2 image taken on December 31, 1993. The difference between the two images illustrates the extraordinary clarity of HST's repaired optics. Figure 1.25 shows a starfield and clear images of individual stars. These images established the fact that Cepheids can be seen in M100. So, astronomers will be able to use the distance determinations we discussed earlier to find a more accurate value for the Hubble Constant. At some point, HST's gaze will be turned to the dynamic core of this beautiful galaxy, where it appears the complex process of stellar formation is taking place.

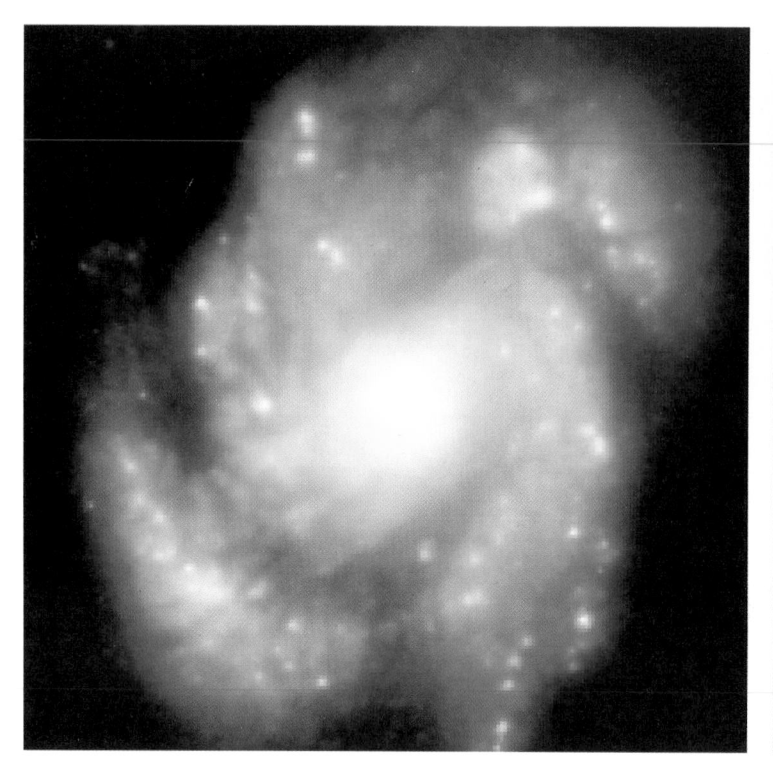

**Figure 1.24.** The central region of M100, taken with WF/PC-1 on 27 November, 1993 (left), and WF/PC-2 on 31 December, 1993 (right). The improved image quality allows HST to cleanly resolve structure as small as 30 light years across. (STScI; NASA; ESA)

Astronomers have long talked about the possibility of black holes hiding at the centers of some galaxies. Confirmation of the existence of these cosmic 'beasts' is one of HST's jobs. Figure 1.26 shows a 'before-and-after COSTAR' sequence of the galaxy NGC 1068 (a Seyfert-type galaxy to be discussed in Chapter 5). HST has looked at this galaxy a few times, and from the data received astronomers know it sports a massive complex of ionized clouds at its heart. These clouds are heated by something at the core of the galaxy, and the most likely suspect is a 100-million-solar-mass black hole. HST will study finer details of these clouds and will enable astronomers to compare the ultraviolet light and visible light from the clouds. Eventually, the resulting data should confirm that this galaxy, like others HST has observed, is home to a black hole.

HST also took a clear look at a star cluster somewhat closer to home: the R136 stellar complex in the 30 Doradus Nebula. This area of the sky is a densely populated region in the Large Magellanic Cloud, the Milky Way's closest galactic neighbor. HST's new vision gives an unprecedented look into the center of the cluster of hot, young stars. Where once HST could only resolve a few hundred stars, WF/PC-2 shows more than 3000 stars crowded together.

One of the most memorable events observed by HST was the collision of Comet Shoemaker-Levy with the planet Jupiter. During an 8-day period from July 16–22, 21 large comet pieces smashed into the atmosphere of Jupiter, creating large dark scars across

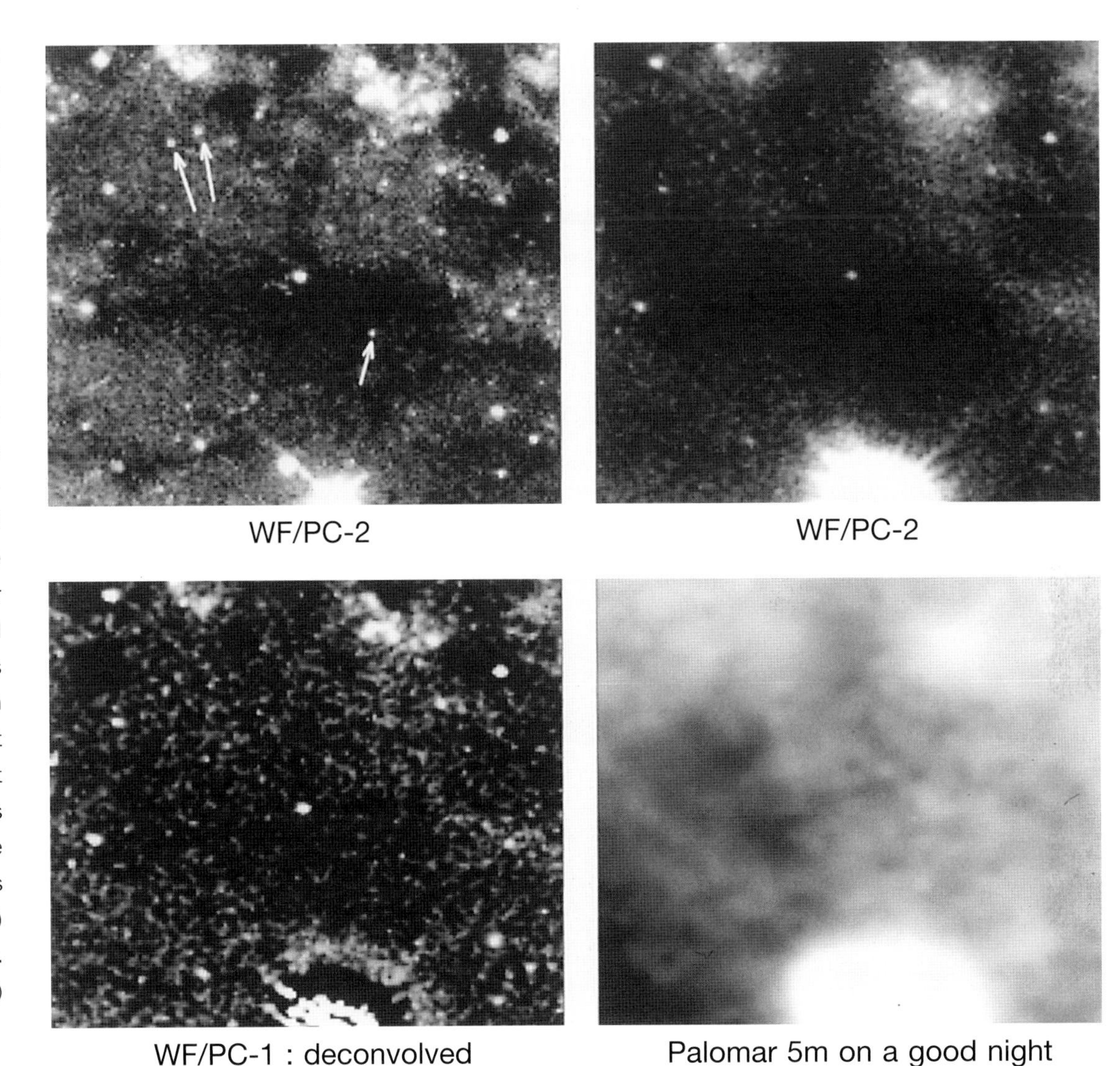


**Figure 1.25.** Four views of a starfield in M100 showing the improvement from a ground-based image to a WF/PC-2 image. The upper left image, taken by WF/PC-2, illustrates the instrument's ability to detect and measure the light from individual faint stars in distant galaxies. The upper right image was taken by WF/PC-1 just a few days prior to the First Servicing Mission. The lower left image is a computer-processed WF/PC-1 image, and the lower right image is the same field seen with the Palomar 5-meter telescope on a good night. The sequence also shows that while deconvolution techniques can sharpen bright features, deconvolution cannot recover faint objects. The arrows in the WF/PC-2 panel indicate stars approximately as bright as the distance indicators (Cepheids) that will be studied in M100. (STScI; NASA; ESA)

Jupiter's lower cloud belts. HST looked at the events as they rotated into view, using its cameras and spectrographs to gain optical and ultraviolet views of the disturbances in the clouds. A team of astronomers continues to study the data from HST in an effort to understand the dynamics of cloud–comet interactions at Jupiter, and we present results from their work in Chapter 3.

The Hubble Space Telescope's work continues, with more than 400 observation programs scheduled each year. The next servicing mission is set for February, 1997. Astronaut crews will make any needed repairs, take out the Goddard High Resolution Spectrograph and the Faint Object Spectrograph and replace them with two 'second-generation instruments' – the Near-Infrared Camera and Multi-Object Spectrograph (NICMOS) and

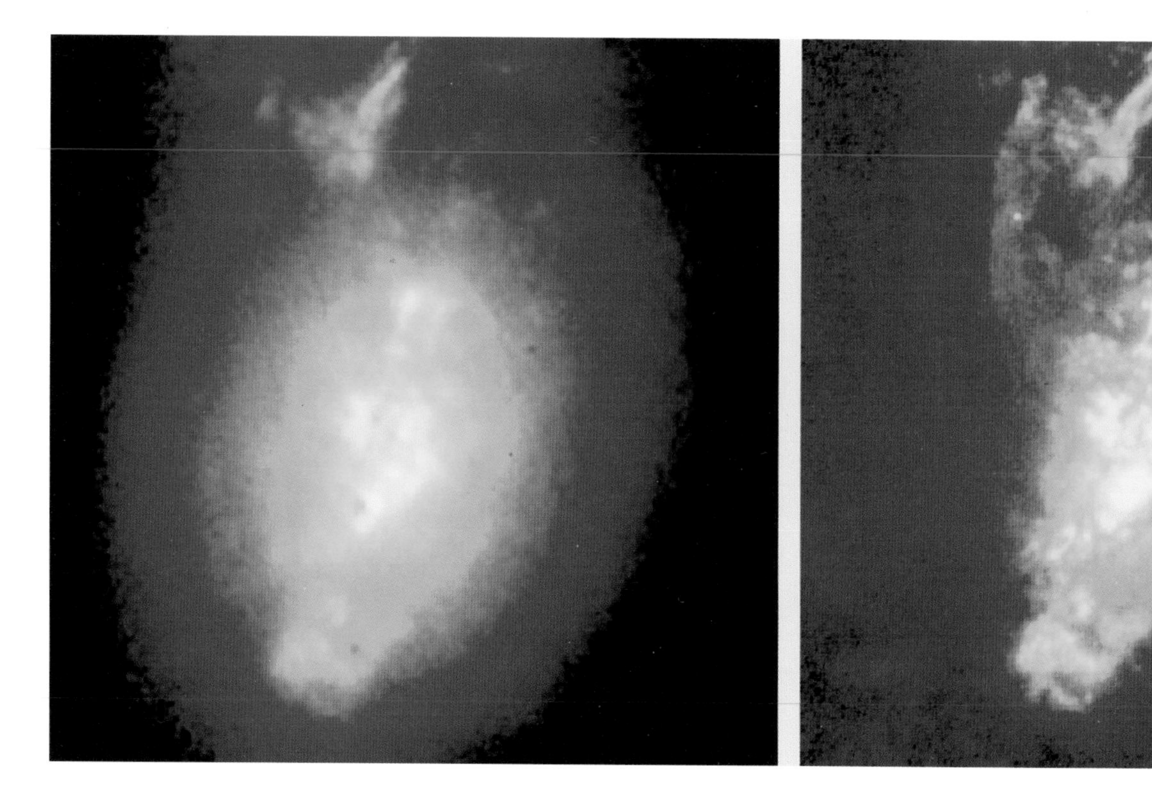

**Figure 1.26.** FOC images of the galaxy NGC 1068, before (left) and after COSTAR (right). With corrective optics, HST was able to detect fine detail in this energetic region of the galaxy. (NASA; ESA)

the Space Telescope Imaging Spectrograph (STIS). Also on the drawing board is an Advanced Camera (AC), which may be installed in HST during a servicing mission in November, 1999. All the instruments will contain optics to correct for the spherical aberration. These replacements are still being designed or constructed, and their exact deployment may be affected by the failure of one or more instruments already on board HST.

The odyssey of Hubble Space Telescope has been a long, exciting and sometimes depressing one. It has gone from being a scientist's dream to an expensive 'Big Science' project in a little over 40 years. During that time, HST has been many things to many people – a multi-billion-dollar boondoggle, a complex engineering project, a failure, a 'techno-turkey', a miracle machine, a job, an obsession, a whipping boy, the future of NASA, the death of NASA, the solver of mysteries and the key to the universe. HST's future is much brighter now than when it was first launched, but it is important to remember that the telescope *did* achieve impressive science even during its first three and a half years of life. Proof of HST's success comes from increased public acceptance of the telescope and its accomplishments

The scientific rewards that HST promised are at hand. As we will see in later chapters, the telescope is busy measuring stellar winds, scanning planetary surfaces, watching supernovae pop off, studying galactic structures, confirming the existence of black holes,

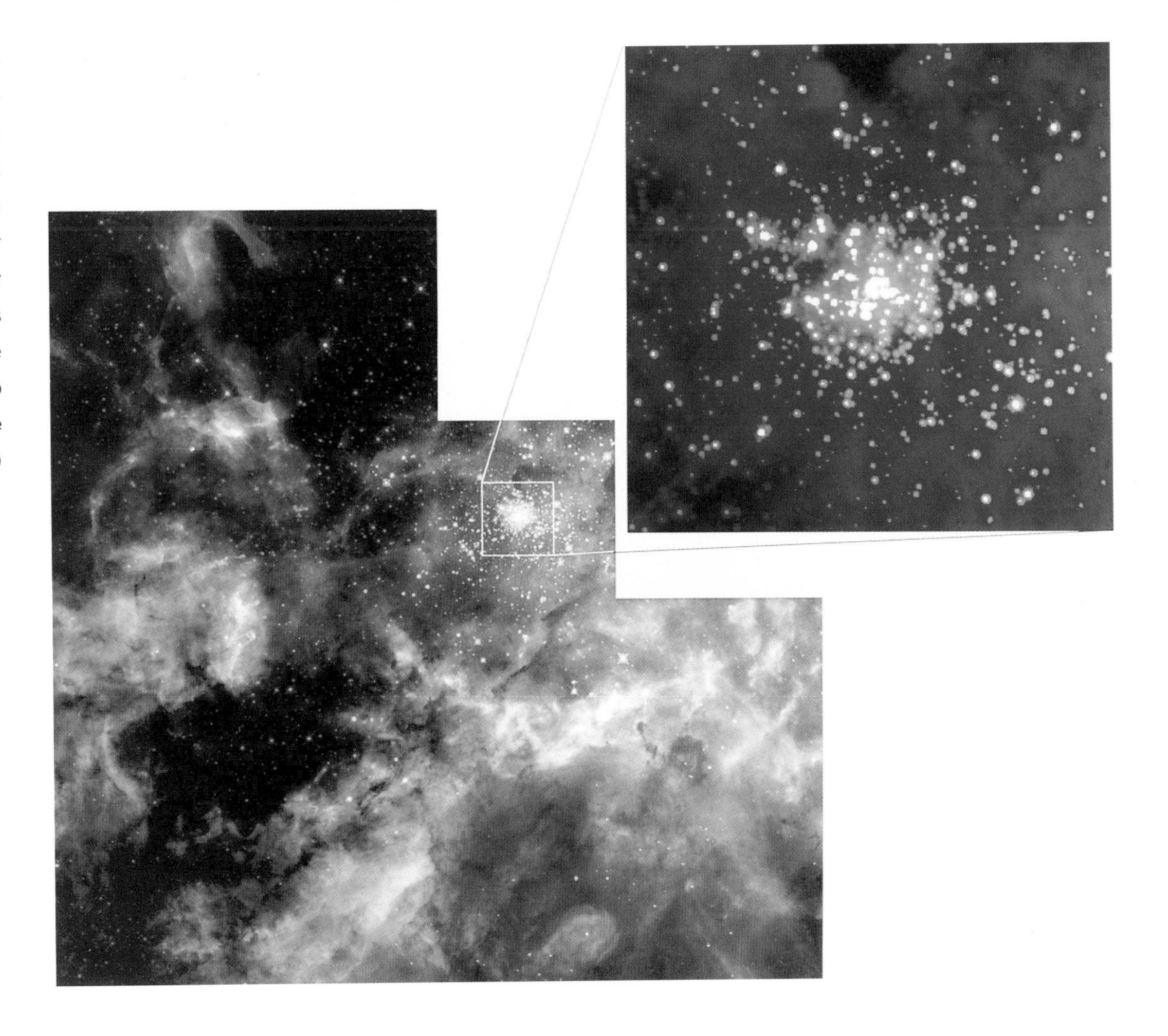

**Figure 1.27.** A WF/PC-2 image of the central region of the 30 Doradus Nebula. The blown-up image shows R136, a dense cluster of hot young stars at the center of 30 Doradus. The cluster is imaged with the Planetary Camera at full resolution. Clouds of gas and dust surrounding the star cluster R136 are being lit up by ultraviolet light from the cluster. (STScI; NASA; ESA)

determining accurate distances to faraway objects and figuring out just what powers quasars.

It is an exciting and rewarding time to be an astronomer. The less tangible rewards are found in the resiliency of the human spirit – of the determination that scientists and engineers and politicians and administrators showed in the face of continuing adversity throughout the program. Ed Weiler, in an interview before the First Servicing Mission, said: 'My greatest reward on this program will be able to get in front of my friends in the media who turned out not to be so friendly and remind them that three and a half years ago when we said we had a problem, we also told them we could fix it by the end of 1993. Three years ago, HST was called a techno-turkey, a disgrace, a dead satellite. But, we worked very hard to get science out early and fast. It is a tribute to everyone's hard work that Hubble has generated more positive press stories in the last two years than all of the NASA satellites combined.'

HST's age of discovery can be likened to Galileo Galilei's astronomical accomplishments. Today, there are hundreds of Galileos, pointing the Space Telescope in all directions, finding mind-boggling things. Because of HST and the people who use it, our understanding of the universe has changed forever.

# 2 Observing the universe

> Light from distant places has made the journey to earth, and it falls on these new eyes of ours, the telescopes, the all-frequency time machines. . . with these new eyes, we can make voyages as dramatic as those of the early navigators, out to the limits of the known universe.
>
> *Michael Rowan-Robinson*

Many years before Hubble Space Telescope's odyssey began, a father and daughter stood outside on a balmy autumn night, looking up at the sky. All day the girl had heard the adults in the house talking about a strange new thing in the sky – something called 'Sputnik'. She begged her father to take her out and show 'Sputnik' to her. Finally, when it was dark enough, the two went out, and he tried to tell his daughter that Sputnik was up there, but that it would be hard to see. All his daughter could see were stars, even though she was not sure what they were. What *he* saw was a little girl who wanted something she could not understand. But they stayed out there for a while that night, looking for a bright, shining point of light, moving across the sky – Earth's first artificial satellite. They never saw it, but it was her introduction to the night sky, and it expanded her universe for ever.

Nearly 40 years have passed since Sputnik raced skyward, and humans have been sending instruments and animals and people into space on a regular basis ever since. Like that of the little girl, our understanding of the cosmos changes with every space mission. Not only is the universe more complex than we ever dreamed, but the instruments we use to study it have evolved from a simple telescope into intricately interlocked systems of optics and electronics.

The complexity of the cosmos is reflected throughout the universe, from the microscopic world of the atom to the macrocosmic domain of the galaxies. To fully appreciate that breathtaking range, we are compelled to transform it into units we can understand. The simplest are, of course, the ones by which we measure our daily lives. We wake up at home, we leave to go to work or school, we travel from one country to another. On a slightly larger scale is the Solar System. From there, we cast inquisitive eyes from our little

**Figure 2.1.** The size and scale of cosmic systems. At the bottom, astronauts are servicing the Hubble Space Telescope. This scene is just a point within the cube containing the Earth, which is just a point within the cube containing the Solar System. In turn, our Solar System is just a point within our stellar neighborhood, which is a point within an arm of our Milky Way Galaxy. The Milky Way is a point within our galactic neighborhood, which finally is just a point in the entire universe. (Dana Berry, STScI)

planetary niche to the stars in our own galaxy. Beyond the Milky Way, other galaxies wheel through the cosmos in clusters. Just at the faintest edge of our astronomical eyesight, we find evidence of the birth of the universe. Of course, what we are really searching for is an understanding of the processes that constantly change the cosmos.

We see the seeds of change, of evolution, every day. An earthquake destroys a city, rains wash away homes and hurricanes devastate coastal areas. Great sprawling cities and highways, dams and bridges change the face of the Earth as well. On the surface of a planet like Mars, we find the evidence of a watery past, completely different from what exists today. In one part of the sky, an immense cloud of gas and dust turns out to be the birthplace of stars, while elsewhere an ominous-looking cloud is all that is left of a supergiant star that died in a violent cataclysm. In the beauty of a nearby galaxy, we see the seeds of its birth.

## Astronomy and light

Light is the Rosetta Stone of astronomy – the guide to understanding the complexity of the universe. Locked within the light that surrounds us is an incredible amount of information about the objects that radiate and reflect it. We can learn a great deal of information about a light source – ranging from its temperature and chemical makeup, to its speed and direction, as it travels through the universe. The history of astronomy itself can be characterized as the history of humans learning how to decode the mysteries of light. Once we knew light was the key to the universe, we developed a vast array of devices to capture and analyze it in all its forms.

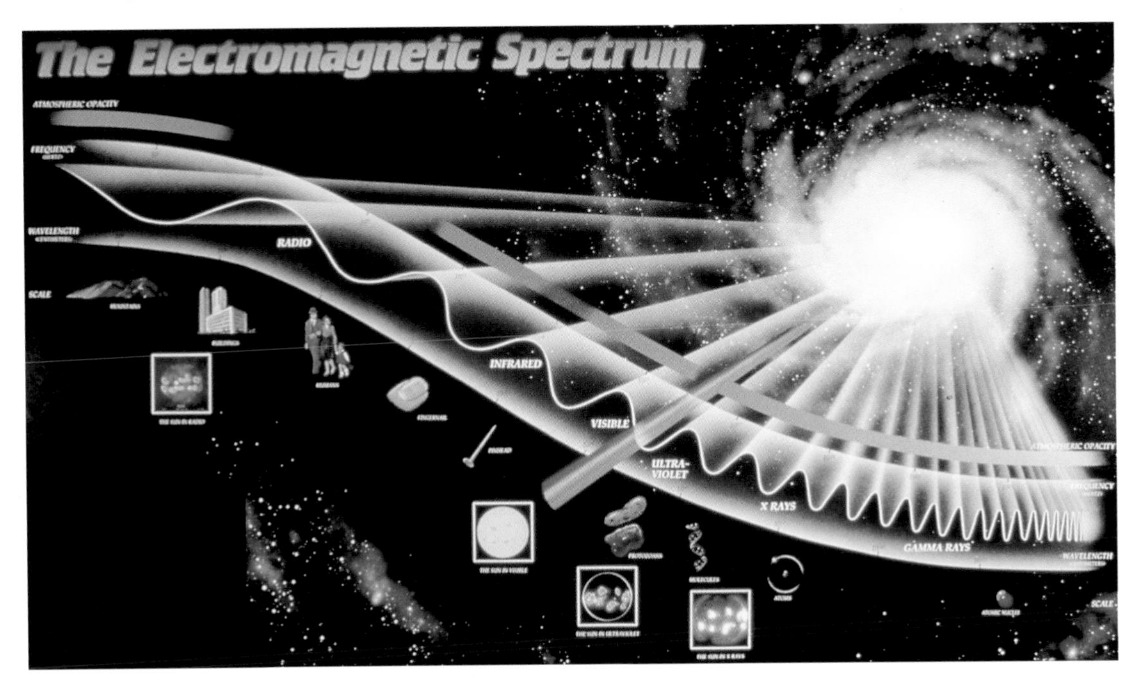


**Figure 2.2.** The electromagnetic spectrum (EMS). The schematic diagram shows the entire spectrum of electromagnetic waves, relevant scales and objects. The atmospheric opacity, which determines what radiation reaches the Earth's surface, is shown above as a silvery band. Visible light and radio waves reach the surface, and most other waves do not. (Dana Berry, STScI)

The light most familiar to us is what we see with our own eyes. This is called 'visible light'. It is part of a larger array that astronomers call 'light', which runs from the shortest wavelengths of about $10^{-8}$ meter to the longest wavelengths of 3 millimeters. Astronomers' 'light' is, in turn, part of a larger array called the electromagnetic spectrum (commonly referred to as the EMS). Astronomers also refer to very small wavelengths of light by another unit – the angstrom, which is equivalent to $10^{-10}$ meter. Thus, the shortest wavelengths for light would be 100 angstroms. The EMS encompasses everything from gamma-rays, X-rays and ultraviolet light through visible wavelengths, infrared and radio waves. To put it into perspective, imagine if all the wavelengths of the electromagnetic spectrum – up to radio wavelengths of 100 meters – could be spread out across an area about the length of an American football field (91.4 meters). Visible light – the light we can see with our eyes – lies in a band between 4000 angstroms and 7000 angstroms. This band is 0.0000003 meter wide – much narrower than one of the blades of grass on that imaginary football field.

Light, however, is not totally defined as a collection of wavelengths. Certainly it can behave as a wave, but it can also act as a particle. Actually, it can act as both a wave and a particle at the same time; the only difference lies in how we observe it. A particle of light is called a *photon*, which can be thought of as a packet of energy. Each photon has an energy value, given in units called *joules*. The photon also has an underlying wave nature – i.e. it exhibits the characteristics of its wavelength. Red light, for example, has a wavelength of about 6500 angstroms; a photon of that red light has about $3 \times 10^{-19}$ joules of energy. The common light bulbs used in our homes radiate yellowish-white light, which consists of photons that are just a little more energetic than red light. When you turn on a 100-watt light bulb, it emits about $10^{20}$ photons every second and dissipates the rest of its energy as heat.

To form some idea of the duality of light, imagine going down to the beach on a warm summer's day. As you sunbathe, photons of visible light from the Sun hit your eyes and enable you to see your surroundings. You also feel infrared wavelengths as warmth from the Sun, and some ultraviolet radiation burns, or tans, your skin. The Sun also emits higher-energy ultraviolet and gamma radiation, but you do not experience it as you lie there on the sand because the Earth's atmosphere screens it out.

Let us now turn that experience to stargazing. You stagger outside with your observational accoutrements – telescope or binoculars, star charts, hot drinks, chocolate bars, blanket, radio or CD player – and start an evening of observation. You are very likely going to be observing objects that radiate in the visible wavelengths of light. If you watch the Moon or the planets, you are looking at reflected light because these objects do not generate their own light. Stars and galaxies provide you with radiated light because they do generate their own photons.

It is not much different for professional astronomers (and the many advanced amateurs out there!) who simply use bigger telescopes and more complex instruments to study the universe. But, instead of limiting their view to visible wavelengths, these astronomers try to study every wavelength of light, using both ground-based and orbital instruments.

## Astronomy: the observational science

The principle behind the array of astronomical instruments in use today is fairly simple: gather as much light as you can from objects in the universe, and analyze it. What sort of light you gather depends on what you are using to gather it. There is no ideal light detector; no instrument that senses all wavelengths of light perfectly. So astronomers use different detectors to study light of various wavelengths.

Telescopes are the most familiar astronomical tools around. At its heart, a telescope is simply a device that gathers electromagnetic radiation – usually in the form of ultraviolet, visible, infrared or radio wavelengths. For 'optical' devices, mirrors at the heart of the instrument focus and reflect light to other sensors, such as film cameras and electronic recorders called CCDs. A CCD (charge-coupled device) is a specialized camera used on a wide variety of telescopes. Within this camera is an extremely sensitive chip that gathers light across the 'picture elements', or pixels, of its surface. The image is 'read out' from the chip and transferred to an electronic storage medium. Like other cameras, the CCD captures an astronomical event at a single fixed point in time. From the images produced by cameras, astronomers can learn about such things as brightness of an object, its shape, location and relationship in space to other nearby objects.

The bigger the telescope you build, the more light you can gather. However, some very practical considerations limit the size of the instrument and its placement. To begin

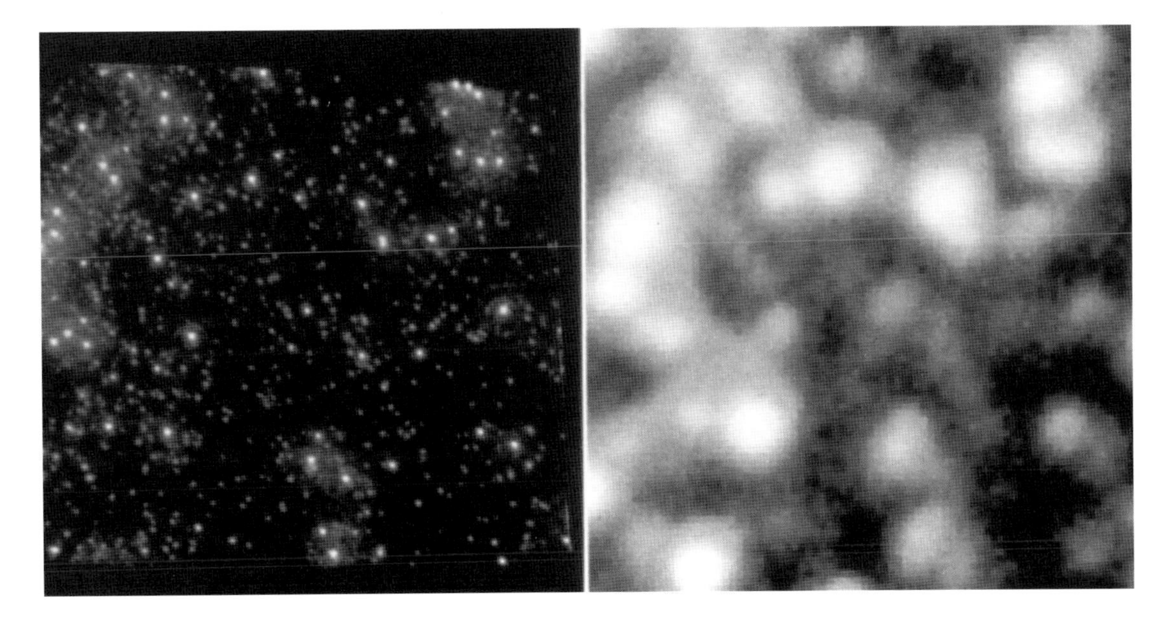

**Figure 2.3.** This pair of images probes the inner regions of the globular cluster M14 (located 70 000 light years away) and contrasts the differences in spatial resolution between HST and ground-based telescopes. The image on the right was taken with the 4-meter telescope at Cerro Tololo Inter-American Observatory in Chile and has a resolution of 1.5 arcseconds. The pre-COSTAR image on the left comes from HST's Faint Object Camera and yields stellar diameters of 0.08 arcsecond. (STScI; CTIO)

with, any telescope mirror has limits on how big it can be. A large mirror can gather more light, giving its users a chance to look at dimmer and more distant objects. However, mirrors cannot be made infinitely large because the structures to support them would have to be infinitely large as well. The shape of a mirror also limits its ability to concentrate incoming light to a tight focus, and large mirrors bend and sag under their own weight.

Observatories function best when they are located away from sources of light, heat and radio wave pollution. Ideally, observatories should also be at high elevations to minimize the effects of Earth's atmosphere on the incoming light. Many of the world's facilities are located on mountain tops (such as the collection of telescopes at Mauna Kea, Hawaii; Cerro Tololo, Chile; and Pic du Midi, France). Others, like the Very Large Array of radio telescopes near Socorro, New Mexico, are located well away from cities in wilderness areas and deserts. The latest twist on 'getting away from it all' involves the use of space-based observatories. Spacecraft – both Earth-orbiting and so-called 'fly-by' missions – have carried a variety of instruments and sensors to study light of wavelengths that are filtered out by the Earth's atmosphere.

Basically, astronomical observations can be divided into three categories: *imaging*, *photometry* and *spectroscopy*. In imaging, the light is recorded on film or by an electronic camera such as the CCD we mentioned earlier. Sometimes a telescope is pointed at a field crowded with objects, such as a star cluster. If we are studying just one star in the cluster, it helps if the instrument can separate one star from another. This is the concept of *spatial* resolution – the ability to distinguish objects in a crowded field from each other and to produce clearly defined images of them.

Photometry is the practice of measuring the intensity of light from an object. A photometer can be thought of as a very sensitive light meter, similar to that used in flash photography. Photometers are used to determine fluctuations in light intensity from a variable star, for example. The determination of intensity of light from stars, and other objects in the universe, is extremely important because from those studies comes the determination of brightnesses, or magnitudes. Quite often an astronomer will refer to an 8th-magnitude star or a 5th-magnitude comet. What the number refers to is the relative brightness of the object compared to other objects. The brightest stars have the smallest numbers, while dimmer stars have larger magnitude numbers. The brightest star in the night-time sky is Sirius, with a visual magnitude of –1.5; Canopus is –0.7; while Betelgeuse is around 0.5. The dimmest stars we can see with the naked eye range around 5th or 6th magnitude. Determining the magnitudes of stars is an important step toward determining their distances.

Different photometers measure different wavelengths, so there are whole sets of instruments sensitive to infrared and ultraviolet light as well as to visible light. Photometry is a very systematic way to classify celestial objects and to observe how their changes of light output affect the way they appear to us. Before it was replaced by COSTAR, HST's High

Speed Photometer worked in this way, recording the brightness of objects at what is called 'high temporal resolution'.

The concept behind high *temporal* (time) resolution is simple. We now know that some objects in the universe change rapidly, and scientists learn more if they can study an object or event as it evolves. Certainly, telescopes can image something many times over the course of an observation, or for a much longer time than the human eye can gaze at an object, but there are limits to just how many 'pictures' can be taken with ordinary film cameras, plates and single CCDs. So, specialized cameras are used to capture an event as it happens over a period of time. The best systems will give us many 'snapshots' of an event taken within fractions of a second.

Spectroscopy takes light and divides it into its component wavelengths, or disperses it. We are all familiar with some everyday types of spectroscopy: white light shining through a prism and sunlight shining through raindrops to form a colorful rainbow, for example. These examples demonstrate very simply that the Sun's light (or light from any white-light source) radiates all the wavelengths in the spectrum.

Spectroscopy answers a variety of questions that might not be answered if we relied solely on undispersed light to study objects: what is the chemical composition of an interstellar gas cloud? the temperature of a star? the chemical makeup of a comet? the velocity of a gas jet at the center of a galaxy? The methods of spectroscopy allow us to answer these

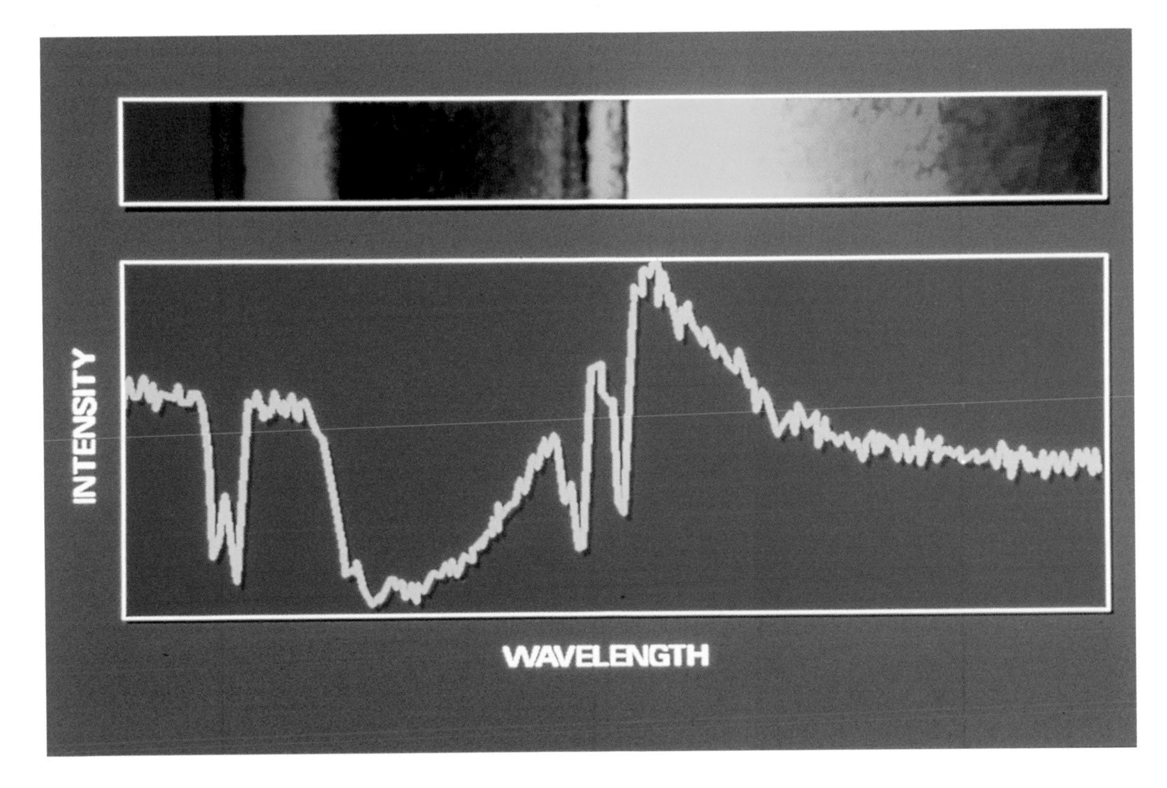


**Figure 2.4.** Half of the data gathered by HST is spectroscopic data. This is what the science data from spectrographs look like. The black lines in the color bar correspond to dips in the graph. The lines create patterns of absorption or emission, which tell astronomers what elements are present in the object being observed. Depending on the position of the lines, astronomers can also tell whether the object is moving away from or toward the Earth, as well as its velocity relative to us. (Dana Berry, STScI)

questions by studying the way in which stars and galaxies, comets and planets emit and absorb light.

The simplest spectrographic tool is the spectroscope, which breaks light down into very fine divisions by wavelength. A prism will work, but most modern astronomical spectrographs – instruments that collect light, break it down and record the results on film or as computer data – use diffraction gratings. These are mirrors scored with very thin lines. As the light shines across the grating, it is broken up into very, very fine wavelength divisions. You can begin to see how a diffraction grating works by looking at a compact disc in sunlight. A very fine, continuous rainbow of colors appears.

Spectroscopy is a very powerful tool in chemistry, where the identifying characteristics of an element can be determined with great precision. It is a fairly simple process in the lab – all you do is apply heat to an element and study the light given off as the element burns. Each element has a very distinctive 'fingerprint', or spectrum. Generally, a spectrum looks like a smooth continuum of color, broken by very bright or dark lines. The story that spectroscopy reveals about an object is found in these lines. The concept of *spectral resolution* applies in spectroscopy – the ability to cleanly separate adjacent features in a spectrum.

The basic rules of spectroscopy were written in the last century by a German chemist named Gustav Kirchhoff. His Laws of Spectral Analysis describe what sort of spectra appear as elements are burned. Kirchhoff's first law states that a hot, high-density gas or an incandescent solid body will radiate across a continuous spectrum. We will see light of all wavelengths in its spectrum. The second law states that a hot, low-density gas will produce what is called an 'emission-line' spectrum, i.e. elements that are abundant in the object will produce very bright lines in its spectrum. (Parts of the Orion Nebula, for example, shine very brightly in emission spectra.) Kirchhoff's third law states that when a source of continuous radiation, such as a star, is viewed through a cooler, low-density gas, an absorption-line spectrum will be produced. The study of absorption spectra is a particularly ingenious way of determining what lies between us and a star. The spectrum of that star will show dropouts where certain wavelengths of light have been absorbed by clouds of material in interstellar space. All an astronomer needs to do is to compare that spectrum to laboratory spectra of elements suspected to exist in interstellar space to make an identification.

## How HST does astronomy

Like its Earth-bound counterparts, the Hubble Space Telescope makes its observations with mirrors and detectors. It operates in just the same way as other telescopes – by gathering light from objects for analysis by science instruments. Because HST's science instruments are sensitive to a wide range of light, it is often said it gives astronomers an 'extended, high-resolution view' of the universe. Since HST was designed to operate in orbit around the Earth, it has a few extra features that ground-based telescopes do not have – solar panels to generate electrical power for everything on board, and magnetometers to detect the Earth's magnetic field. However, it has other features that Earth-bound astronomers would recognize – its own dedicated computer, gyros and sensors to help position it and lock on to guide stars (similar in spirit to the setting circles and finders used to help position smaller telescopes) and a communication link to Earth (just like the telecommunications lines between facilities on the ground). Like most Earth-bound facilities, it can even cover its aperture when it needs to – just like capping the telescope tube or covering a mirror here.

The heart of the telescope is the mirror system. HST's mirrors are part of the Optical Telescope Assembly. It is a special Richey-Chrêtien version of a standard Cassegrain-type telescope. In a Cassegrain telescope, light from an object enters the telescope tube and bounces from a primary mirror to a secondary mirror, which sends the light back through a hole in the primary mirror and focuses it onto an imaginary surface, the focal plane. The HST system was built to provide light in the focal plane of the telescope that is close to the limit specified by the laws of physics – or, as the telescope scientists call it, the 'diffraction limit'.

The diffraction limit is a physical property of light that depends on the size of the primary mirror and the wavelength of the incoming light. As we discussed in Chapter 1, when light hits the primary and secondary mirrors on HST, it is supposed to be focused down

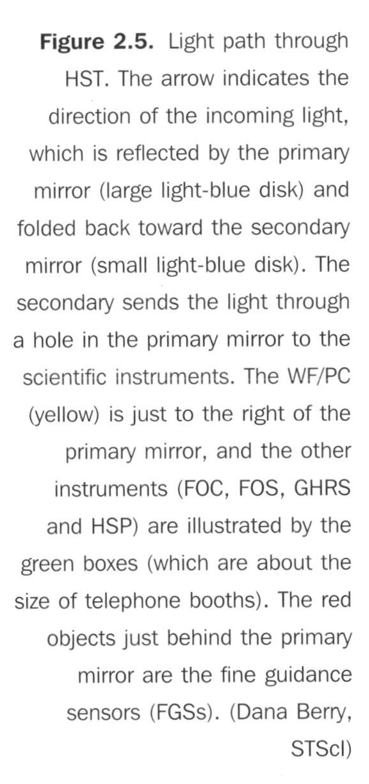

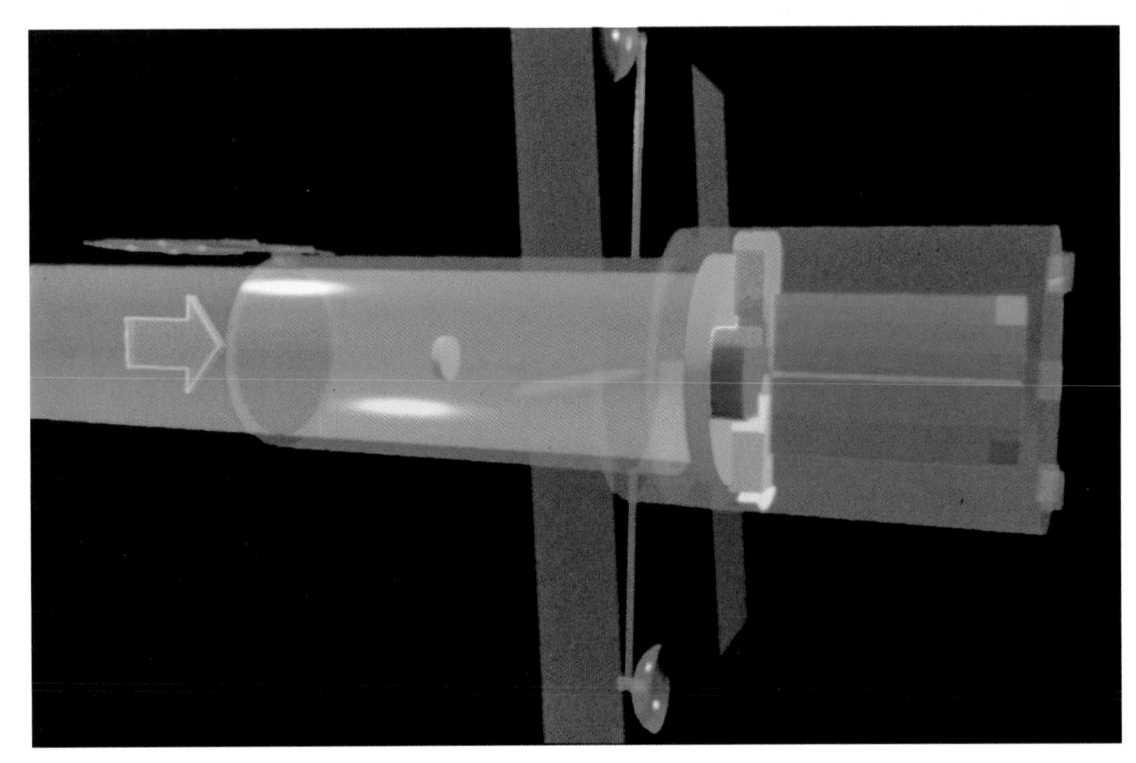

**Figure 2.5.** Light path through HST. The arrow indicates the direction of the incoming light, which is reflected by the primary mirror (large light-blue disk) and folded back toward the secondary mirror (small light-blue disk). The secondary sends the light through a hole in the primary mirror to the scientific instruments. The WF/PC (yellow) is just to the right of the primary mirror, and the other instruments (FOC, FOS, GHRS and HSP) are illustrated by the green boxes (which are about the size of telephone booths). The red objects just behind the primary mirror are the fine guidance sensors (FGSs). (Dana Berry, STScI)

to a small dot of encircled light energy on the focal plane. Theoretically, HST's mirrors could concentrate around 85% of the incoming light into this tiny circle. This is the diffraction limit of the mirror, and it is supposed to guarantee delivery of the highest possible quality image to the science instruments. As everyone found out in 1990, the mirrors spread the light out. The corrective optics installed during the First Servicing Mission worked around the error by focusing 84% of the light back into the circle.

## The HST science instruments

HST was originally designed with five science instruments. The best-known is the Wide Field and Planetary Camera (WF/PC). This is a radial instrument, which means that it is mounted radially, inserted through the side of the telescope behind the mirror. The other instruments are mounted parallel to the long axis of the spacecraft behind the mirror, and

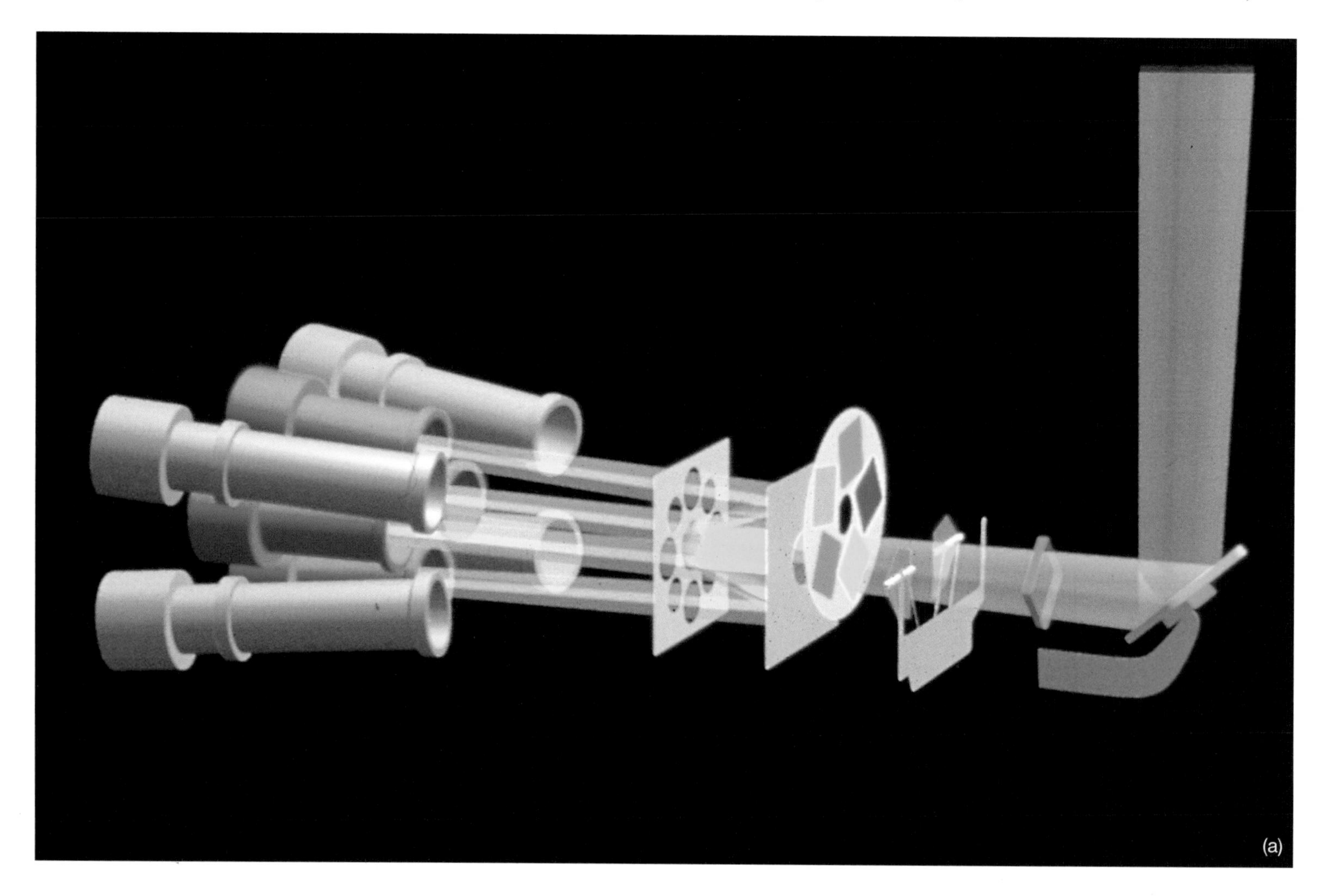

are called axial instruments. The Faint Object Camera (FOC) does visible imaging, but is sensitive to ultraviolet light as well. The Faint Object Spectrograph (FOS) operates over a wide wavelength range, and the Goddard High Resolution Spectrograph (GHRS) is a purely ultraviolet-sensitive instrument. The High Speed Photometer (HSP) was HST's 'light meter'. Although not built as science instruments, the Fine Guidance Sensors (FGSs) serve as star trackers, and also perform astrometry – the science of accurately measuring stellar positions. During the First Servicing Mission, WF/PC-2 replaced WF/PC-1, and the High Speed Photometer was removed to make room for the COSTAR system. Because the HSP gathered data during the three years it was on orbit, it will continue to be a part of several discussions throughout the book. None of the other instruments was removed during the First Servicing Mission.

## The Wide Field and Planetary Cameras

All of HST's instruments overlap each other to some degree in function and wavelength range, allowing scientists to collect the maximum data relating to an object. During the first three years on orbit, WF/PC-1 was the most-used instrument on the spacecraft, and WF/PC-2, with its own built-in spherical aberration correction system, is continuing that tradition.

Both Wide Field and Planetary Cameras were built by the NASA Jet Propulsion Laboratory in Pasadena, California. WF/PC-1 was designed under the direction of the Investigation Definition Team headed by California Institute of Technology professor James Westphal. Ironically, at first he was not very interested in involving himself with HST. 'Jim Gunn came into my office one day in 1976,' Westphal recalled. 'He said, "We've

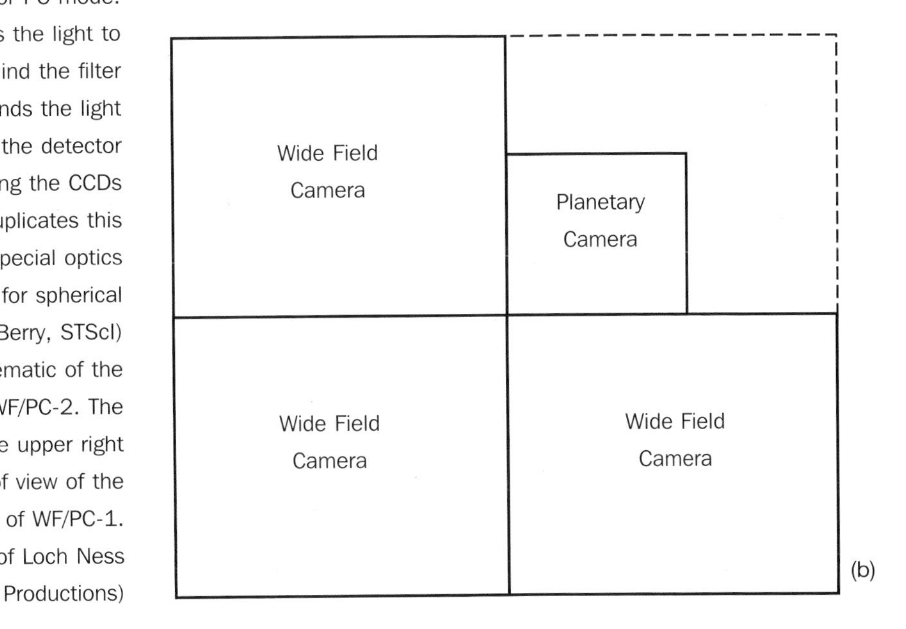


**Figure 2.6.** (a, facing page) Light path through WF/PC-1. Light from the secondary mirror enters at the top and is sent towards the main instrument by the pick-off mirror. The light passes through the entrance aperture and then encounters the shutter, which determines the exposure length. The light then passes through the filter wheel and falls on the pyramid; the orientation of the pyramid determines whether the camera is in WF or PC mode. The pyramid sends the light to relay mirrors (behind the filter wheel) which sends the light through the baffle to the detector assemblies containing the CCDs (left). WF/PC-2 duplicates this operation, but has special optics installed to correct for spherical aberration. (Dana Berry, STScI) (b, this page) A schematic of the 'field' seen by WF/PC-2. The dashed line in the upper right indicates the field of view of the 'WF' portion of WF/PC-1. (T. Kuzniar; courtesy of Loch Ness Productions)

got to build a wide field camera for the space telescope." I looked at him and said, "Jim, you're crazy, that's not our style of thing! My God, I don't want anything to do with that cast of thousands and millions of dollars and bureaucracy and all the thrashing around and all the traveling back and forth across the country!"'

Gunn prevailed on Westphal to look into the possibilities. They did some quick figuring and agreed that the project was possible. Still, Westphal did not think that he wanted to be involved. He recalled Gunn saying, '"If we don't do it, they won't do it right. It's the truth." And I said, "Yes, I know it's the bloody truth, but we should never say that in public."'

Westphal later changed his mind, and became the Principal Investigator for the team that won the right to design and build WF/PC-1. Because NASA expected WF/PC-1 to deliver images and data of wide appeal to the public, its replacement in case of failure was a high priority. Well before WF/PC-1 was placed in orbit, NASA provided funds for a 'clone', which became WF/PC-2. Jet Propulsion Laboratory scientist John Trauger headed up the team that designed and built WF/PC-2.

To do the kind of work that everyone wanted to do with the WF/PCs, the team designed the instrument to operate in two modes – wide field mode or the higher spatial resolution planetary camera mode. Theoretically, both WF/PC-1 and WF/PC-2 are sensitive to wavelengths of light ranging from about 1200 angstroms (in the ultraviolet) to just under 11 000 angstroms (in the infrared). This range also encompasses visual wavelengths. The WF/PCs have been responsible for some of the most spectacular images released from HST. The spatial resolution of both WF/PCs is approximately 0.05 to 0.1 arcsecond.

**Figure 2.7.** A sample WF/PC-2 image, the central region of the remote cluster of galaxies, CL 0939+4713. The extraordinary resolution of WF/PC-2 allows a clear view of the shapes and structure of galaxies as they were when the universe was two-thirds of its present age. This cluster is discussed further in Chapter 5. (Alan Dressler, Carnegie Institution; NASA)

WF/PC-1 used two sets of four CCDs as its main light receptors, giving a full picture with an array of $1600 \times 1600$ pixels across the four CCDs. When data were received, the four CCD images were electronically 'pasted together'. WF/PC-1 performed magnificent science, despite the effects of spherical aberration, a loss of ultraviolet sensitivity, and the development of less-sensitive spots on the chips called 'measles'. The replacement instrument, WF/PC-2, uses four CCDs at two different magnifications. There are three Wide Field camera fields on the new instrument and one planetary camera field, which is why some images from the instrument appear to be 'chevron-shaped'. Nonetheless, WF/PC-2 allows its users to see stars as faint as 28th magnitude during long-term exposures

## The Faint Object Camera

The Faint Object Camera (FOC) was built for HST by Dornier Corporation and was funded by the European Space Agency. The Investigation Definition Team responsible for its design was chaired by H.C. van de Hulst of the Leiden Observatory in the Netherlands. The FOC is part of the European contribution to the spacecraft, and its operations are overseen at the Space Telescope Science Institute by Duccio Macchetto, who has been with the project for many years, and acts as Principal Investigator for the instrument.

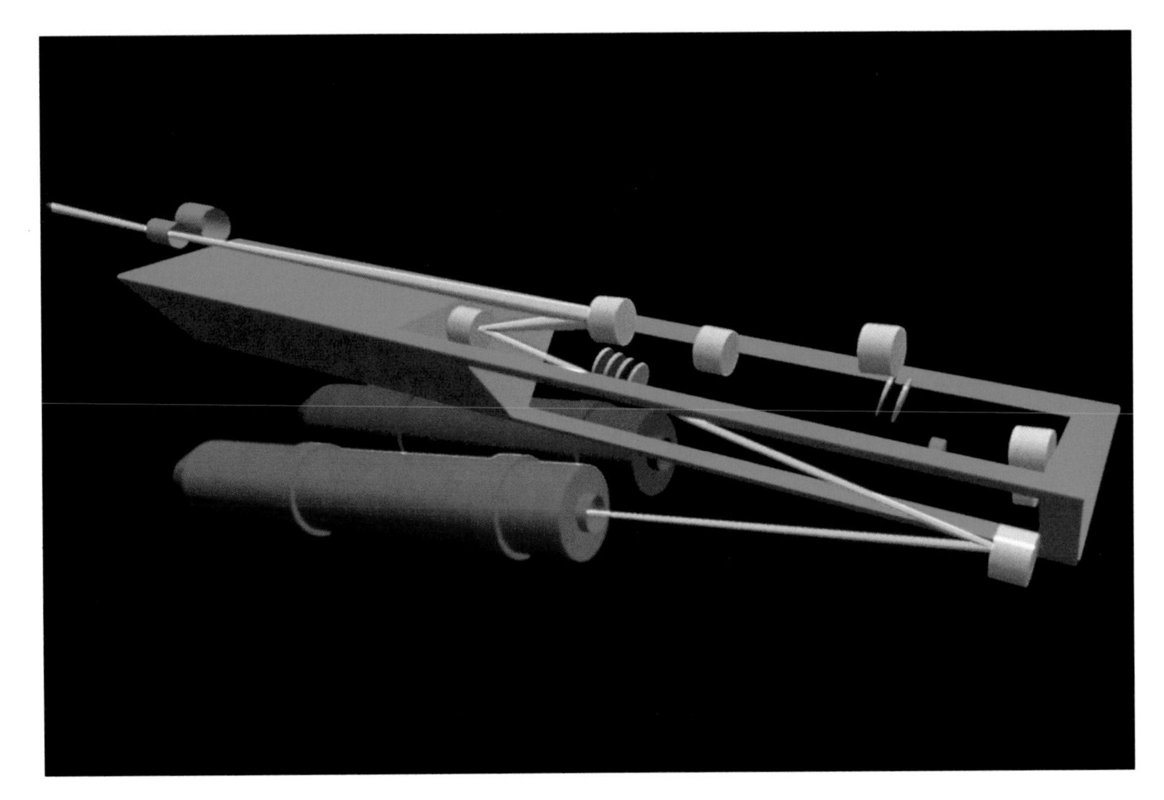

**Figure 2.8.** Light path through the Faint Object Camera. Light from the secondary mirror passes through the entrance aperture to strike FOC primary and secondary mirrors. The light then passes through filter wheels (the multi-colored disks) and is refocused onto the detectors (red), an image-intensification device and television-type camera tube. The optical elements not involved in the light path shown are an independent (but similar) mode of the FOC that ends in the other (green) detector. (Dana Berry, STScI)

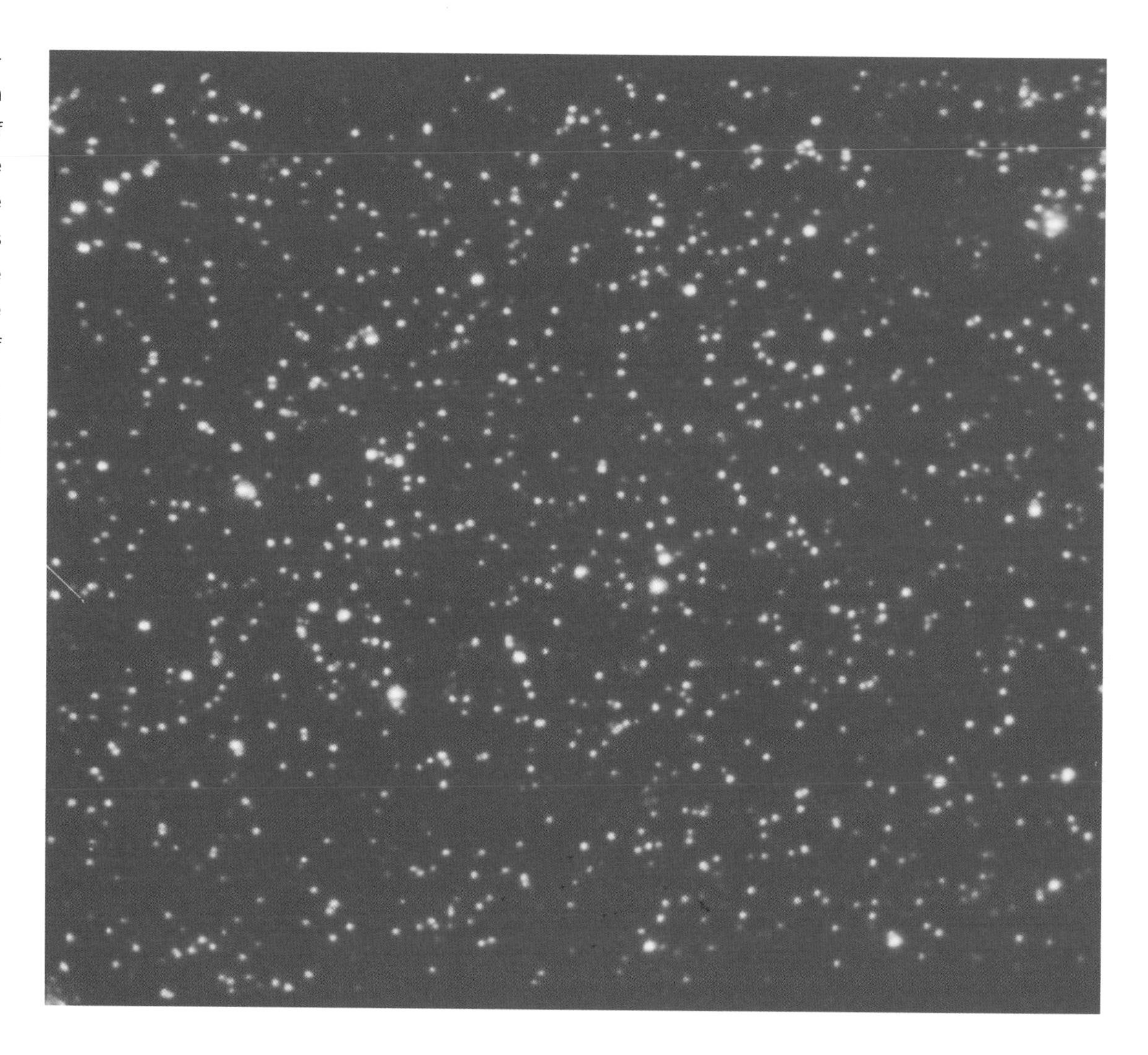

**Figure 2.9.** A sample post-COSTAR FOC image taken on January 10, 1994, of the core of globular cluster 47 Tucanae. The image is 14 arcseconds on a side and shows individual star images that are crisp and clean. The brightness of the stars can be accurately measured; some of the stars may be white dwarfs. (R. Jedrzejewski, Space Telescope Science Institute; NASA; ESA)

The FOC 'sees' light in the range of 1150 to 6500 angstroms. It has two complete detector systems with image intensifiers to gather as much light as possible from dim, distant objects. Its maximum field of view is a square measuring about 22 arcseconds on each side. The FOC is designed to 'oversample' the image and thus provide the highest possible spatial resolution, even better than 0.05 arcsecond. The image is produced on a phosphor screen, and that screen is scanned by a specialized television camera. If the FOC looks at anything brighter than 21st magnitude, filters must be used to dim the light so that the detectors do not become saturated

As with other instruments on HST, the FOC suffered from the effects of spherical aberration and lost its ability to suppress the light from bright objects to observe nearby dim ones. The FOC also has problems with the electronics in one of its observing modes.

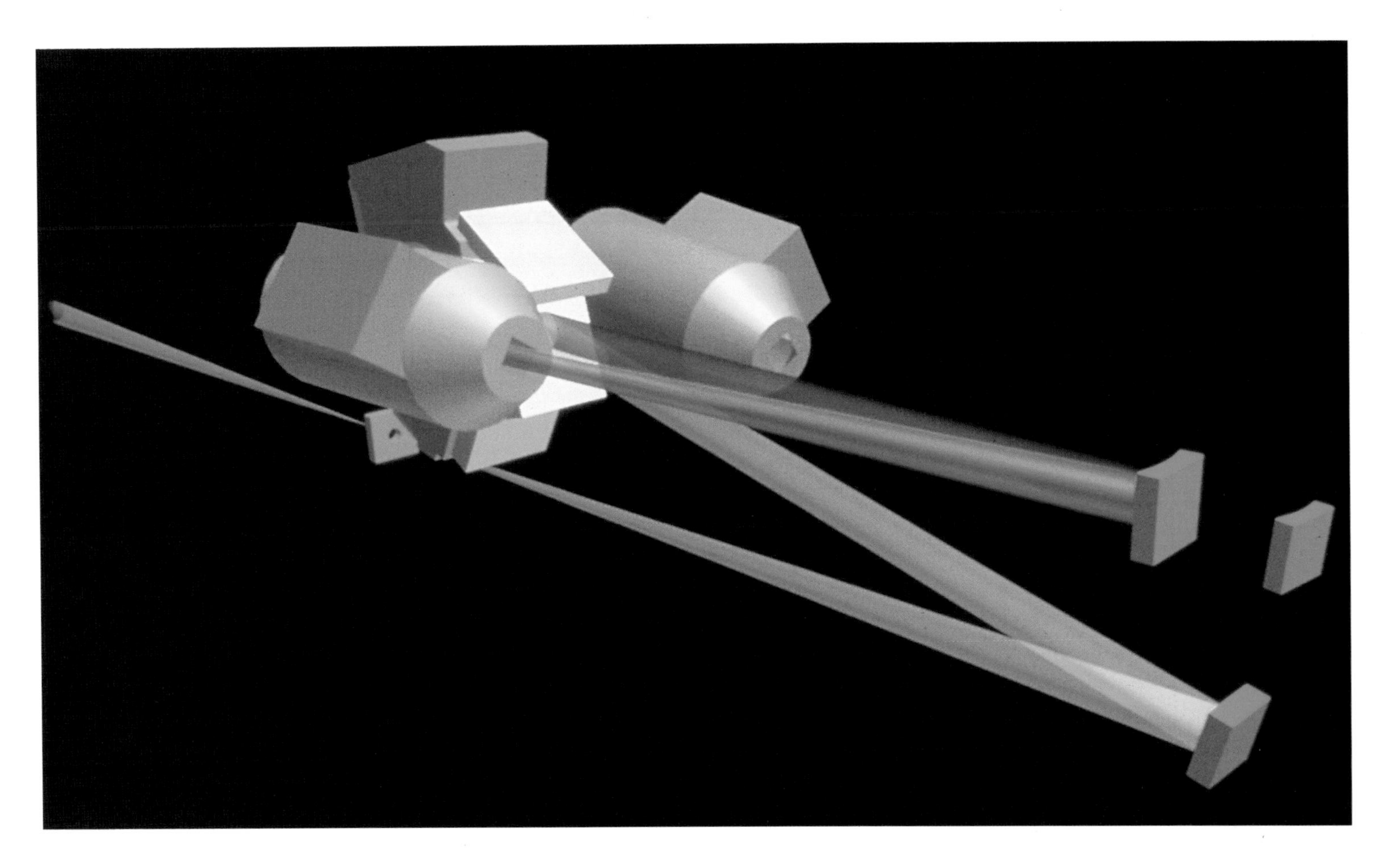

**Figure 2.10.** Light path through the Goddard High Resolution Spectrograph. Light from the secondary mirror enters from the left and is sent by the collimating mirror (lower right) to the carousel (between the two detectors). The carousel rotates to bring different gratings into the light path to determine the wavelength to be observed and the resolving power. In the mode shown, the dispersed light (with colors) from the carousel goes to a mirror and then to the Digicon detector. For the highest resolving power, the light from the carousel goes to another grating (called the cross disperser, and the element not used in the diagram) and then to the Digicon detector. (Dana Berry, STScI)

## The Goddard High Resolution Spectrograph

Because spectroscopy is such an integral part of modern astronomy, HST has two spectrographs – the Goddard High Resolution Spectrograph and the Faint Object Spectrograph. These instruments overlap each other to some extent. The FOS takes the 'big picture' and the GHRS zeros in on finer detail. Each of the spectrographs performs science that cannot be done from the ground because ultraviolet light is absorbed by our atmosphere.

The Goddard High Resolution Spectrograph (GHRS) was built by Ball Aerospace Systems Division, Boulder, Colorado, under the direction of an Investigation Definition Team headed by John C. Brandt (who was at Goddard Space Flight Center at the time). When Brandt moved to the University of Colorado in 1987, Goddard scientist Sara Heap was named as Co-Principal Investigator.

GHRS contains an optical system, support electronics and a structural system. The way it works is quite simple: ultraviolet light enters the spectrograph through one of two apertures and is sent to a rotating carousel by a collimator mirror. The carousel contains gratings that spread the ultraviolet light out. Depending on the gratings chosen, a scientist can select certain wavelengths of ultraviolet light for specialized study. GHRS is sensitive to light between 1150 and 3200 angstroms.

The GHRS uses two 512-pixel Digicon detectors and operates in three resolution modes: low, medium and high. The design specifications at low resolution call for measurement of a spectral feature 1 angstrom wide. At medium resolution, it should be able to measure features 0.06 angstroms wide, and, at high resolution, it should measure features 0.02 angstroms wide. GHRS can deliver data with a very high signal-to-noise ratio. The challenge is to collect the maximum amount of information about an object with a minimum of 'noise' from the system.

As with some of the other instruments on the telescope, the GHRS developed a few problems. A year or so into operation, a power supply began working intermittently, and, for safety reasons, some modes of the GHRS were not used after discovery of the problem. On the last day of the First Servicing Mission, a special electronics package was strapped and electrically connected to the GHRS, which returned the instrument to full working order.

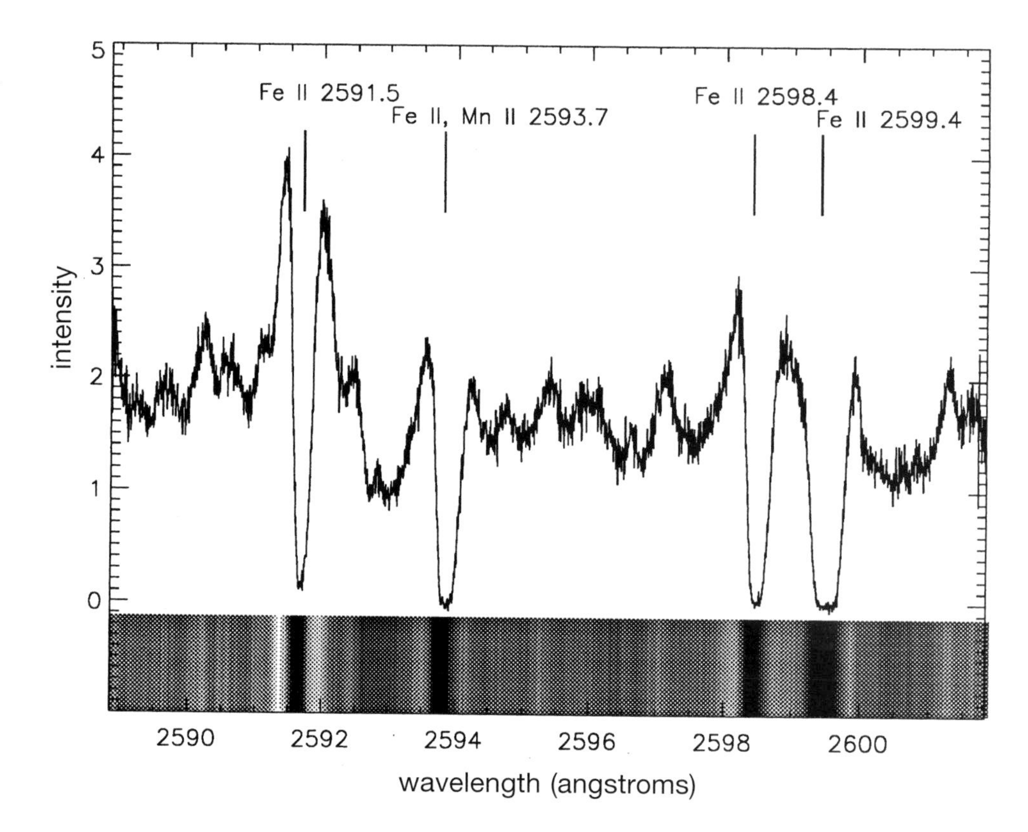


**Figure 2.11.** Sample GHRS observations, spectrum of Betelgeuse (Alpha Orionis), the bright reddish star in the shoulder of Orion. The spectrum shows the bright and dark regions of the star's spectrum (below) and how the same information appears in the graphical form used by astronomers (above). Some identifications, Fe for iron and Mn for manganese, are shown. (Martin Snow, University of Colorado and GHRS Science Team)

## The Faint Object Spectrograph

The Faint Object Spectrograph (FOS) studies fainter objects than can be detected by the GHRS. The FOS was built by Martin Marietta Astronautics Group of Denver, Colorado, under the direction of an Investigation Definition Team headed by Richard Harms (who was Vice-president of Science Applications, Incorporated, at the time). The FOS was designed to be sensitive in the wavelength range of 1100 to 8000 angstroms, more extensive than the GHRS range. There is a tradeoff involved with this wider wavelength range and its ability to study fainter objects. The FOS studies fainter objects at a lower spectral resolution of light than does the GHRS, but the GHRS looks 'in depth' at brighter objects in a limited wavelength range of ultraviolet light and at higher spectral resolution.

The FOS operates in two resolution modes: low and high. In low resolution mode, it can image 26th magnitude objects in 1 hour exposures and can trace features 15–20 angstroms wide. In high-resolution mode, it can achieve 22nd magnitude in a 1 hour exposure, and can trace features 3–4 angstroms wide.

Like other instruments, the FOS took a 'hit' from the spherical aberration, losing about 3 magnitudes in observing capability. One FOS mirror is also oxidized slightly (meaning that it was exposed to oxygen at some point), which reduced some of its ultraviolet sensitivity. This happened directly as a result of the launch delay from 1983 to 1990.

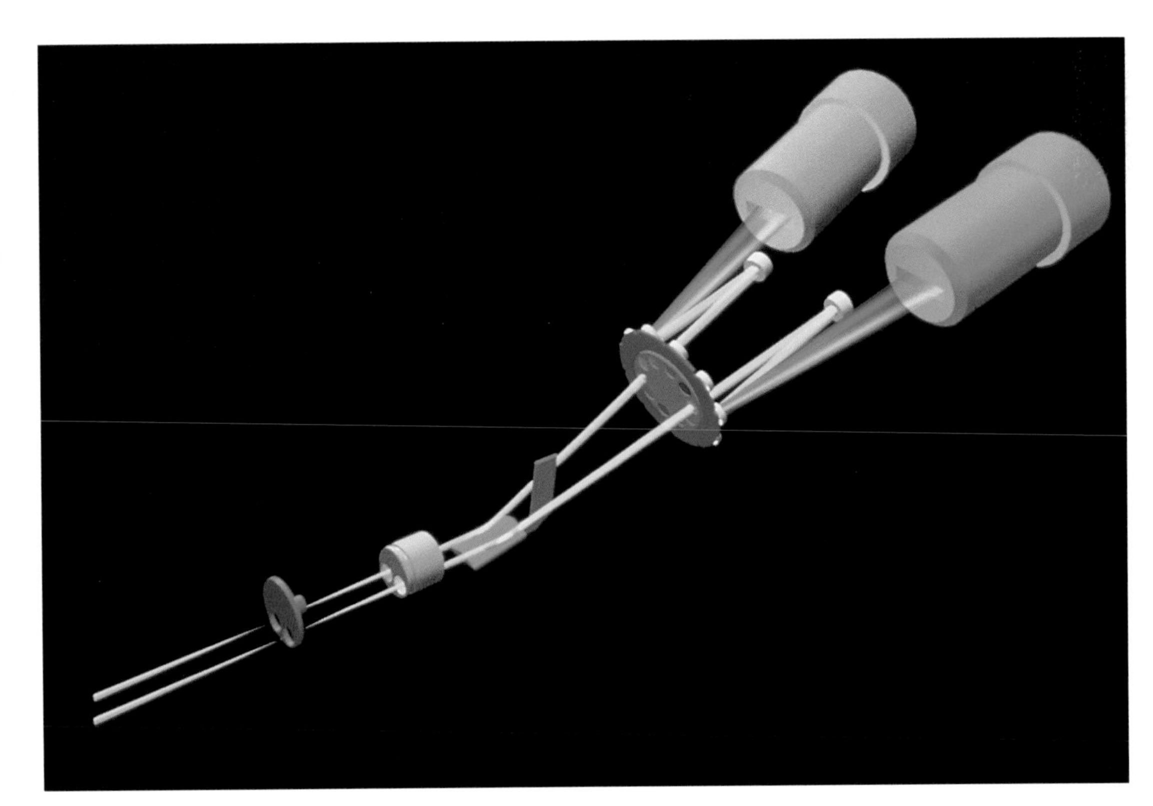

**Figure 2.12.** Light path through the Faint Object Spectrograph. Light from the secondary mirror passes (from left) through an entrance port and through an aperture assembly. After this assembly, the light reflects off a mirror (light blue) and goes to the filter/grating wheel. Filters block out the undesirable wavelengths when the light passes through, and mirrors send the light back to the gratings mounted on the wheel. The dispersed light then goes to the Digicon detectors (red and blue). (Dana Berry, STScI)

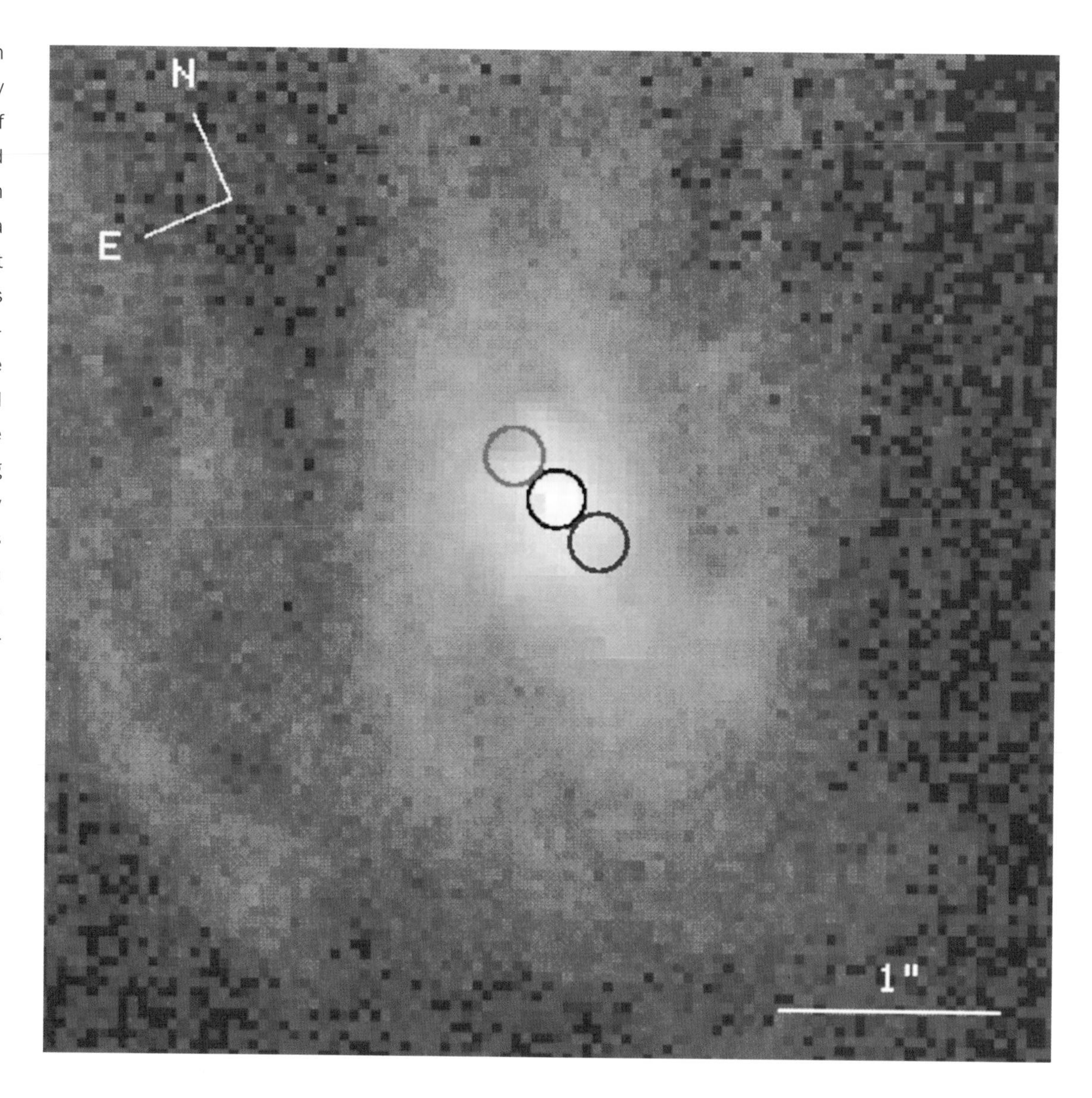


**Figure 2.13.** An FOS spectrum of the center of active galaxy M87. The locations and size of the regions above are color coded (red, black, blue) to the plot on p. 71. The red-labeled area clearly shows redshifted light coming from the region that is moving away from us. The blue-labeled spectrum shows a blue shift, indicating light from material that is moving toward us. The velocity of the material moving away from us is approximately 550 kilometers per second. This sort of evidence is important in establishing the existence of a massive black hole at the center of M87. (This is discussed in more detail in Chapter 5.) (R.J. Harms, RJH Scientific, Inc.)

The spectrograph's designers knew that the mirror's coating could oxidize, but, during the course of the delay, that was apparently forgotten, and the degraded wavelength sensitivity was not discovered until after the telescope was on orbit.

## The High Speed Photometer

The High Speed Photometer (HSP) was built at the University of Wisconsin, under the direction of Professor Robert Bless and his HSP Investigation Definition Team. According to Bless, the photometer was designed to measure high speed fluctuations of light from some of the most energetic objects in the universe: 'The point of the instrument was to do

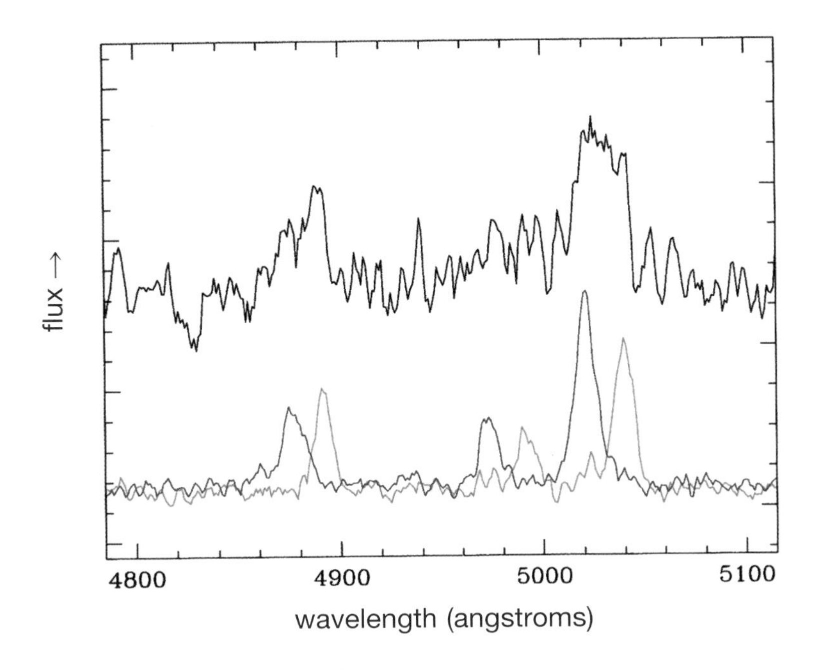


*very accurate* photometry – High Speed measurements of light fluctuations up to 100 000 samples per second,' Bless said. 'It had to be very precise. Any jitter would be reflected immediately in our data.'

The HSP was able to measure changes in light every 1/50 000 second. During the first three years of the mission, the HSP functioned flawlessly, but, because of the

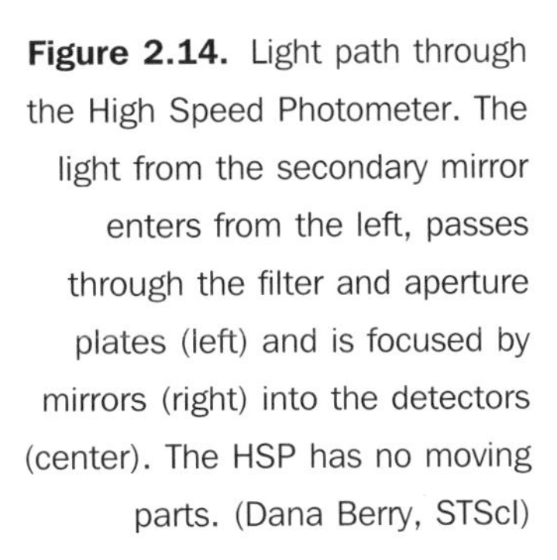

**Figure 2.14.** Light path through the High Speed Photometer. The light from the secondary mirror enters from the left, passes through the filter and aperture plates (left) and is focused by mirrors (right) into the detectors (center). The HSP has no moving parts. (Dana Berry, STScI)

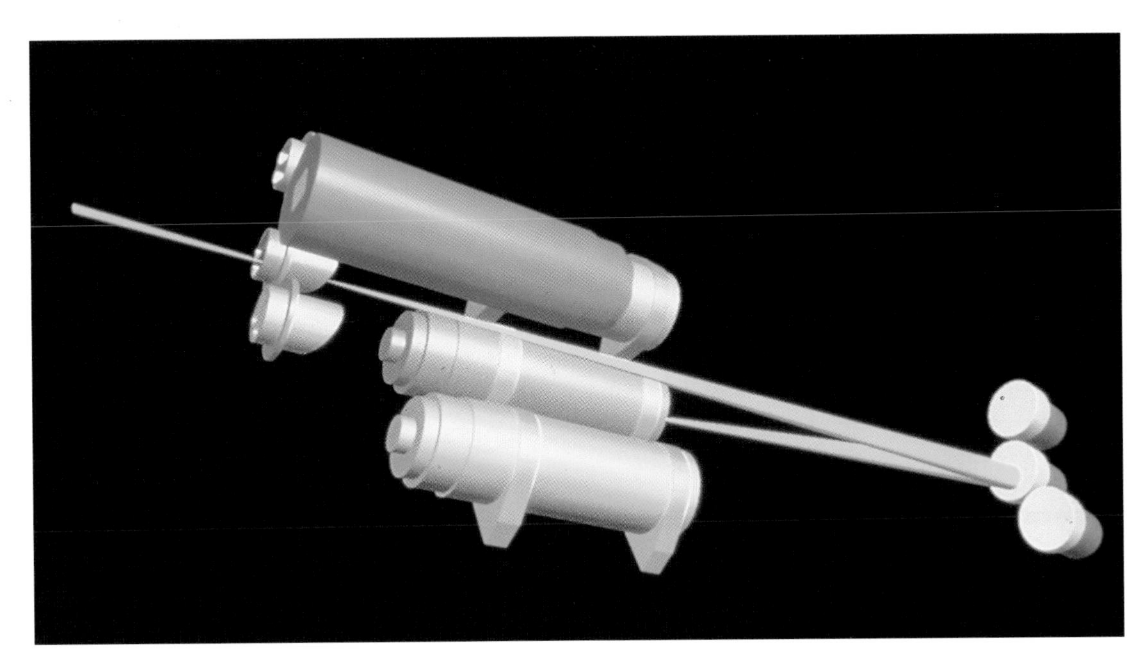

spherical aberration and jitter, its science program was reduced. The spherical aberration in particular was quite damaging to HSP's observational capability because stellar images simply overflowed the aperture. Most precise work with such a large image was impossible.

Bless knew that as soon as the spherical aberration was announced HSP would be the 'fall guy' for the corrective optics. 'I think it's fair to say that we were more affected by the spherical aberration than anybody else,' he said. 'In the case of the spectrographs, for example, they could not work in the crowded fields, but that's a small part of their program and basically they had to integrate for a longer time. On the cameras you could still get good qualitative data.'

That was not so for the telescope's only university-built instrument. According to Bless, the mirror problems were just the tip of the iceberg. 'For us, the combination of the huge image and the jitter was just devastating,' he explained. 'Our most-used aperture was 1 arcsecond across, and when you had an image of a couple tenths of an arcsecond within that, it wouldn't matter if there had been only a little motion or jitter.'

Ultimately, the photometer was removed from HST and replaced with COSTAR – a disappointing end to a promising program for Bless and his team-mates. That end was made all the more bittersweet when they actually received some usable data just prior to the instrument's removal from the telescope.

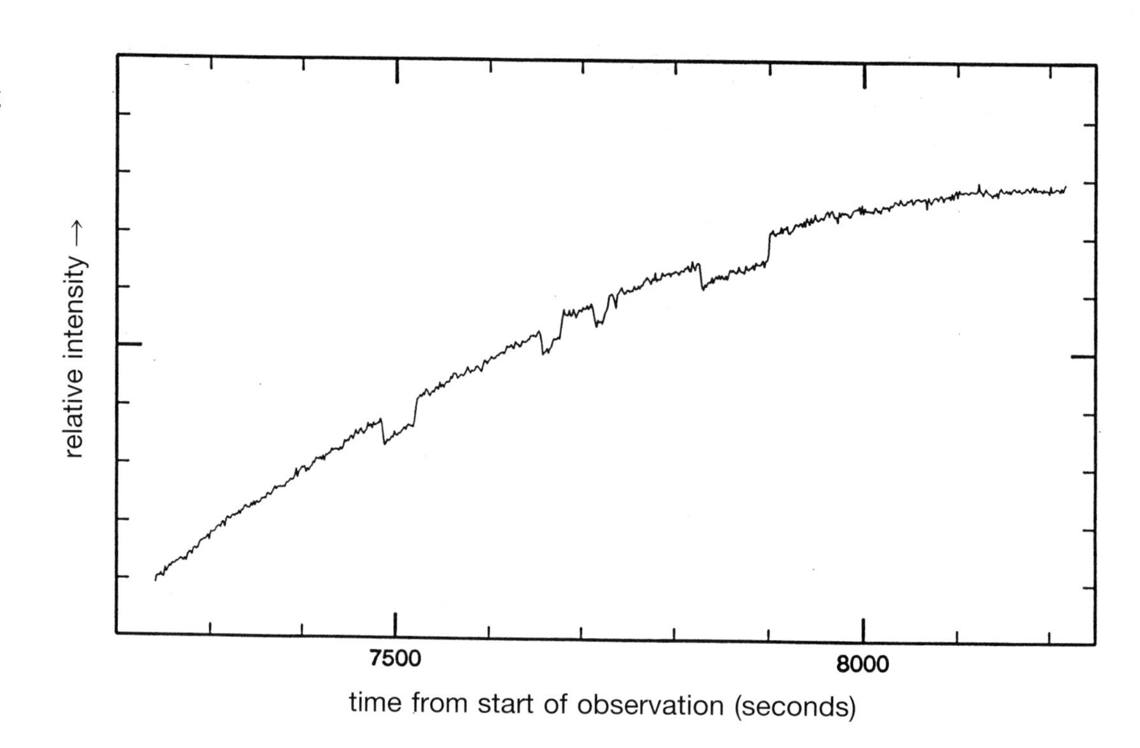


**Figure 2.15.** Sample HSP data, taken while observing a star being occulted by the rings of Saturn. The star moved from the middle into the outer Crepe ring, and several obvious features are shown. See Figure 3.16 and accompanying discussion. (Robert Bless, University of Wisconsin and the HSP Team)

## The Fine Guidance Sensors

The Fine Guidance Sensors (FGSs) were built by Perkin-Elmer, now Hughes-Danbury Optical Systems. They function as HST's sixth scientific instrument, particularly FGS3, and they provide star positions that are ten times more precise than those from ground-based measurements. The Earth's atmosphere blurs star images taken from the ground, and consequently astronomers cannot measure their positions too accurately. High above the atmosphere, however, the seeing is fine, and from there astrometry can be done exceedingly well.

According to Astrometry Team Principal Investigator and University of Texas researcher Bill Jeffreys, HST's astrometric studies are the latest chapter in a long history of astrometry: 'There's a lot akin to what we're doing with the space telescope and the kind of astrometry which has been done for maybe a hundred years or so – photographic astrometry.'

Photographic astrometry is actually a kind of relative astrometry. An astronomer takes a picture of the sky and measures the positions of the stars on the plate relative to other stars. If the stars have moved, the scientist must determine the direction of motion of the star. If one star is closer than the background stars, the astronomer can calculate the parallax (how much it has shifted against the background of stars during the course of an Earth year), and from this can derive the star's distance.

HST's Fine Guidance Sensors are ideal space-based astrometry instruments, and they are being used along with the European Space Agency's Hipparcos satellite to measure positions and parallaxes of stars. Because spacecraft time is so limited, and because it takes months to achieve parallaxes with HST, Jeffreys and his team have to concentrate on objects of particular interest, either for their astrophysical characteristics, or for their function as fundamental distance calibrators. What the team loses in numbers of parallaxes, it more than makes up for in terms of accuracy of parallaxes. 'We have about a half a dozen stars in the Hyades that we're going to try and wring the best parallax out that we can get,' Jeffreys said.

Hipparcos, on the other hand, is doing a faster survey, looking at 120 000 stars out to 11th magnitude. One of the goals for astrometry with HST is to obtain reliable parallax measurements for very faint stars and such difficult-to-measure stars as binaries and Cepheid variables. According to Jeffreys, that sort of work has been slow going for the team until relatively recently for two reasons. The first was that the telescope's jitter made it hard to measure angular distances accurately; since the First Servicing Mission, however, that problem has gone away. The other reason is that taking astrometric measurements is a lengthy process. Jeffreys explained, 'It takes a lot of spacecraft time to get a parallax with HST. The number of parallaxes we're going to be able to get is very small. It will not significantly expand the number of parallaxes that are currently known. Currently, I think there's something on the order of a thousand or so good parallaxes.'

The FGSs work by knowing approximately where to expect a guide star to be and then acquiring the star in their apertures. The astronomer designing an HST observation plan has to know the positions of guide stars in advance. Early on in the mission, no database existed that contained more than about half a million stars. So, a catalog was created containing the stars that could be used as guide stars by the HST. One of the Astrometry Team's early tasks was to advise NASA on how to create that database. They built a prototype system in Texas, and then delivered the design for the Guide Star system now in use by HST's controllers.

## COSTAR

The spherical aberration of HST's main mirror has been 'fixed' by an ingenious optical instrument called the Corrective Optics Space Telescope Axial Replacement, or COSTAR. This instrument deploys five pairs of coin-sized mirrors in front of the Faint Object Camera, the Faint Object Spectrograph and the Goddard High Resolution Spectrograph. The mirrors refocus the incoming light for these axial instruments, correcting for the spherical aberration.

As we mentioned in Chapter 1, COSTAR was built for NASA by Ball Aerospace Systems Division, under the direction of Ball Program Manager John Hettlinger and Space

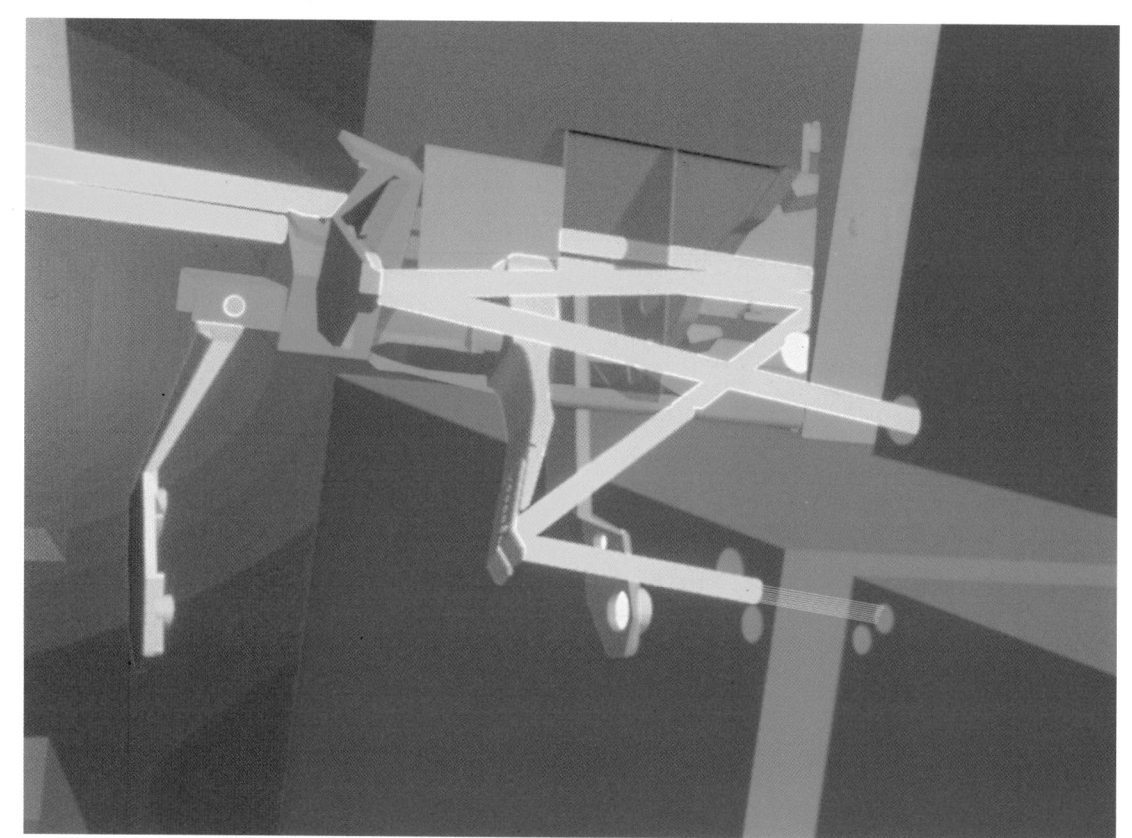

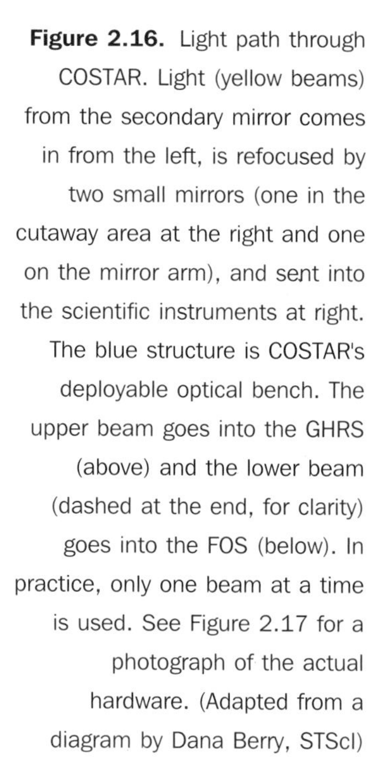

**Figure 2.16.** Light path through COSTAR. Light (yellow beams) from the secondary mirror comes in from the left, is refocused by two small mirrors (one in the cutaway area at the right and one on the mirror arm), and sent into the scientific instruments at right. The blue structure is COSTAR's deployable optical bench. The upper beam goes into the GHRS (above) and the lower beam (dashed at the end, for clarity) goes into the FOS (below). In practice, only one beam at a time is used. See Figure 2.17 for a photograph of the actual hardware. (Adapted from a diagram by Dana Berry, STScI)

**Figure 2.17.** COSTAR during a test. The time-lapse image shows the second GHRS mirror arm going from the stowed to the fully deployed position. (Ball Aerospace Corporation, Aerospace and Communications Group)

Telescope Science Institute's Holland Ford. Ford and others described the engineering stakes as being very high. Not only was NASA sporting a black eye from the discovery of spherical aberration, but anyone who tried and failed to repair HST would share in the disgrace. Fortunately, Ball put an exceptional team to work on the problem, including one of their best optical designers, the late Murk Bottema. Due to the hard work put in by this team, COSTAR did not fail. 'By putting all the light into a very small number of pixels,' Ford said, 'we can do in orbit the equivalent [using the FOC] of spotting a firefly in Tokyo from Washington, D.C. The resolution enabled by COSTAR is so good that if you used it to look at two fireflies nine feet apart in Tokyo, you could tell they were two fireflies from Washington.'

## Observing with HST

The process of using HST for an observation is more complex than any other observing process in the history of astronomy. Hundreds of users have been assigned and have used

HST time, but to do this they worked their way through what many have termed the 'most incredibly complex' process of proposal and observation allocation ever seen.

With ground-based facilities, if observers want observation time they apply to the observatory and describe the observations they want to make. A group called a Time Allocation Committee (TAC) reviews the request and either approves or rejects it. If it is accepted, the scientist can travel to the observatory, set up the equipment, with the help of a night assistant, and do the observations. Then the observer takes home photographs, computer disks, videotapes – whatever storage medium is needed. Final data analysis takes place at the observer's 'home' institution. Often, the observer stays home and monitors the observations via computer linkup, and has the data delivered.

In principle, the system is similar for HST, but there are some major differences – observations with an observatory in low Earth orbit can be an involved process. The orbital period produces an observatory 'day' of a little over 90 minutes. Because of this, the observations are often chopped into small segments. As a consequence of this difficult

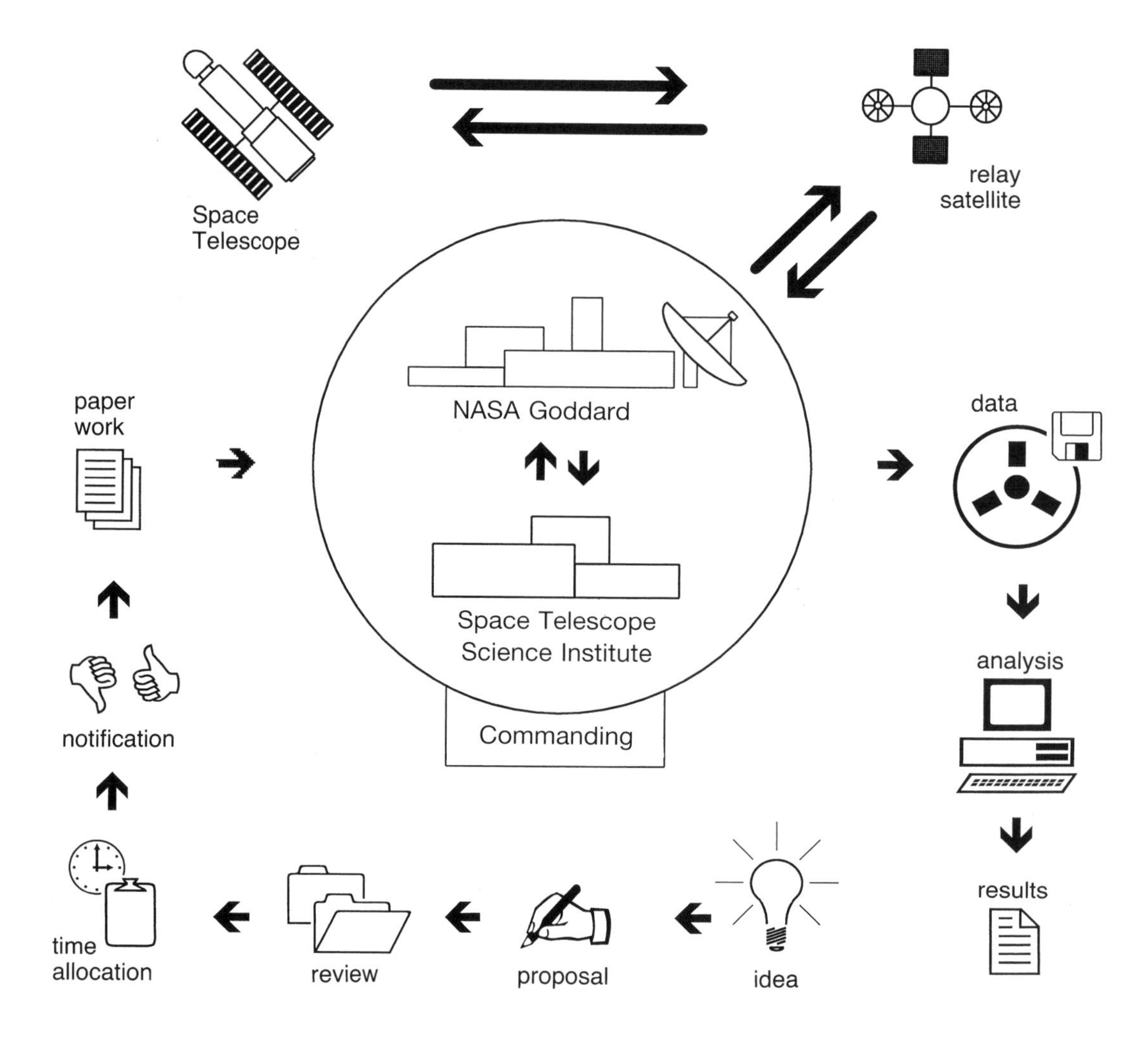


**Figure 2.18.** Flow of an HST Scientific Project, from original idea (starting at the light bulb) to published results. (Adapted from a diagram supplied by the STScI)

observing environment, the ground system must command many spacecraft functions, but these are not under the direct control of the observer.

An observer still applies for time on HST, but it is done in several steps. First, the observer fills out what is called a Phase 1 form, describing the object to be observed, the instruments needed and the scientific justification for the observation. This file is sent to the Space Telescope Science Institute via electronic mail. The observer then follows this up with an application in writing. An HST Time Allocation Committee meets yearly to consider these proposals for using HST, and sends its recommendations to the Institute Director for final consideration.

The Phase 1 form is submitted to a group called a 'discipline TAC', which comprises the scientist's peers in the field. If, for example, an astronomer wants to observe planetary nebulae, the discipline TAC that examines the proposal will have most or all of its membership from scientists who have done work with planetary nebulae. The TAC is very large, because currently there are more than 800 requests for time in any given cycle of observations, and only about 300 requests are granted. Thus, a large field of TAC members is needed to choose the successful proposals.

If a candidate's proposal is accepted, the next step is to fill out a Phase 2 form. This is also computerized, but requires an incredible amount of detail on exactly how the HST is to be used. The observer refers to a set of manuals that specify the operating modes of the different HST instruments. In current operations, the observer has access to a graphical representation of the proposed program's structure and its impact on the spacecraft's schedule.

In the telescope's early days, a number of observations failed because the pointing was not specific enough to capture an object in the correct apertures. Now the observer receives help in specifying more accurate pointing information from the Space Telescope Science Institute, which maintains responsibility for accepting and implementing the science programs proposed by observers.

There are several categories of observers on HST: one group consists of the so-called 'guaranteed time observers', who were promised a certain amount of observation time in return for their work on developing the various instruments on board the telescope. There are also general observers, who apply for time in yearly competitions, and amateur observers, who apply for time through a special TAC set up by Jet Propulsion Laboratory's Steve Edberg and Space Telescope Science Institute's Peter Stockman to judge amateur proposals. Amateur observers receive observation time apart from that allocated to the guaranteed time observers and general observers, from a fund known as the 'Director's Discretionary Time'.

As part of program implementation, the Institute maintained what was called a User Support Branch, which helped observers to get their proposals written and implemented. The proposal passed through several other groups, who converted the observer's request into a form the HST software could understand. Eventually, the information made it to a system of computer hardware and software called the Science Operations Ground Systems (SOGS), designed to help the Institute plan, schedule and perform observations.

After the Phase 2 proposal is filled out and accepted, it is submitted to the Institute via computer. It is integrated into the Science Mission Specifications (SMSs) by SOGS and placed into the schedule of observations for its intended cycle.

The SMSs comprise a pipeline of programmed observations for HST. Once the SMSs have been sent to the Goddard Space Flight Center, where the control center for HST is located, it is very difficult to make changes. You could think of the SMSs as a set of 'executable' files on a computer, telling the telescope where to point, how long to point, which instruments to use, what filters and apertures to use, and so on.

In a typical command load, HST will be looking at something – Mars, for example. The next observation, which may or may not be planned to take advantage of the current slew position, comes up and tells HST to slew to the Orion Nebula. The program will contain guide star information for HST's sensors to find, and a command to start slewing to a position near where the guide stars are. If all pointing parameters are satisfied, i.e. if the object is not closer than 50 degrees to the Sun, or too close to the Earth's limb, or the Moon, and if the telescope is not close to (or going through) a region of strong radiation called the South Atlantic Anomaly, then HST can proceed to lock on to its guide stars and start the search for the object in question.

When HST is in position and the instruments are ready to go, it is time to do the observation. In what some people have termed a frustrating 'sit and wait' mode, the telescope then waits for 10 to 20 minutes before it starts the observation. (This wait time is based on a combination of two factors: the maximum expected time to set up the observation, and where the telescope is in its orbit, i.e. there must be sufficient time before the next interruption.) Some observations take only a few minutes; others take several hours. Since HST is in orbit, hours-long observations continue over many orbits. Each time the telescope moves out of sight of the object, it loses its lock on that object. As soon as HST comes back around the Earth, it must relocate its target and continue the observation. If the observer has chosen an object near the poles of the Earth, an area called the 'continuous viewing zone,' then the telescope can observe for longer periods of time.

At the end of the observation, data are saved by a tape recorder on board the telescope. At some point the data are sent to Earth, where they are received at a ground tracking station at White Sands, New Mexico. The data are then relayed through Goddard to the Space Telescope Science Institute, where they are usually released to the observer within a few hours or days. In some cases, however, for unknown reasons, observers have waited much longer.

Unlike a ground-based observatory, very few scientists sit around Space Telescope Science Institute watching their data come down from the telescope. Usually the astronomers are aware of their scheduled observation times, but wait out the observations at their home institutions. However, some observers do come to the Institute, and they

**Figure 2.19.** Part of the Science Support Center at the Space Telescope Science Institute. (STScI)

often work with people whose job it is to monitor the observations and occasionally take action during real-time observations.

Once the observation is concluded, the received data are archived immediately. A set of analyses called 'pipeline reduction' is performed, which presents the data in a form the scientist can use for analysis, and it is also done to spot any serious problems with the data that might require more than the usual amount of work.

The final analysis of the data is done by the observer who proposed the work (or often enough by a graduate student). Early on, this often included the 'deconvolution' process, in which a mathematical set of reductions that partially compensated for the effects of spherical aberration was applied to the data. Now that the spherical aberration has been corrected, deconvolution can still be used to improve the resolution on a given image, but it may not always be necessary.

The users have proprietary rights to the data for one year after receiving it, but very often a truly spectacular set of images will be sent through to the Space Telescope Science Institute Office of Public Outreach for immediate release to the media, with the permission and participation of the observer.

At the end of the user's proprietary year, all data sets are, theoretically, available to any other researcher who wishes to use them. This proprietary period has often been the subject of debate. Some outside the science community feel that the information should be released right away, since – as some of them put it – 'the taxpayers paid for it'. Scientists within the community point out that, in return for researching the data needed, devising the analysis tools to understand the data, and going through the lengthy process of submitting a proposal, researchers should be entitled to some period of time to review their data. It is an ongoing debate, but most scientists think that the one-year period is adequate and that it does not impede the flow of scientific information.

The last step in the process of conducting research with HST is publication of data results – usually in a peer-reviewed journal such as *Astrophysical Journal*, *Astronomical Journal*, *Astronomy and Astrophysics*, *MNRAS*, *Nature*, *Planetary and Space Science*, *Science*, and *Icarus*.

Observation with HST is a lengthy process. For most observers, more than a year will pass between the time of first proposal submission and the time the observation is actually executed. Contrast this with the time frame of several months needed to complete an observation at a ground-based observatory, and one can see why observing with HST is not a simple undertaking.

Historically, the system for time applications, allocations and user support has not always functioned perfectly. For the first few years, many observers complained about the strict constraints on scheduling, planning, observing and other problems that can and do foul up even the best-written observing proposals. Some recent examples have ranged from lost observing coordinates to incorrect record-keeping of observing time. These and many other problems are frustrating to users who have been granted precious time on the world's premier space-based instrument.

Despite the difficulties involved in the observation process and the inherent frustrations of applying for time and using the telescope, increasing numbers of astronomers vie for time on HST. They follow it through, and, if all goes well, they are rewarded with information that at least one scientist frankly characterized as 'wonderful data, but sometimes painful to get'.

Within the community, experienced observers cannot help but compare the differences in operation between HST's hierarchy and that found at any ground-based institution. According to Ray Weymann of the Observatories of the Carnegie Institution of Washington, the procedure may or may not be worse than a ground-based facility: 'It's a complex procedure,' he says. 'People who are very familiar with the details say that if you know the ins and outs, if you're very close to Space Telescope, there are ways of doing things which can save you a lot of trouble.'

Weymann has cited guide star selection for observations as one example where knowing someone at the Institute could come in handy: 'You may just be told "there are no suitable guide stars and we can't schedule your observation". But, there is also a certain amount of the squeaky wheel getting the grease, and if you say "Hey this is really important", make a big fuss about it, then people will pay attention to you and do it. If you don't know those things, then you may find yourself with observations that fail for one reason or another, or observations that simply don't get done. I don't think the situation is any worse or any better than any other facility. If you would talk to people at Kitt Peak National Observatory, or the users of that facility, you would hear similar stories.'

Other scientists complained early on about not receiving good service from the User Support Branch of the Institute when they ask for help in filling out the proposal paperwork. For some, the problems came when they did not receive their data. University of Colorado scientist Ted Snow recounted a story of having run an observation on HST early in the mission, and then never receiving the data. 'During that observation some of the data were lost in transmission,' he said. 'We noticed the data were missing and we made inquiries. Nothing happened for the longest time. We didn't even get an answer about what could be done. I finally had to write a formal request to redo the observation because it was determined the data were just lost.'

It is important to realize that many of these problems can and should be categorized as 'growing pains'. In response to user concerns and budgetary constraints, the Institute organized a complete overhaul of the way it does business with the telescope's users. Some of this overhaul was long overdue, but some of it is due to a change in philosophy prompted by Robert Williams when he took over the Director's job in 1993. 'I think it's really important that the Institute respond to the user community and at times that takes a Herculean effort,' he said. 'There's a lot of pressure on the Institute to produce science, and it's an expensive project. There are a lot of people who feel that other space projects have failed because of the cost of HST.'

The overhaul resulted in the melding of five Institute branches, including the old User Support Branch, into a new division called 'Project to Re-Engineer Space Telescope Observing', or PRESTO. Headed by Mark Johnston, PRESTO has substantially reduced the paper-work hassles for observers and has streamlined the way proposals are encoded for execution on the telescope. According to Johnston, since PRESTO's implementation, observing efficiencies have been increased, the lead time for changes in observations during the SMS phase has been cut from eight to four weeks, and, most importantly, users have access to a liaison person from the PRESTO office during the entire proposal and observation process.

'The steps we've taken to improve this part of HST operations are all in the right direction,' Johnston explained. 'There's still a way to go to make user support and the whole system work as smoothly as it can. It's a lot more user friendly now, and if you ask people to compare their experience in previous cycles with the current system, you'll get a definite view that things have improved. It's been satisfying to see a lot of the problems from the early years rectified now, five years after launch.'

For all of the problems that scientists recount, there are as many people who have had few problems with their observations, received their data and got as much help as they needed analyzing it. One special group of observers has been the small but constantly changing group of amateur observers chosen to do science slightly out of the mainstream of what professionals might do.

They are selected first by a coalition of amateur organizations around the United States. The program was instituted by former Space Telescope Science Institute director Riccardo Giacconi, who felt a need to connect HST with the public in some way. 'Through the amateur program, I think we were quite successful with bringing HST science to the community,' he said. 'I remember that one of the first amateurs was a biology teacher, and he was telling us that the interest in his school in biology just shot way up when the kids learned he was involved with HST. Each one of the amateurs has been a carrier of culture, of the interest in astronomy, and that's basically what we wanted.'

More than two dozen amateurs have won time on HST, and the Director's office continues to make time available for the program. At another level, the Institute funds a series of 'Hubble Fellowships' for deserving scientists – usually post-doctoral researchers.

Quite aside from the HST user community, there remains a high public interest in the telescope's operations. The Space Telescope Science Institute maintains an active Office of Public Outreach, which supplies press announcements of HST discoveries to members of the media. The educational community is served by the same office, which makes materials available to teachers on request.

There is no doubt that HST has taken its place among the premier observatories of the world. From observations of faraway quasars, to comet crashes in our own cosmic backyard, scientists have come to recognize the value of the telescope. To be sure, it is still very much a political animal. In an age when governments are looking for any way possible to cut spending, HST might even be termed a 'Big Science' luxury. The question that everyone can ask is whether using the telescope is a worthwhile endeavor. Even a casual look at the science it has done during its first five years on orbit tells us that it is.

# 3 HST and the Solar System

> I have observed four planets, neither known nor observed by any one of the astronomers before my time, which have their orbits around a certain bright star [Jupiter], one of those previously known, like Venus or Mercury round the Sun, and are sometimes in front of it, sometimes behind it, though they never depart from it beyond certain limits. All of which facts were discovered and observed a few days ago by the help of a telescope devised by me, through God's grace first enlightening my mind.
>
> *Galileo Galilei*

> I think I've found a squashed comet!
>
> *Carolyn Shoemaker*

On a dark hillside overlooking Padua, Italy, in the year 1610, astronomer Galileo Galilei studied the sky through a crude telescope. He was gazing eagerly at a bright object that had tantalized his curiosity for years. Much to his surprise, he found that this object – the planet Jupiter – had four companions that seemed to circle it much as he had observed Mercury and Venus to circle the Sun. Over a series of nights, he watched Jupiter's tiny companions move in an intricate dance near the planet, and he concluded that they were a family of moons in Jovian orbit. To him they seemed to be an analog of the Solar System. In time, Galileo focused his gaze on Saturn, and saw something around that planet as well. Perhaps, he reasoned, they were moons as well, although through his simple instrument they looked more like ears. While he often speculated what these things were, he never did resolve the mystery of these Saturnian companions.

Galileo was not the first to notice movement in the heavens. During the centuries before his observations, others had studied the heavens, and noted the actions of so-called 'wanderers'. These point-like objects seemed to move at will across the sky. People went to great efforts to chart and predict them. While earlier astronomers, such as Aristotle and Ptolemy, knew of at least five planets by their motions, it did not occur to anyone that these wan-

**Figure 3.1.** Silhouette of Galileo. (Artwork by T. Kuzniar, courtesy Loch Ness Productions)

derers – these planets – were worlds in their own right. Galileo's journal entry tagging these planets as other worlds changed for ever humanity's views of the Solar System.

Nearly 400 years after Galileo's observations, on a warm July evening in 1994, Jupiter was the focus of world-wide attention. For one group of astronomers crowded into a control room at the Space Telescope Science Institute, it was the end of a feverish year of preparations. They were waiting and watching as 21 fragments of Comet Shoemaker-Levy 9 crashed into the swirling clouds of Jupiter. It was an historic first for planetary scientists – the chance to see just what sort of changes would take place when pieces of a comet interacted with a gas giant planet.

The astronomers – part of an extended network of scientists around the world – would not actually see the pieces of Shoemaker-Levy 9 plunge into the Jovian cloudtops. Instead, everyone gathered at the Institute had pinned their hopes on the HST – with its newly repaired optics and sensitive spectrographs – to capture both visible and ultraviolet views of the aftermath of the collisions. At the same time, ground-based astronomers used infrared instruments to measure the energy given off as each comet fragment was consumed in a huge fireball. It was an event unprecedented in astronomy, and it marked the coming of age of the Hubble Space Telescope as one of the great Solar System observatories, as we will discuss in detail at the end of this chapter.

## Pre-HST exploration

Four decades of active space exploration have changed our view of the Solar System drastically. Like explorers of earlier centuries, robotic spacecraft have scouted out the territory, reporting back a tantalizing array of first impressions. Through the electronic eyes and ears of our spacecraft, we have seen an incredible variety of planetary surfaces, atmospheric structures and moons scattered throughout the Solar System.

In the 1960s, the first explorers went to the Moon – these were spacecraft with names like Ranger and Rover. Mars was the next target, studied by a series of Soviet Mars probes and the US Mariner craft. They prepared the way for the Viking landers in the mid-1970s. One Mariner craft studied Mercury; two Pioneer Venus spacecraft, a series of Venera spacecraft and Magellan went to Venus; Pioneer 11 and 12, and Voyager 1 and 2, went on fly-by missions that included all the outer planets except Pluto; the Galileo spacecraft has looked at Venus and Earth, and is studying Jupiter; and Ulysses – the solar–polar mission – is examining the solar wind from various vantage points around the Sun.

Astronomers have even received their first close-up studies of asteroids and comets, probing Gaspra and Ida (and discovering Ida's small moon Dactyl), as well as comets Halley, Giacobini-Zinner and Grigg-Skjellerup. Recent Earth-approaching asteroids and the Shoemaker-Levy 9 impacts on Jupiter have spurred scientists to mount a systematic search for small bodies in the inner Solar System. Others are concentrating on the population of small bodies out among the gas giant planets and beyond. It has been a magnificent four decades of exploration along the shores of the cosmic ocean, and, as astronomer Carl Sagan wrote in *Cosmos*, 'we have waded a little out to sea, enough to dampen our toes, or at most, wet our ankles'.

So, you might ask, what is left to study in the Solar System? In a word: everything! What we have gained over these past decades of planetary exploration is a mere nodding acquaintance with the Solar System and some interesting theories on how it may have formed. What we lack is the depth of understanding that prolonged study can give us. While we have been able to categorize the forces that shape solid bodies in the Solar System – cratering, tectonism and volcanism – we are still learning about the dynamics of large-planet atmospheres, the plasma physics of comet tails, and the true makeup of the many families of asteroids that inhabit the spaces between the planets.

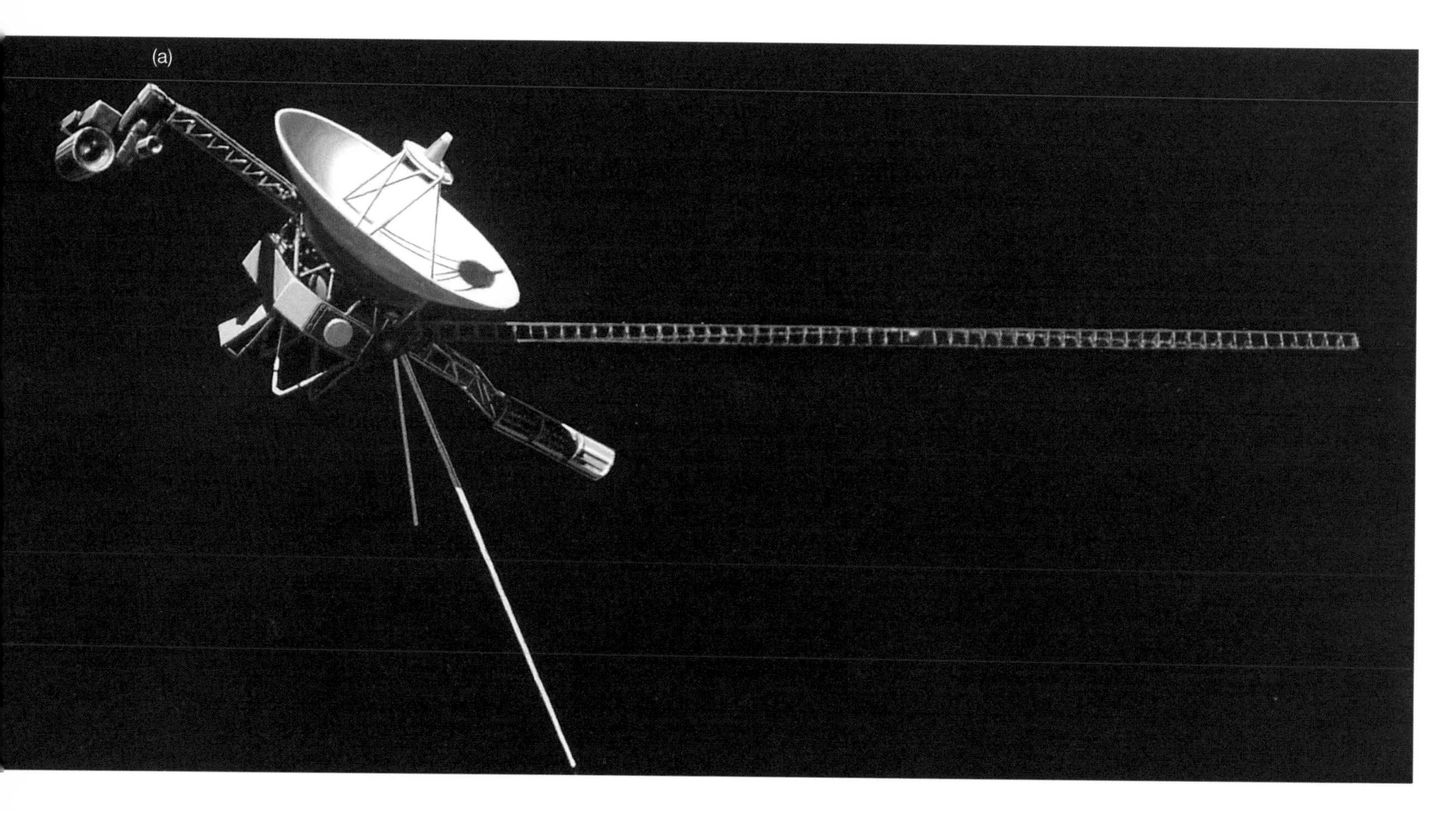


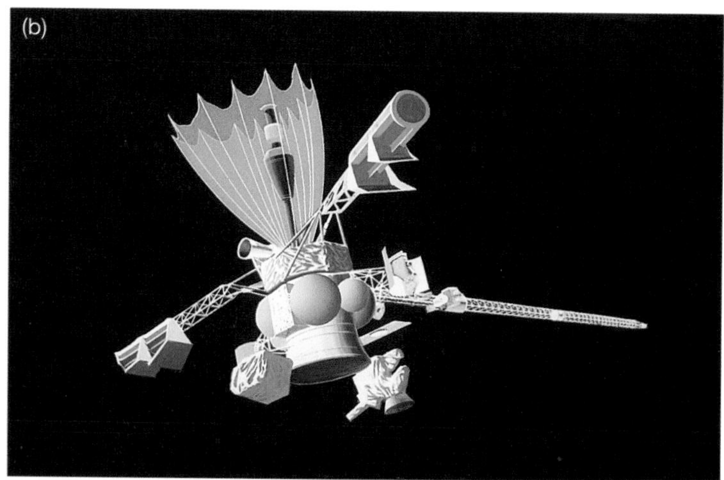

**Figure 3.2.** (a). Voyager; (b) Galileo; (c) Magellan. (Artwork by T. Kuzniar, courtesy Loch Ness Productions)

The earliest cosmologies held that the Earth was at the center of the Solar System, if not the Universe itself. That idea was replaced by the concept of a Sun-centered universe, which was in turn replaced by the current theory of an expanding universe with no center. Our thoughts on the origin of the Solar System itself have gone through similar changes. Not many years after Galileo Galilei's observations, philosopher Rene Descartes tackled the genesis of the Sun and planets. His was a systematic and scientific approach based on precise observations rather than on Biblical suppositions or unscientific 'theories'. Descartes theorized that vortices formed in a primordial gas cloud, and were responsible for the formation of the planets. His early theory did not stand up for very long, and the next person to try and explain the origin and evolution of the Solar System was a French nobleman – the Comte de Buffon. He suggested that the planets formed when a comet plowed into the Sun, the impact causing it to eject material that later became the planets. Buffon was the first of many proponents of what we now call 'catastrophe' theories of Solar System evolution. Some of these theories postulated the collisions of stars as the seed activity for planetary formation, and that the action of one star merely passing by another would be enough to start the planetary birth process.

In the 18th century, philosopher Immanuel Kant advanced a theory that the Sun and planets formed in one continuous process. The idea was fine as far as it went, but it needed a bit more definition. Mathematician Pierre Simon de la Place took it one step farther. *He* saw the birth place of the Solar System as a spinning gas nebula, and this theory is the prototype of the creation story astronomers tell today. The Solar System's birth place was most likely a cloud of hydrogen and helium gas, plus a good supply of heavier elements ejected from the explosion of massive stars (supernovae). The cloud began to contract, possibly triggered by shock waves from a nearby supernova. The waves of contraction compressed the material in the primordial nebula, and the compression caused the cloud to heat. Fluctuations may have set the nebula spinning, slowly at first but speeding up later as gravitational attraction began to pull clumps of material together. As the spinning continued, the nebula flattened out and took on the orbital characteristics of the proto-Solar System.

The bigger agglomerations attracted more and more material to themselves. The largest clump became the proto-Sun. Dust grains trapped in the nebula began to stick together to form planetesimals, which collided and merged with one another to become the protoplanets.

At some point, the compressional heat in the proto-Sun was high enough to trigger nuclear reactions. When that happened, the proto-Sun became a star, flaring fiercely in its newborn youth. Winds from the newly born Sun blew the remaining gas cloud away from the inner Solar System, leaving only the Sun and four rocky bodies that were to become airless Mercury and atmosphere-rich Venus, Earth and Mars. The larger planets, with their higher masses and gravities, managed to hold on to hydrogen and helium atmospheres, becoming Jupiter, Saturn, Uranus and Neptune. The leftovers of this birth process very likely became the comets and asteroids, and quite possibly the many moons that orbit the planets.

This modern theory of Solar System formation is far from perfect. It leaves many unanswered questions about the rate of formation and the origin of the primordial cloud. The theory does not completely explain the existence of the asteroid belt between Mars and Jupiter, or the size and extent of the Kuiper Belt and Oort Cloud of cometary nuclei in the outer Solar System. Still, it is the best theory we have. The only way to prove it right or wrong is to continue our Solar System exploration efforts, and to look for planetary systems around other stars.

Researchers have classified the worlds of the Solar System in a number of ways: terrestrial worlds and gas giants; planets with atmospheres and those without; worlds with moons and those without; comets, asteroids, near-earth asteroids, proto-comets in the outer belts, Sun-orbiting comets, Jupiter-orbiting comets, and a host of others.

The Earth is the benchmark against which we compare all other planets. Earth's atmosphere, for example, seems to be unique in the Solar System. Mercury has no air, and Venus has a poisonous carbon dioxide and sulfur dioxide atmosphere. Mars has a thin carbon dioxide atmosphere, but there is evidence it may have had a heavier blanket of gases in the distant past. We know that our own atmosphere was mainly carbon dioxide in the early days of the planet, and that it has been modified by volcanic outgassing and other processes. However, Venus and Mars also show evidence of volcanism, so what is the difference? Certainly the presence of life and water on the Earth make some difference in the evolution of Earth's atmosphere, but how much? And how do we know for sure?

To answer these and many other questions, we continue to study the other planets. One useful way to carry out these studies is through the use of orbiting spacecraft such as Magellan for Venus, and fly-by spacecraft such as Pioneer and Voyager. Except for the long-term missions (the Viking Mars studies or the Galileo explorations of Jupiter), most of the probes sent on planetary voyages have been fly-bys on fixed trajectories – capturing brief moments in the life of a planet before moving on. What we receive are planetary 'snapshots' – images of worlds frozen in an instant of time. Unfortunately, there are only a few new planetary probes being built to be sent out on voyages of discovery, and for long-term studies snapshots are not enough. The worlds of the Solar System are dynamic places. We need to make more observations of the planets over a long period of time before we can truly say we understand them.

Until any new missions are launched, however, HST can produce good quality planetary images. The main advantage that HST has over ground-based and fly-by type missions to the planets is that it can look at an object over a long period of time through broadband filters and can carry out high-resolution spectroscopy of the other members of the Solar System.

Observing planets with HST is a tricky business. Ironically, these objects – which are Earth's closest 'cosmic neighbors' – present the telescope with its greatest observing challenges. Like any Earth-based telescope, HST has to track its moving targets. Compared to stars and galaxies, however, planets move relatively rapidly against the backdrop of the sky. Scientists using HST must always account for these motions when they plan their observations, and, of course, the telescope's own orbital motions have to be added into the observing calculations. The current method of acquiring and tracking Solar System bodies with HST involves a combination of 'ambush mode', guide star hand-offs (where the telescope uses a succession of guide stars to help keep it pointed accurately as it orbits the Earth) and a slow scan rate that matches the movement of the object.

## Hubble's dynamic Solar System

### Mars

People have always been fascinated by Mars. It stands out in the night sky as a reddish-tinged, unwavering point of light. If we look at it through an amateur telescope (with, say, an 8-inch instrument or larger aperture), we can just make out faint markings on the Martian surface. Over the course of a Martian year, we can watch the polar caps grow and shrink with the change of seasons. We know Mars has an active atmosphere because planet-wide storms sometimes blanket the surface with dust. For clearer views of the planet, we need larger telescopes. However, even the largest ground-based instruments cannot resolve small details on the Martian surface.

The most meticulous explorations of Mars were conducted in the 1970s, when a series of spacecraft from the USA and the then Soviet Union imaged the planet. Two Viking landers settled onto the surface of the Red Planet in 1976 and returned a steady stream of data about atmospheric and surface conditions. Recent attempts at exploration with two Phobos craft and a Mars Observer mission ended in disaster as contact with all three probes was lost. More Mars missions, such as the Mars Global Surveyor, are being planned, which will continue the studies begun more than 25 years ago.

Why all the interest in Mars? Certainly the planet seems to resonate in our collective unconscious as a place to explore. Some Mars enthusiasts point to the planet as a future home for Earth colonists. Others see in Mars a chance to study a planet similar to the Earth in many ways. A great deal of scientific interest in the Red Planet focuses on its atmospheric conditions and surface characteristics. Understanding what happens on Mars and comparing it to the conditions we experience on Earth is a strong motivation for Mars research. For those who want Mars to be a second home to humanity, a knowledge of climate and terrain would be essential. To form the most accurate picture of the changes on Mars, however, we need to monitor the planet over many years.

One of the big questions in Mars studies is: did it form with the same amount of water as the Earth did, or did it form with less? If we assume that it started out with the same amount of water as the Earth, then something must have happened long ago to cause Mars to lose most of its water. The alternative is to assume that it started out with much less and

**Figure 3.3.** A color composite of images of the Syrtis Major region on Mars, taken with WF/PC-1. The image was so striking that it was reproduced on the May 1991 cover of *LIFE* magazine. (Philip James, University of Toledo; NASA)

then try to find out why. Either way, we are faced with the certain knowledge that some amount of water once flowed on Mars because we see evidence of it all over the surface, for example the channels with tributaries spread across the dusty plains, looking remarkably similar to dry riverbeds on Earth. It is possible that some small fraction of the original water supply is locked underground in permafrost, but it seems certain that a great deal of it escaped into space. The question is, how do we go about proving that this happened?

One possible way is to look in the upper atmosphere for the presence of *deuterium*, an isotope of hydrogen. To form deuterium, you take the nucleus of a hydrogen atom (which contains one proton) and add a neutron to it. Since hydrogen is the main element in water (water is two atoms of hydrogen and one of oxygen ($H_2O$)), you might expect to find some deuterium in the water in the upper atmosphere.

Imagine the Earth were to lose massive amounts of water to space in a runaway evaporation cycle. The water, in the form of vapor or gas, would rise high up into the atmosphere. There it would break apart – dissociate – into hydrogen and oxygen. Hydrogen, being very light, would escape into space rapidly, whereas any deuterium, which is heavier, would escape slowly, and much would be left behind. Now assume that millions of years go by as this process takes place. A group of aliens comes by, exploring the Solar System. They find this dry and desiccated planet, pockmarked with craters. It looks as if it used to have water, because its surface is criss-crossed with fossilized channels, dry riverbeds and flood plains. The big question is: why is it that way?

To find out, they start to measure the atmospheric water content. They find a little water vapor and lots of deuterium. That deuterium content tells them that a large amount of hydrogen, from water, escaped into space over time, leaving behind the deuterium. The more deuterium found in the upper atmosphere, the more hydrogen must have escaped. If our aliens calculate a deuterium-to-hydrogen ratio, and work backwards from that ratio, they might deduce that this desert planet once supported huge oceans and teemed with life. Could the same be done for Mars?

For many years, planetary scientists *have* searched the upper atmosphere of Mars in the same manner as our hypothetical aliens, looking for the spectral signature of deuterium. They have looked at the data gathered by all of the missions to Mars in an effort to gain some understanding of the planet. However, Mars exploration did not stop when the Viking 1 spacecraft stopped sending data in 1982; in the years since then, planetary scientists have studied the planet from ground-based observatories, from NASA's Kuiper Airborne Observatory, and by analyzing the makeup of a peculiar set of meteorites found on Earth that are thought to have come from Mars. All of this study has allowed planetary scientists to make some very sophisticated estimates of the existence of water, deuterium, ozone and other gases in the Martian atmosphere. For example, based on theoretical models of hydrogen escape, planetary scientists have come up with a ratio of deuterium to hydrogen on Mars which is five times that of the Earth's. Some estimates of water loss point to Mars losing a layer of water which was somewhere between 3 and 27 meters deep! Clearly there is a lot of work to be done when we resume our *in situ* studies of this planet.

In May, 1991, a team led by French astronomer J.L. Bertaux of the Service d'Aéronomie du CNRS, used the Goddard High Resolution Spectrograph to obtain a short exposure spectrum of the Martian atmosphere. The team reported a possible detection of deuterium in the form of a D Lyman-alpha line in the spectrum of Mars, and have used it to establish an upper limit for the abundance of deuterium on Mars. Using the spectrographs to study the Martian atmosphere shows that such observations are possible, and the story of these types of observations remains very much a work in progress. Other observations have concentrated on searching the upper atmosphere for deuterium and comparing the abundances to amounts suspected to exist in the lower atmosphere. If these observations can be continued and expanded, the next step will be to apply the data to determining the timing of the events that triggered the catastrophic loss of water on Mars.

Other changes in the Martian atmosphere are the main interest of a group of researchers who are devoting their HST time to a multi-year study of Mars. Philip James of the University of Toledo, Leonard Martin of Lowell Observatory, the University of Colorado's Steven Lee, and Todd Clancy of the Space Science Institute in Colorado, planned a series of observations to measure the changes of albedo (reflectivity) of various surface features on Mars, as well as to study the opacity of the Martian atmosphere and to take measurements of the seasonal and interannual variations in the ozone distribution. Apparently, the creation of water vapor is part of a chain of chemical reactions that remove ozone from the Martian atmosphere. HST's ability to image Mars in the ultraviolet allows the team members to measure the absorption of ozone, which, as it does on Earth, plays a role in the upper atmosphere. Of particular interest is the part ozone plays in atmospheric changes over the Martian polar regions. The team also wants to learn more about how the polar ice caps on Mars change with the seasons. The preliminary results of the Mars team's work suggest that the amount of ozone in the Martian atmosphere changes seasonally with latitude. There is also evidence for strong ozone absorption in the north polar region of Mars during late winter.

In the first striking images released by James and his team, some features familiar to Mars-watchers show up in sharp detail. The large dark area shaped like a shark's fin is the Syrtis Major Planitia. The surface material here is probably a coarse, dark-colored sand. Prevailing winds blow the dust into the brighter-looking Arabia Planitia to the west. To the east of Syrtis lies Isidis Planitia, an impact basin formed more than two billion years ago when a meteor slammed into Mars.

The planet-wide dust storms that so fascinated observers back in the 1970s do not seem to be recurring with any regularity. Steve Lee has been monitoring the region around Syrtis Major, which is one of the most variable regional albedo features on the planet. (Albedos change when winds move dust and sand around on the surface.) Wind transport may be responsible for many of the variable albedo features seen on Mars. By early 1994, Lee and his colleagues had seen no major dust storms in their HST data. Yet, if Viking data were to be believed, they should have seen major changes on the Martian surface if a dust storm *had* occurred.

In at least 12 separate observations of the region, little variability was seen. The team continues to wait for a storm that HST can observe. Ironically enough, it appears that a storm did blanket the planet during early 1994, but, as with many another project, the team did not have any HST observations scheduled. Still, their work has resulted in remarkable photographs showing seasonal changes on the surface of Mars as well as in its atmosphere. HST's high-resolution images also showed craters on the Martian surface – something that has arguably never been done using ground-based observatories.

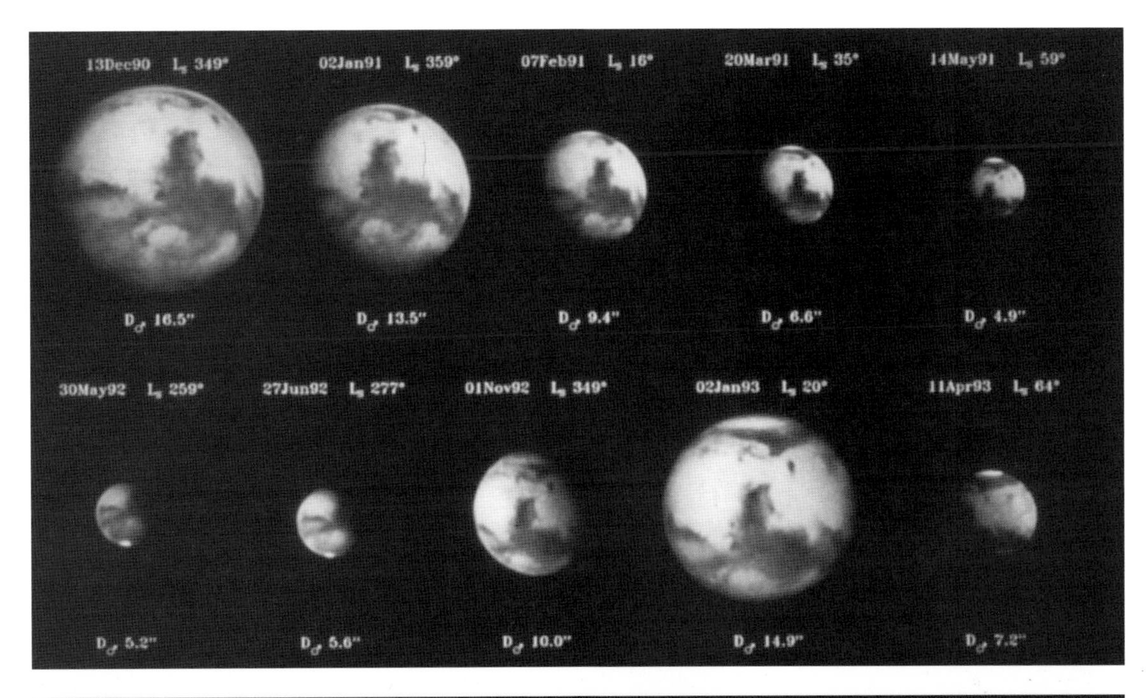


**Figure 3.4.** A sequence of Mars images taken with WF/PC-1. Indicated on each are the date, the longitude of the sub-solar point ($L_S$) and the apparent diameter of Mars ($D_♂$) in arcseconds. (Steve Lee, Laboratory for Atmospheric and Space Physics, University of Colorado)

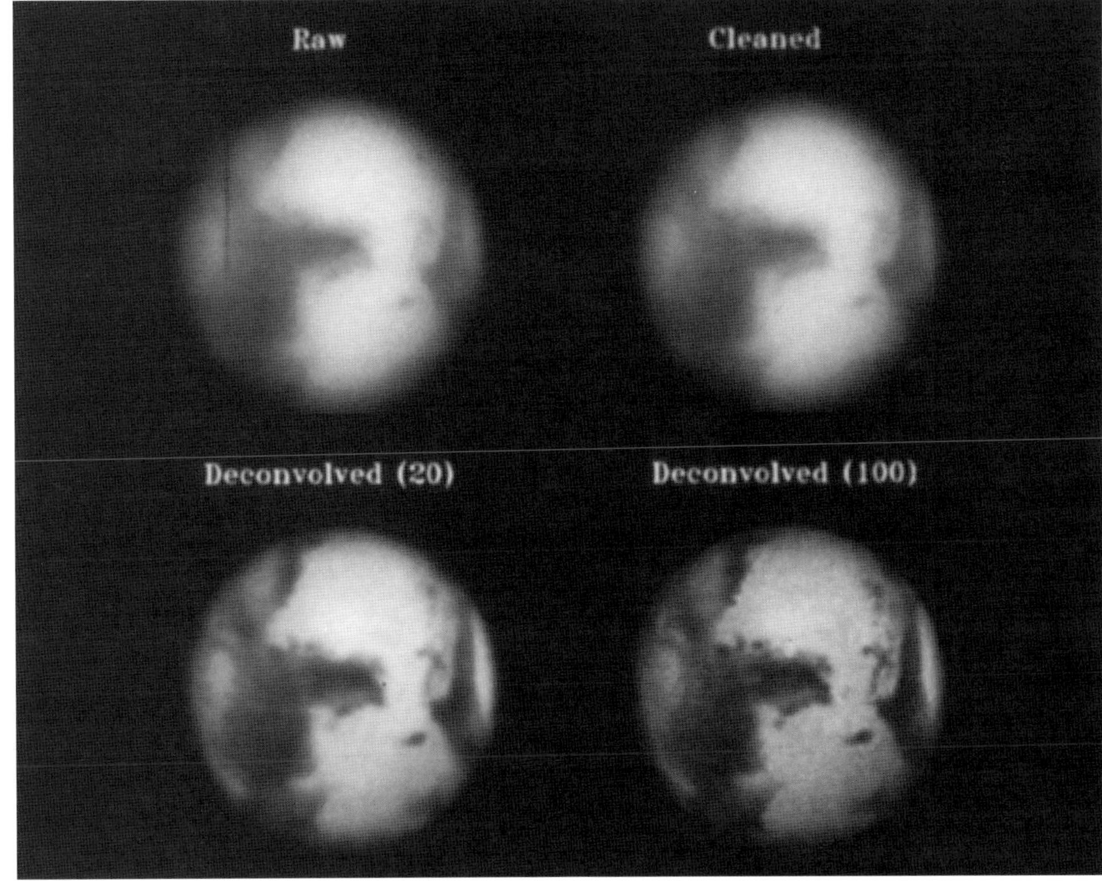


**Figure 3.5.** Steps in the deconvolution of a Mars image. See the text for a discussion. (Steve Lee, Laboratory for Atmospheric and Space Physics, University of Colorado)

**Figure 3.6.** The three faces of Mars. These HST views – taken on February 25, 1995 – provide the most detailed complete global coverage of Mars ever seen from the Earth. Mars was at a distance of 103 million kilometers, and HST was able to resolve features as small as 50 kilometers across. To the surprise of the researchers, Mars appears cloudier than seen in previous years. This means that the planet's atmosphere has become cooler and drier because water vapor has 'frozen out' to form ice-crystal clouds. (Philip James, University of Toledo; Steve Lee, Laboratory for Atmospheric and Space Physics, University of Colorado; NASA)

HST's Mars imagery provides an excellent opportunity to illustrate one of the techniques used to correct for the effects of spherical aberration. In Figure 3.5, the 'raw' image has obvious blemishes – the vertical line, for example. These blemishes are removed to make the 'cleaned' image. Then the deconvolution procedure is used to improve the image. The technique is applied repeatedly to each 'improved' version of the image; 20 times and 100 times, in this case. By contrast, because they were taken with the refurbished HST, the images in Figure 3.6 needed no deconvolution to make them presentable.

## Venus

Early in 1995, University of Colorado planetary scientist Larry Esposito used HST as a sort of interplanetary weather satellite to study Venus. His observations focused on the thick blanket of clouds that covers the heavily volcanic surface of Earth's so-called 'sister planet'. You might think that Venus is an unlikely target for HST observations, and for much of its orbit it is. This is because the planet always appears very close to the Sun, and HST is not allowed to let its gaze wander too close to the Sun. For a few weeks each orbit, however, when Venus is at its farthest from the Sun (as seen from Earth), researchers can

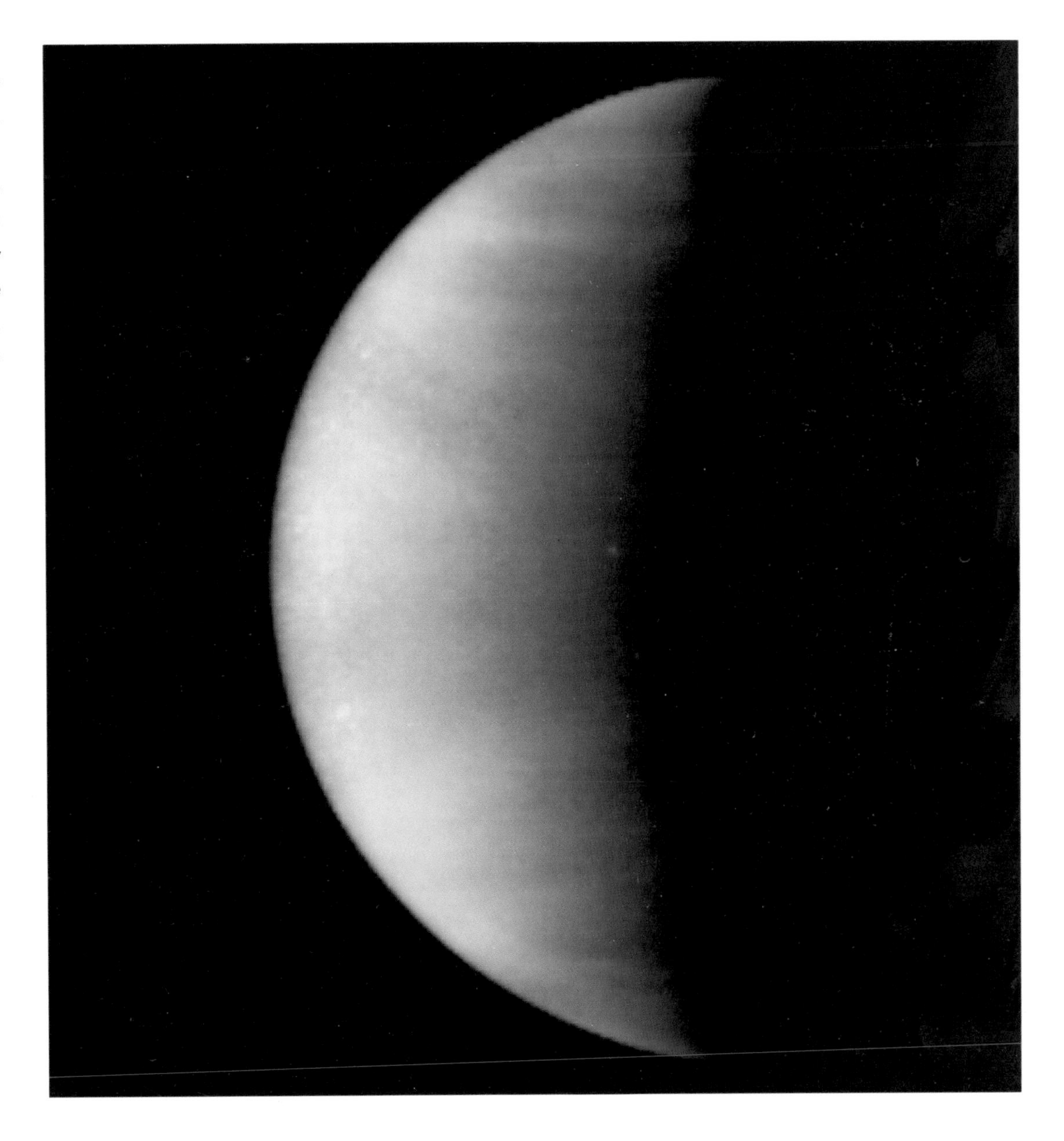

**Figure 3.7.** This WF/PC-2 image of Venus was taken on January 24, 1995, when Venus was 70.6 million kilometers from Earth. False color has been applied to the image to enhance the cloud features. (L. Esposito, Laboratory for Atmospheric and Space Physics, University of Colorado, Boulder; NASA)

risk sneaking a peek at the planet. On January 24, 1995, HST took imaging and spectroscopic observations aimed at studying the cloud patterns on Venus. Figure 3.7 is HST's view of Venus.

At ultraviolet wavelengths, specific patterns in the Venus cloud-tops stand out – particularly a large, horizontal Y-shaped feature near the equator. Other spacecraft, sent to study Venus in the past, reported this peculiar formation, which may be a sort of 'wave' in the atmosphere, analogous to high- and low-pressure cells in Earth's atmosphere. The polar regions of the cloud deck appear bright, and could be covered with a haze of small particles. Dark regions in the clouds show the location of increased sulfur dioxide levels near the tops of the clouds.

The heavy cloud blanket on Venus is very different from Earth clouds, which largely comprise water vapor. On Venus, the clouds consist mostly of sulfuric acid, and it seems that the atmosphere was deluged with a sort of 'sulfuric acid rain' triggered by a volcanic eruption in the late 1970s. It appears that the atmosphere is still recovering from that eruption (there is much less sulfur dioxide in the dark regions now than there was in the early 1980s, for example), and researchers plan to use HST to study the Venus atmosphere the next time the planet is far enough away from the Sun to allow a safe view.

## The Jupiter system

Jupiter presents a treasure trove for study by HST – as the spectacular events surrounding the impact of comet Shoemaker-Levy 9 in July, 1994, demonstrated. Aside from cometary impacts (which we will discuss later in the chapter), some very rewarding research programs focus on the turbulent atmosphere of the planet and the interaction of that atmosphere with Jupiter's magnetic field.

Jupiter is a frequent target of a number of ground-based observatories. From the spectacular Voyager 1 and 2 images, most of us are familiar with Jupiter's moons, its cloud belts and zones and the Great Red Spot – which has been part of the Jovian 'cloudscape' for at least 300 years.

However, studying Jupiter with HST reveals a world very different from the gas giant revealed by the Voyagers in 1979. For one thing, the Great Red Spot continually changes its color. Also, the cloud belts constantly evolve and mutate, and multitudes of smaller storms have raced across the face of the planet and disappeared.

As useful as the Voyager work is for showing the scale and extent of features in the Jovian atmosphere, it is hard to get a feel for Jupiter's true nature from a series of 'snapshots'. HST's contribution to the study of Jupiter is an ability to deliver high-resolution images of the planet every 1.5 hours, to take ultraviolet spectra of the planet, and to give a more consistent feel for what the planet's upper atmosphere does over the long term.

One of the astronomers who uses HST to look at Jupiter is Reta Beebe, a scientist at New Mexico State University in Las Cruces, New Mexico. Among other things, she studies wind displacements in the cloud zones on Jupiter. Beebe and her fellow observers obtained images of Jupiter using HST's Wide Field and Planetary Camera, and are doing a long-term series of observations of the planet before Galileo sends a probe into the atmosphere in mid-summer, 1995.

HST's spectrographs are continually scanning Jupiter, gathering data on the aerosol hazes in the upper atmosphere of the planet and probing the limb darkening over certain latitudes. While HST cannot see Jupiter's magnetic field, it can and does study the effects

**Figure 3.8.** The first true-color image of Jupiter obtained by HST. Red, green and blue light WF/PC-1 images on 28 May, 1991, were combined to produce this view. The Great Red Spot is visible in the lower right quadrant. (STScI; NASA; ESA)

of the field on particles in the upper atmosphere. The Jovian magnetic field is the strongest and most extensive of any of the planets in the Solar System. Jupiter's magnetosphere is generated by the planet's fluid, electrically conducting interior, and it extends out about 2 million kilometers, well past the orbit of its moon Callisto.

This magnetosphere traps electrons along lines of magnetic force, as well as sulfur ions from the tiny moon Io. Radiation from these electrons results in radio emissions, and aurorae occur over Jupiter's polar regions when the ions slam into the atmosphere. These aurorae were first seen in ultraviolet imaging from the Voyager 1 spacecraft in 1979. Since

**Figure 3.9.** An image of Jupiter taken on May 18, 1994, by WF/PC-2. Jupiter was 670 million kilometers from Earth at the time the image was taken. The dark spot on the disk of the planet is the shadow of the moon Io, seen in the upper center. Even from this distance, HST was able to resolve some surface details of the tiny moon. (H.A. Weaver and T.E. Smith, STScI; J.T. Trauger and R.W. Evans, NASA–Jet Propulsion Laboratory; NASA)

then, astronomers have studied the phenomenon in an effort to understand the fine details of Jupiter's electromagnetic environment.

University of Texas astronomer Larry Trafton used HST's Faint Object Camera and the Goddard High Resolution Spectrograph to look at Jupiter's northern polar region, while the Ulysses spacecraft simultaneously measured Jupiter's X-ray and trapped-particle emissions.

Trafton looked at the ultraviolet light output of one small region within an aurora. To do that, the instruments were 'tuned' to observe hydrogen gas ($H_2$) emissions. When $H_2$

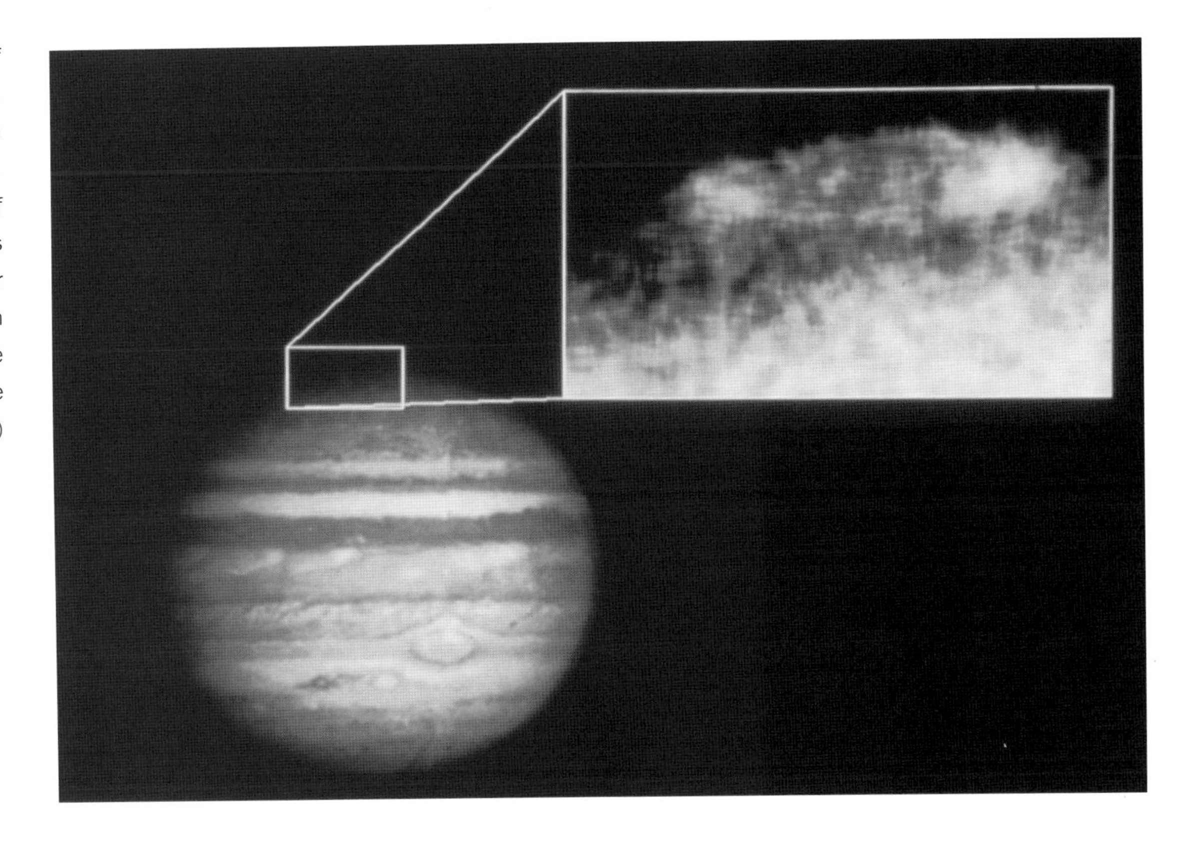

**Figure 3.10.** The FOC image of Jupiter taken on 8 February, 1992, shows the aurora in ultraviolet light (1600 angstroms). The box on the WF/PC-1 image of Jupiter shows the aurora's location. See Figure 3.35 for another example of Jovian aurorae. (John Caldwell, Institute for Space and Terrestrial Science and York University; NASA; ESA)

molecules are excited by interaction with a magnetic field, they radiate ultraviolet light. By focusing the spectrographs on the auroral regions and measuring the strength of the emissions, Trafton and his colleagues found that the gas heated to a temperature of 1200 kelvin (about 927 °C) in the aurora. The next step in the process is to determine the total amount of ultraviolet light emitted. From there, Trafton and others can move on to determining the total $H_2$ emission across an *entire* aurora. Figure 3.10 is an HST image of the Jovian aurora.

In 1992, Trafton, as well as a team led by York University (Canada) astronomer John Caldwell, turned HST's spectrographic attention to the ultraviolet emissions of sulfur dioxide found in another part of the Jovian system – the tiny moon Io. This rocky, silicate-rich moon is under continual study by both professional and amateur astronomers, using ground-based facilities along with HST. Before Voyager, scientists suspected the presence of some kind of volcanic activity on Io, and indeed both spacecraft showed Io to be an incredibly active world. At least eight eruptions were seen by Voyager 1, and Voyager 2 imaged six volcanoes. Squeezed into an orbital resonance between Jupiter, Ganymede and Callisto, Io is caught in strong gravitational fields, leading to flexing, which causes heating and melting. Eventually, the 'melt' (molten interior) reaches the surface in the form of sulfur flows and plumes, which send fountains of sulfur dioxide up through Io's tenuous atmosphere. It is the precise amount of this sulfur dioxide that Trafton and his colleagues want to determine with their HST observations.

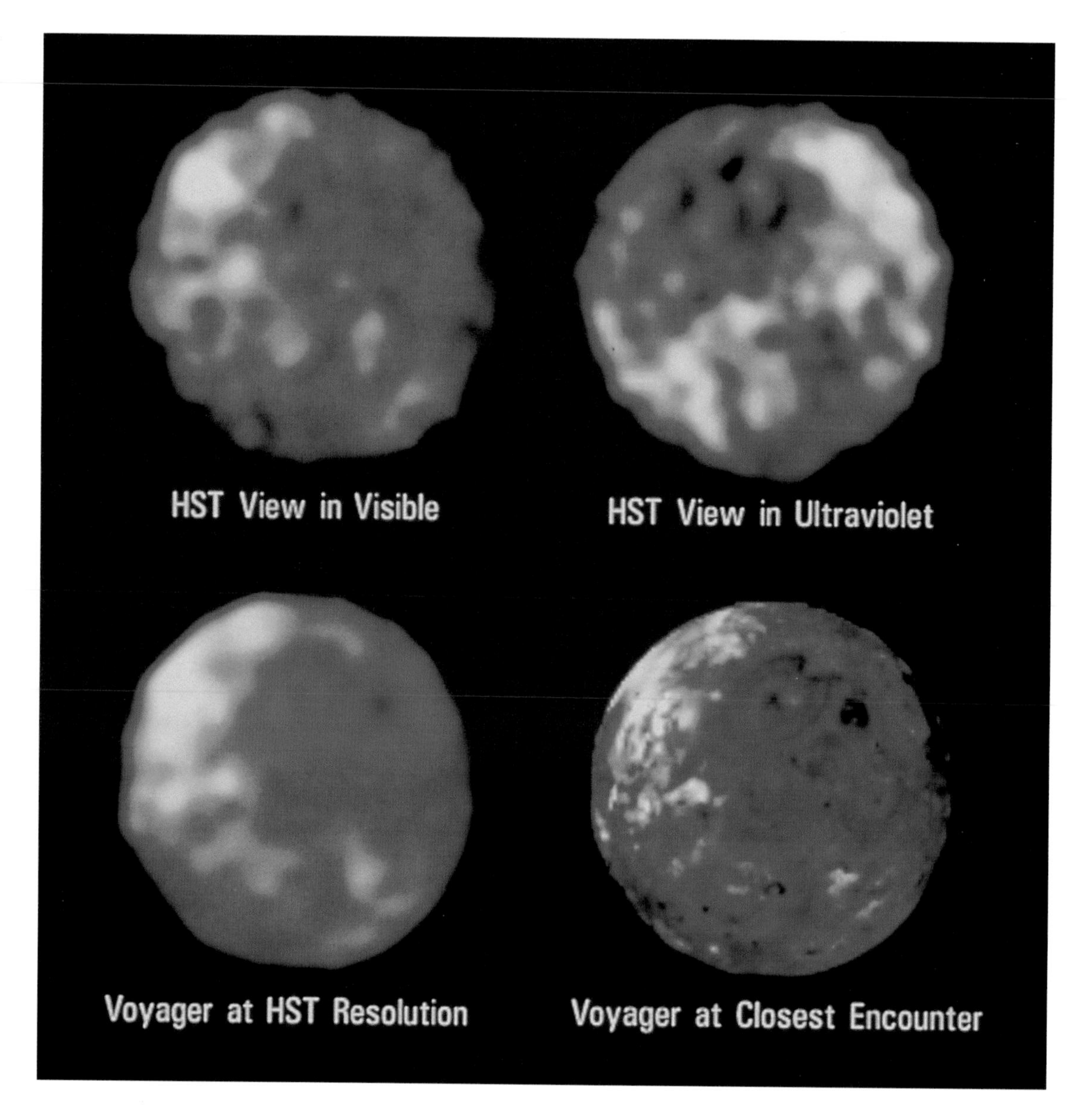


**Figure 3.11.** FOC and Voyager images of Io (Francesco Paresce, ESA and STScI; Paola Sartoretti, University of Padova; NASA)

Along with Trafton's atmospheric sulfur dioxide studies, other observers, led by Space Telescope Science Institute's Melissa McGrath, have studied Io's atmosphere with the Faint Object Spectrograph. That instrument was also programmed to look for sulfur and oxygen emissions from Io's surface. The results show that Io's atmosphere is three times smaller than previously thought. After Voyager passed by in 1979, planetary scientists theorized that Io's atmosphere was a result of emissions from volcanoes and evaporation of surface frost in sunlit areas. They also established an upper limit for the extent of the atmosphere at 5 Io diameters. Now, it is thought to be an average of 1.5 Io diameters, thicker over known volcanic regions and frost deposits, and thinner in other places. Some of Io's sulfur falls back onto the surface as a sort of 'snow' and some of it feeds into a giant ring of high-temperature gas surrounding Jupiter called the Io Plasma Torus. Ionized sulfur in the torus is one possible source of the particles that cause the Jovian aurorae. During HST's observations, however, oxygen emissions from the torus itself were discovered. It turns out that oxygen is twice as abundant in the torus as sulfur.

Some of the first Faint Object Camera images of Io's surface were taken by Francesco Paresce of the Space Telescope Science Institute and Paola Sartoretti of the University of Padova, Italy. Comparing these images with those from Voyager shows dramatically that the moon really has not changed very much in the years since the Voyager encounter, despite continual volcanic activity.

In Figure 3.11, the upper right image is an ultraviolet scan. Notice that it looks very different from the visible light image on the left. This is probably because Io is covered with sulfur dioxide frost, which absorbs ultraviolet light and causes those areas to look dark. The same areas look bright in visible light.

Io orbits Jupiter once every Earth day, and for half the orbit it is hidden behind Jupiter, out of sunlight. Astronomers have noticed that sometimes Io looks about 10 to 15% brighter when it emerges from Jupiter's shadow. This phenomenon is called Io post-eclipse brightening, and was the subject of early HST observations by Jim Secosky, one of HST's amateur observers. Secosky, a high-school biology teacher, said he ran across some material on Io's brightening when he was researching a paper on water in the Solar System. He wanted to track down a cause for the brightening and to determine whether evaporation of sulfur dioxide frost was a possible explanation. His HST program first ran in the spring of 1992, but his results were ambiguous, possibly due to timing problems with the observation. He was not able to observe Io exactly after eclipse, although he had three snapshots of the limb of Jupiter to try and spot the phenomenon. Still, Io does seem to brighten from time to time as it comes out of Jupiter's shadow, and Secosky now thinks that the brightening may be driven by sporadic volcanic activity.

Because Io is the Solar System's most dynamic moon, its evolution will be the impetus for more HST studies in the future. These include a continuing series of ultraviolet atmospheric observations, surface mapping and studies of the complex interactions between Io and Jupiter.

## Saturn

The gas giant Saturn, with its glittering ring system, has long been a favorite of both amateur and professional astronomers. So it should come as no surprise that Saturn was selected as one of the first planetary objects to be studied by HST. In fact, Saturn presented HST with its first planetary challenge only four months after launch.

On August 26, 1990, HST took what many called a 'Voyager quality' image of the planet. Figure 3.13 shows Saturn as it would appear from the ground if it were only twice as far

**Figure 3.12.** Voyager 2 image of Saturn. Compare with the HST images in Figures 3.13 and 3.15. (NASA)

away from us as the Moon. In reality, Saturn is $1.4 \times 10^9$ kilometers away – or about 2000 times farther away than the Moon. HST imaged the northern hemisphere of Saturn, showing the banded structure in the planet's upper cloud decks. The famous 'Cassini Division' (the largest dark gap) in the rings is clearly visible, as is a thin division near the edge of the ring system. This is the Encke Division, which had never been photographed from the Earth. The faint innermost 'crepe ring' is also shown.

About one month after HST's first image of Saturn was taken, a pair of amateur observers noticed a large white spot developing in Saturn's northern equatorial region. The date was September 25, 1990. They announced their find, and shortly afterward Reta Beebe made a series of ground-based observations. She reported that the storm was expanding eastward at about 400 meters per second, smearing out across the planet's cloud tops. Immediately the HST observing community clamored for time on the telescope to observe

**Figure 3.13.** WF/PC-1 view of Saturn reconstructed from red, green and blue light images taken on 26 August, 1990. Image deconvolution has been applied. (NASA)

this 'target of opportunity'. Everyone knew the telescope was still in orbital verification mode, and that procedures to track a moving target were barely functional, but Beebe, along with members of the Wide Field and Planetary Camera Team, convinced the Institute Director (Riccardo Giacconi) that this was an event worthy of special attention.

On November 9, 1990, after a hurried few weeks of planning, HST was aimed at Saturn and took a series of time-lapse images. The storm turned out to be so dynamic and complex that observers pleaded for more and more time to catch every nuance before it faded away. Life became very interesting very quickly for the observers and the moving targets planning group; they had to rush to put together a second observing run even before the first was completed. A request for a second observation came in only a week before Saturn would be too close to the Sun to be observed with HST. If the planners missed the opportunity, it would be 100 days before Saturn emerged from behind the Sun, and the storm might be gone. The decision was made to proceed, and everybody had just over a day to deliver the observation parameters and spacecraft commands to Goddard for execution.

Two problems threatened to complicate matters for the Saturn observers. First, the telescope gathered so much data that there was some danger of overwriting the data buffers

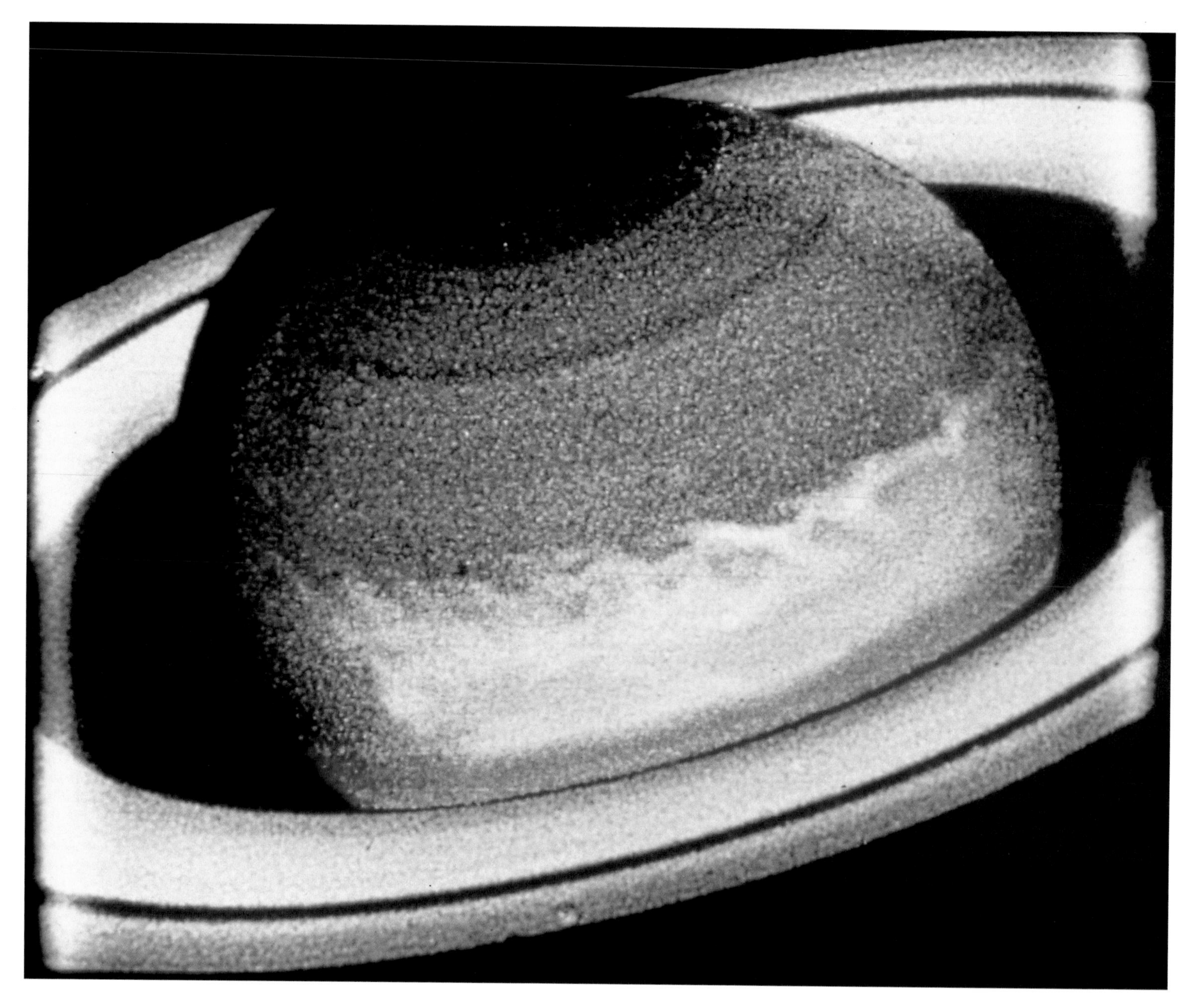

**Figure 3.14.** An HST color enhanced image of Saturn's great white spot, taken a month after the spot's discovery. The turbulent upper atmosphere of Saturn had already smeared the spot's clouds across the planet. (STScI; NASA)

on the spacecraft. Secondly, it also happened that the US military was running a secret shuttle mission during the week of the Saturn observations. This put a heavy load on the Tracking and Data Relay Satellite System, which, in turn, threatened the flow of HST data. Fortunately, controllers managed to work out the difficulties.

While the storm in the upper cloud belts came as a surprise to many people, it is actually a type of event that occurs relatively frequently – about once a generation. Planetary scientists first characterized it as Saturn's answer to Jupiter's Red Spot, but there are major

**Figure 3.15.** This image of Saturn, taken on 1 December, 1994, shows a rare storm that appeared in September 1994. The arrowhead-shaped storm was generated by an upwelling of warmer gas from deep within Saturn's clouds. (Reta Beebe, New Mexico State University; D. Gilmore, L. Bergeron, STScI; NASA)

differences between the two storms. On Saturn, high winds make such large-scale atmospheric events very short-lived phenomena. The event turned out to be less like a storm and more like an eruption from deep below the Saturnian cloud decks. One scientist was quoted in *Science* as saying, 'Saturn burped'. Indeed, the 'Great White Spot of 1990' has now been explained as an upwelling plume of ammonia ice crystals that burst out through the cloud tops. To see it so shortly after HST's launch and deployment was truly serendipitous.

HST observations of Saturn continue in an effort to gauge the long-term weather patterns in the clouds. In late 1994, Reta Beebe and Space Telescope Science Institute's Glenn Bergen imaged Saturn and spotted yet another storm. Like the 1990 storm, this one is primarily a white cloud of ammonia ice crystals that formed when warm gases flowed from underneath the colder, upper cloud decks. It now seems that these storms occur in a cycle, usually appearing during northern hemisphere summer. Saturn's high-speed winds, clocked up to 500 meters per second, are spreading out the storm of 1994. As with the 1990 white spot, HST will continue to track the storm's progress as long as Saturn is accessible to the telescope.

Few sights in the Solar System compare to the beauty of Saturn's rings. They have been known to astronomers since Galileo's time. They so puzzled him at first that he was at a

**Figure 3.16.** Rings of Saturn and the HSP. The path of the star behind the rings is shown. The dots indicate 5-minute intervals for the entire period of observation (20 hours). Data were taken in the intervals marked by the green dots and were not taken due to the South Atlantic Anomaly (red dots) or due to Earth occultations (violet dots). See Figure 2.15 for a sample of the HSP data taken for Saturn's rings. Analysis of the data revealed occultations by 43 different features. (Robert Bless, University of Wisconsin and the HSP Team)

loss for words to describe these 'handles' on Saturn. Indeed, the true nature of the rings kept astronomers conjecturing for years. Were they solid or made up of particles? Did they rotate? If so, how fast? If they were solid, how did they manage to stay together as they rotated?

Today we know that the rings are vast collections of particles, stretching some 200 000 kilometers across, but only a few kilometers thick. Thanks to the Voyager 2 spacecraft, we know that they sport an amazingly complex structure. During the August 1981 encounter, Voyager observed a star move behind the rings and recorded light intensity fluctuations as the starlight passed through. The study of those fluctuations (called 'occultation work'), coupled with some very detailed imagery, revealed an amazing array of complex phenomena in the rings: gaps, moonlets, spiral density waves and narrow ringlets. However, Voyager was only one spacecraft, and the next spacecraft to go out Saturn's way – the Cassini probe – will not leave Earth until late 1997 for an arrival in 2004. Until that mission flies, scientists need a way to carry out occultation work from Earth.

For an occultation to be successful, the telescope, the rings and the star they will occult have to be aligned perfectly. Using HST before the servicing mission, such things were possible, but incredibly difficult. The High Speed Photometer was perfectly designed to do high-resolution occultations from Earth orbit, but, of course, it was limited by the spherical aberration and jitter problems. The problems were daunting: the position of the planet had to be known precisely, the location of the star also had to be quite accurately known so that HST could follow it, and the position and stability of the telescope itself was important. Despite the complexity of the task, the HSP team went ahead with an occultation in October, 1991.

Figure 3.16 shows the path of the occulted star behind the rings. The most striking result from this experiment was that 43 different ring features could be identified using the HSP's data, despite the problems with spherical aberration and jitter. It is significant that the observation worked as well as it did, and even though the HSP was removed from HST in 1993, the work proves that successful occultations can be done from Earth-orbiting spacecraft.

Before leaving the Saturnian system, we should make a stop at Titan. Titan is one of

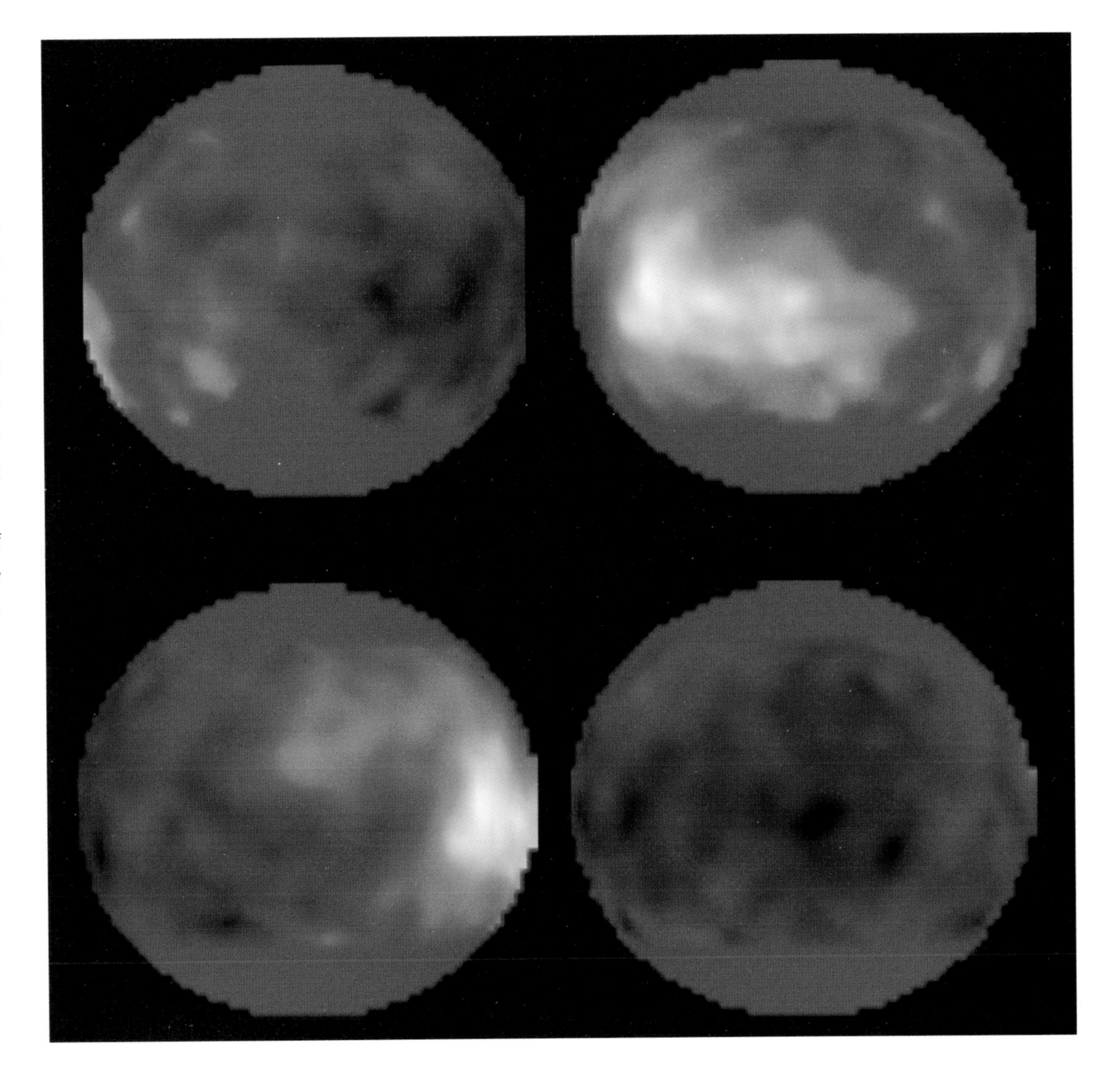

**Figure 3.17.** These four global projections of Saturn's largest moon Titan were assembled from 14 images taken by WF/PC-2 between October 4 and 18, 1994. The upper left image is the hemisphere which faces Saturn. The upper right image is the leading hemisphere and shows the brightest region. The lower left image is the hemisphere which faces away from Saturn, and the lower right image is the trailing hemisphere. (Peter H. Smith, University of Arizona Lunar and Planetary Laboratory; NASA)

the more intriguing moons orbiting Saturn, and is larger than the planet Mercury. Voyager showed it to be a world shrouded with methane clouds, with photochemical smogs that float high in its atmosphere of primarily nitrogen and methane. Its present temperature, less than –178 °C, keeps methane and ethane in gaseous and liquid states, but is cold enough to make water ice as hard as rock.

The thick clouds and haze prevented Voyager from seeing down to Titan's surface, and scientists had to wait until HST's Wide Field and Planetary Camera-2 was installed to take a near-infrared look under the clouds. Led by University of Arizona's Peter Smith and Mark Lemmon, a group of HST observers looked at Titan during a complete 16-day rotation of the moon. They took advantage of the fact that Titan's smog layer is transparent to near-infrared light and mapped surface features according to how much light they reflected. The image resolution of this technique allowed the researchers to see objects as small as 576 kilometers across. One large bright area appeared to be just over 4000 kilometers wide (about the size of Australia) and represents some sort of solid surface area. Other bright and dark regions could be continents, oceans, craters, or different kinds of surface features. Continuing analysis of the HST data will yield new information on small-scale surface features and atmospheric winds. HST's Titan information will be useful to the planners of the Cassini mission to Saturn, which will release the probe Huygens to study the atmosphere and surface of Saturn's cloud-shrouded largest moon.

**Figure 3.18.** A Voyager 2 image of Uranus, taken on January 17, 1986. (NASA–Jet Propulsion Laboratory)

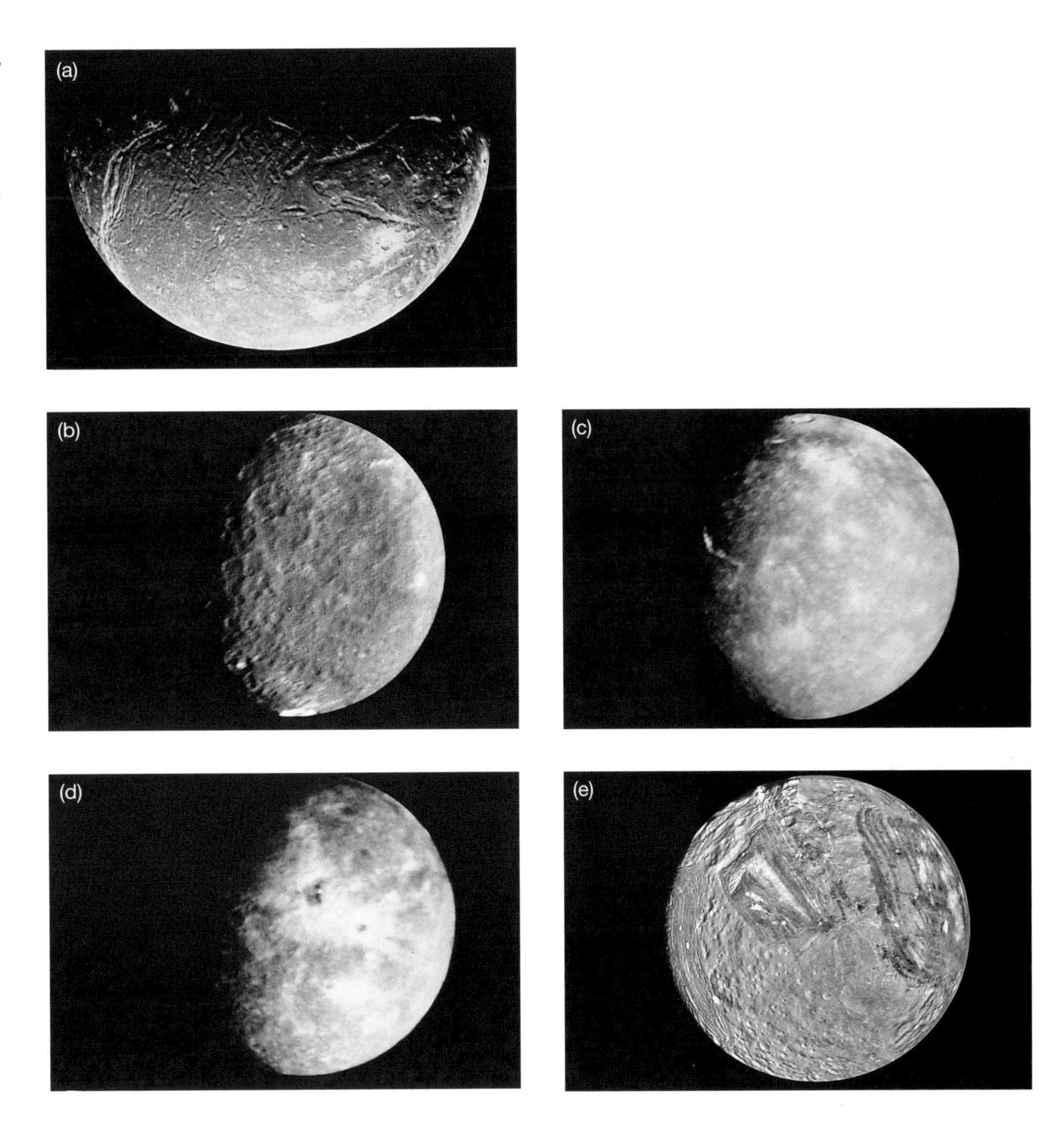


**Figure 3.19.** Voyager 2 images of Uranus's five largest moons: (a) Ariel, (b) Umbriel, (c) Titania, (d) Oberon and (e) Miranda. (NASA–Jet Propulsion Laboratory)

## Uranus and Neptune

In 1986, Voyager 2 flew past the gas giant planet Uranus, and gave planetary scientists their first close-up look at this enigmatic place. It appeared as a featureless blue ball, with a mainly hydrogen atmosphere. Voyager's ultraviolet cameras revealed cloud features and high-altitude methane hazes floating over the planet's sunward-facing southern pole. Voyager also imaged several of the moons of Uranus. These little worlds showed unusual variation in their surface features, ranging from the darkened ice of Umbriel to the wildly variable terrain of Miranda.

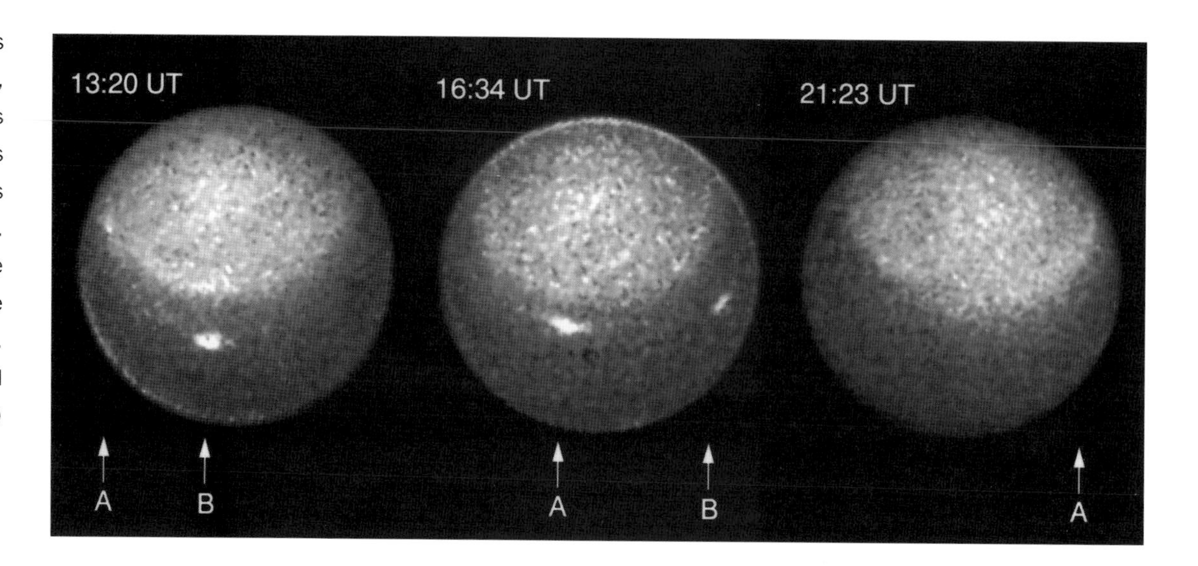


**Figure 3.20.** Images of Uranus taken by WF/PC-2 on 14 August, 1994, showing the planet's rotation and high-altitude clouds (marked A and B). Uranus rotates once every 7 hours, 14 minutes. These atmospheric details were previously seen only by the Voyager 2 spacecraft in 1986. (Kenneth Seidelmann, US Naval Observatory; NASA)

Eight years after Voyager 2's fly-by, a team of astronomers led by Ken Seidelman of the US Naval Observatory began a program of Uranus observations with HST. The first image was taken on August 14, 1994, when the planet was 2.8 billion kilometers from Earth. Clouds, rings and moons appear in sharp detail. The on-going observation program will focus on determining the composition and precise orbits of the moons and rings, and on detecting seasonal changes in the high-altitude hazes that blanket the planet.

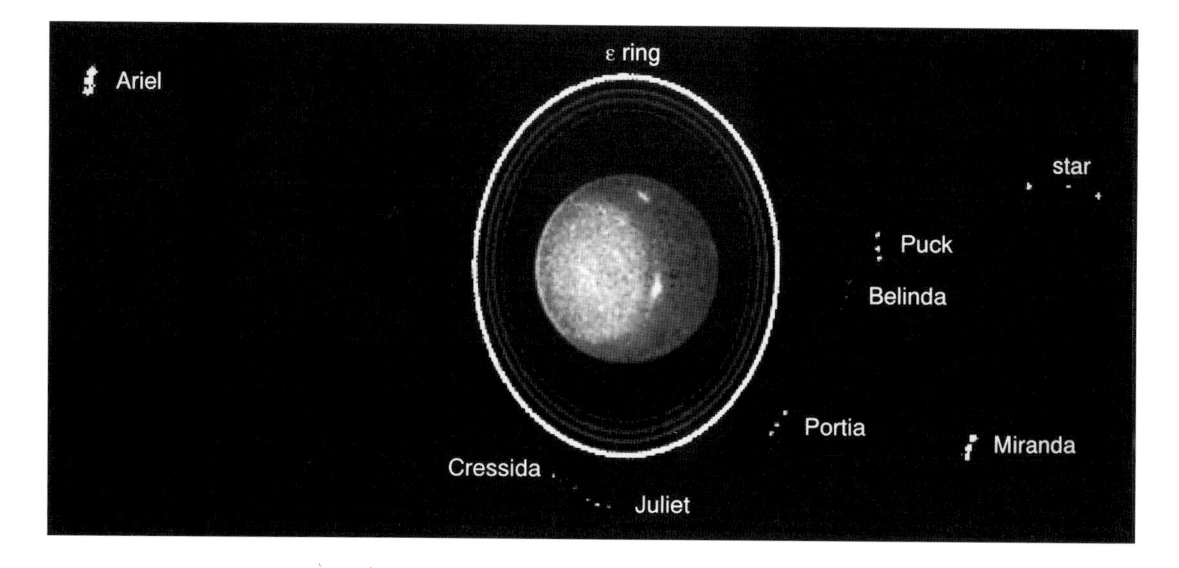


**Figure 3.21.** The moons and rings of Uranus as imaged by WF/PC-2. The picture is a composite of three images taken about 6 minutes apart. Because of their rapid motion, the satellites show up as a string of three dots. Since the Voyager 2 fly-by in 1986 (which discovered these inner satellites), none of them has been observed further. (Kenneth Seidelmann, US Naval Observatory; NASA)

Neptune is the farthest gas giant planet observed by HST, and it appears to have changed in appearance since Voyager 2 visited in 1989. Similar to Uranus in atmospheric composition (hydrogen, helium and methane), Neptune nonetheless distinguished itself by sporting a set of storms that were quickly dubbed the 'Great Dark Spot' and 'Dark Spot 2' by Voyager

**Figure 3.22.** Voyager 2 image of Neptune showing the Great Dark Spot. (NASA)

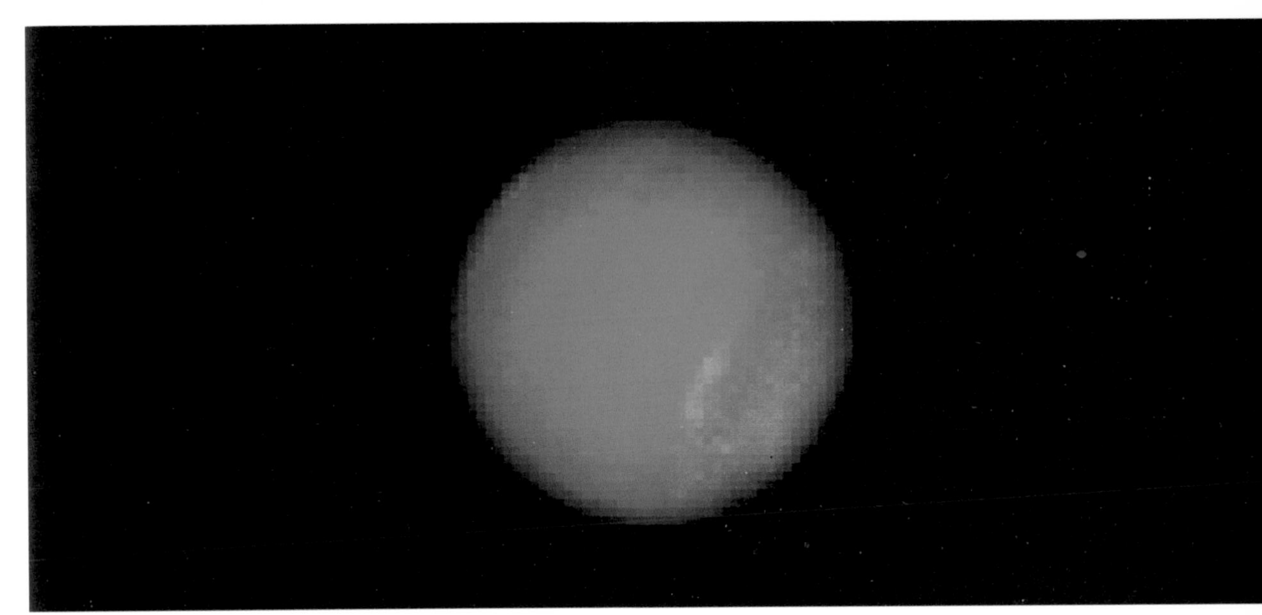

**Figure 3.23.** True-color picture of Neptune constructed from red, green and blue WF/PC-2 images. The blue color of this planet is beautifully shown. (David Crisp and John Trauger, NASA–Jet Propulsion Laboratory and the WF/PC-2 Science Team; Heidi Hammel, MIT)

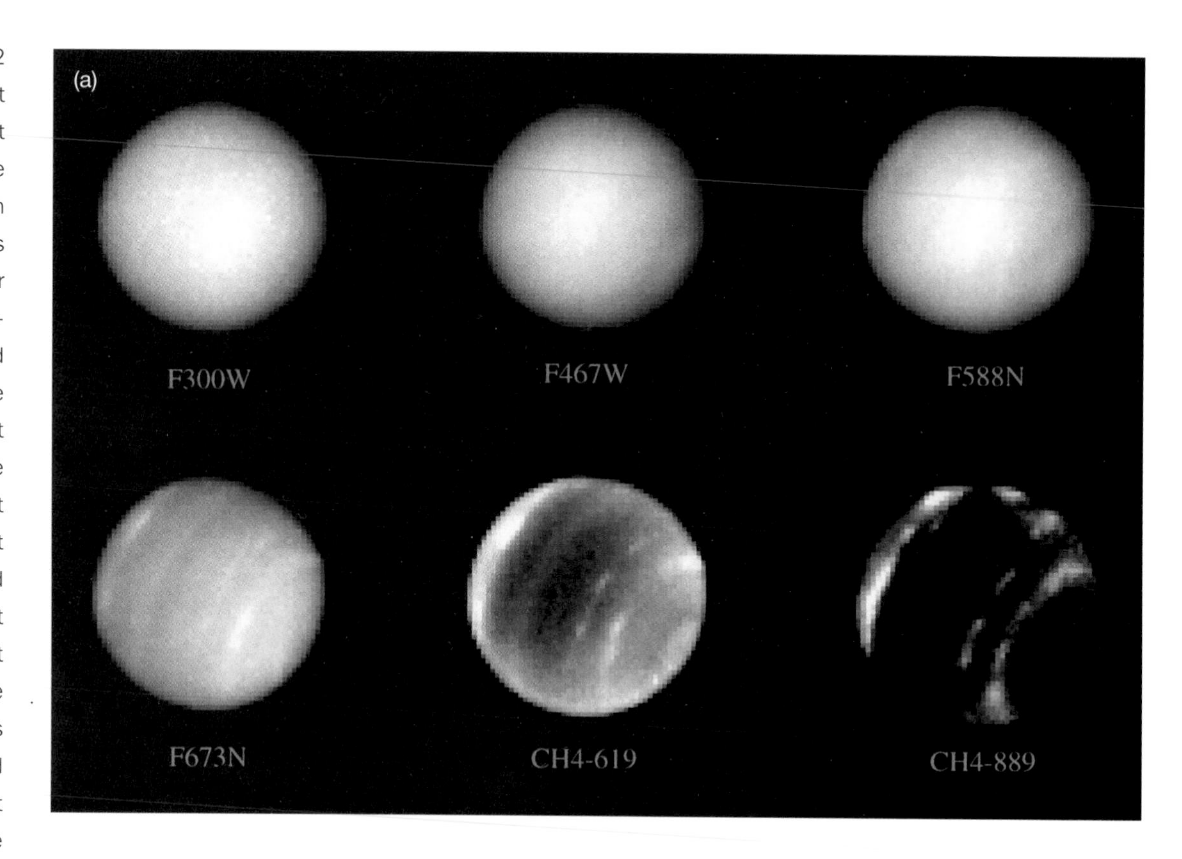


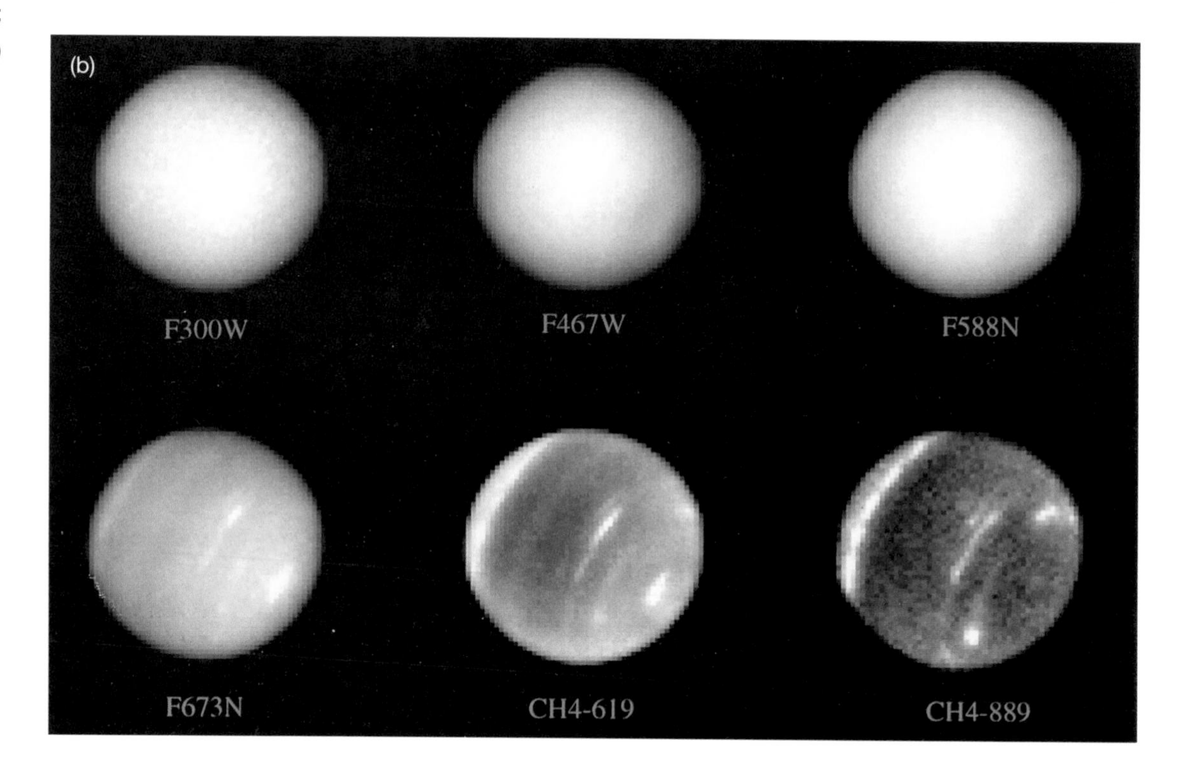


**Figure 3.24.** (a) WF/PC-2 images of Neptune (first hemisphere) in different wavelength bands, taken on June 28–29, 1994. The wavelength in angstroms is given by ten times the number shown and the filter is either wide-band (W) or narrow-band (N). The images marked CH4 isolate light from methane absorption bands. The bright clouds are most obvious in these filters. (b) Same as in (a) but Neptune has rotated by about 180 degrees and the second hemisphere is shown. The Great Dark Spot, which covered almost 40 degrees of longitude at the time of the Voyager 2 fly-by, has disappeared. (David Crisp and John Trauger, NASA–Jet Propulsion Laboratory and the WF/PC-2 Science Team; Heidi Hammel, MIT)

scientists. These storms raced across the planet's disk, revealing wind speeds as high as 325 meters per second. Along with the storms, a series of high-altitude, cirrus-like clouds set Neptune apart from its distant neighbor.

Two groups of scientists have zeroed in on Neptune, using the WF/PC-2 to obtain high-resolution images of the planet. A team led by John Trauger, Principal Investigator for WF/PC-2, and including Jet Propulsion Laboratory astronomer David Crisp and Massachusetts Institute of Technology astronomer Heidi Hammel, acquired the first high-resolution images in June, 1994. Because HST can resolve the disk of Neptune at least as well as ground-based instruments can resolve Jupiter, it is in an ideal position to monitor atmospheric changes on the much more distant planet. Ultraviolet images show some of the same features in Neptune's upper atmosphere that Voyager saw. The bright clouds are probably well above the main cloud deck and the methane layer that causes Neptune to look so blue. According to Heidi Hammel, the biggest discovery in the HST data is that the Great Dark Spot and Dark Spot 2 have disappeared. The disappearance is not completely understood, but it does emphasize the changing nature of the Neptunian atmosphere.

A more comprehensive program of Neptune studies was executed late in 1994 by Heidi Hammel and Wesley Lockwood of Lowell Observatory. Surprisingly enough, in the images from that program, Neptune was sporting a new dark spot in its northern hemisphere. This spot is a mirror image of the Great Dark Spot first mapped by Voyager 2 in 1989, and, according to Hammel, the latest HST image shows that Neptune has changed radically since the Voyager visit. 'New features like this indicate that with Neptune's extraordinary dynamics, the planet can look completely different in just a few weeks,' she explained.

## Pluto

Pluto is the ninth planet in the Solar System, and so far it is the only one that has not been visited by a spacecraft. It is considered a 'weird' little place, and – with its companion Charon – is often referred to as a double planet.

Pluto probably deserves its deviant, peculiar reputation. It orbits at $5.9 \times 10^9$ kilometers from the Sun in a tilted, highly elliptical orbit. Right now it is inside the orbit of Neptune, temporarily making it the eighth planet out from the Sun. Pluto's 'year' is 249 Earth-years long, so only a small fraction of one Pluto 'year' has passed since Clyde Tombaugh discovered the planet in 1930.

Because Pluto is so distant, nobody knew that it had a companion until 1978, when Charon was discovered by James Christy at the US Naval Observatory. The two worlds orbit each other at a distance of 19 000 kilometers, and, from our viewpoint, some 5 billion kilometers away, it has been nearly impossible to tell the two apart, much less determine the two worlds' orbital periods around each other. It was not until HST looked at the pair in 1990 that astronomers were able to clearly separate both components of this 'double planet'. Figure 3.25b is the first clear image of the Pluto/Charon system ever made.

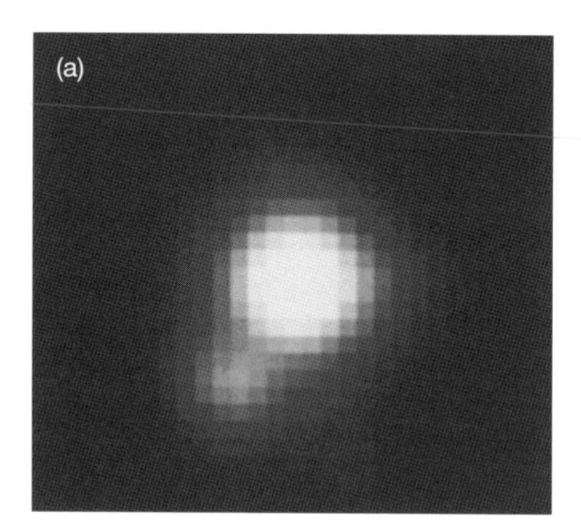


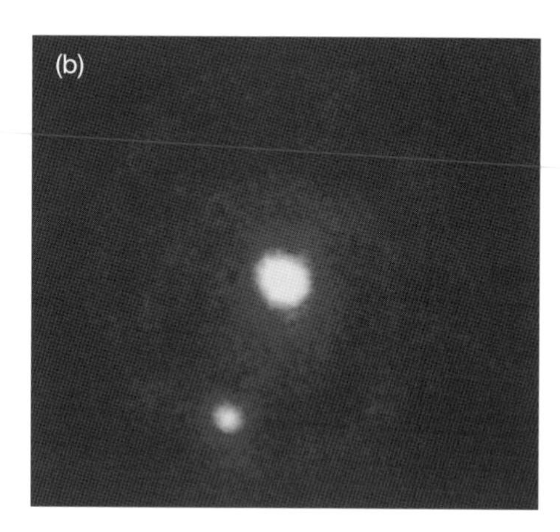


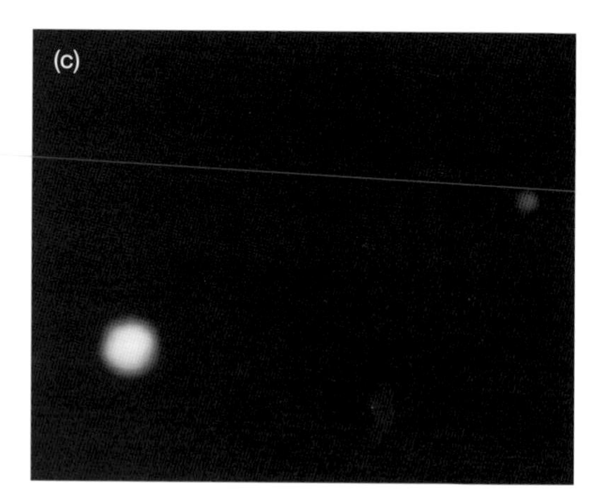


**Figure 3.25.** (a) The best ground-based image of Pluto and Charon. (b) FOC image (pre-COSTAR) of Pluto and Charon. Pluto and Charon were separated by 0.9 arcsecond. (c) FOC image (post-COSTAR) shows these bodies as distinct and sharp disks. Their diameters can be measured to about 1% accuracy and are 2320 kilometers for Pluto and 1270 kilometers for Charon. (R. Albrecht, ESA/ESO Space Telescope European Coordinating Facility, and NASA)

Figure 3.25a is the best ground-based image, taken from the Canada–France–Hawaii telescope atop Mauna Kea. The angular separation between Pluto and Charon is 0.9 arcsecond. Now that precise measurements of the orbits can be established with HST observations, astronomers can refine their estimates of the masses and densities of these two distant worlds. Coupled with ground-based studies of the pair, astronomers think Pluto is about 60% rock, whereas Charon is mostly ice.

In 1994, HST observed Pluto and Charon again, when Pluto was 4.4 billion kilometers from Earth. In the new image (Figure 3.25c), the pair appear as disks. At least one set of Pluto/Charon observations shows some striking differences between the two. For one thing, Pluto is 2320 kilometers across and looks to have smooth ice covering its rocky interior. Charon, on the other hand, looks bluer than Pluto and is only about 1270 kilometers across, about half the diameter of Pluto. Pointing to these differences, some astronomers question whether the pair evolved together or may have been thrown together by cosmic circumstance when the Solar System was forming.

It is possible that the pair was created, along with other similar 'planetary embryos', in the outer fringes of the primordial solar nebula. The other embryos were either used up in the creation of the gas giant planets, or expelled out of the cloud. Pluto and Charon may be the only large leftovers. Since they are more unalike than alike, possible future missions such as the Pluto Express may reveal more about their origins than HST's observations can. Until then, continuing observations of Pluto with HST will supply mission planners with a wealth of information as they decide how best to observe Pluto and Charon up close.

## HST and comets

Comets are a perennial favorite of skywatchers. These icy bodies spend much of their time locked in the deep-freeze of the outer Solar System – the Kuiper Belt and Oort Cloud. Because they formed at the same time as the Solar System, they are also tracers of our past.

They contain chemical mixtures prevalent when the planets were forming, some 4.5 billion years ago. It is only when comets are tugged from their orbits in the outer Solar System into the relatively warmer climes of the inner Solar System that they are heated. They sublimate and, as they do, they give off the water vapor, carbon dioxide, carbon monoxide and many other gases that provide clues to their makeup.

Since they are rapidly moving objects, comets present special challenges to the tracking capability of the HST, but the scientific returns can be magnificent. HST has had the good fortune to observe several comets since its launch. They can be imaged with the Wide Field Camera, and they can be studied in ultraviolet light by the spectrographs to determine their chemical compositions, the structures of their comae, and the activity of their dust and plasma tails.

The first comet to be studied by HST was Comet Levy 1990c. Space Telescope Science Institute scientist Hal Weaver, in collaboration with University of Maryland astronomer Michael A'Hearn and Institut d'Astrophysique of Belgium astronomer Claude Arpigny, was particularly interested in an arc of dust that had been spotted in the tail by a ground-based observatory. They used HST's Wide Field Camera, in conjunction with the European Southern Observatory in Chile, to search for that arc, and came up with a series of images showing the dust.

Astronomers are using HST's spectrographs to compile an inventory of the chemicals that make up comets. For example, Arpigny and Weaver worked on a series of observations that looked at 'Cameron band' emissions in carbon monoxide from Comet Hartley 2. These emissions are probably produced when sunlight photodissociates (breaks down) carbon dioxide to carbon monoxide. Measuring these emissions is an important indicator of how much carbon dioxide exists in a comet.

## The Great Comet Crash of 1994

Of course, the most spectacular comet studied by HST was Shoemaker-Levy 9 – the famous 'string of pearls' comet discovered by the team of Carolyn and Gene Shoemaker and David Levy. For years, the trio of observers had been conducting standard searches for comets and asteroids from Palomar Mountain in California. On the night of their discovery, they were ready to give up observing because it was starting to cloud up. They decided to use up some defective film before stopping work for the night. Later on, Carolyn Shoemaker was looking at the film through an instrument called a stereo microscope.

In David Levy's words, it was an electrifying experience. 'Suddenly she [Carolyn] sat up straight in her chair and looked more intently [at the image on the instrument],' he said. 'Then she looked up at us and she said, "I think I've found a squashed comet!"'

The other two took turns looking through the stereo microscope and each saw what had so surprised Carolyn – a bar of cometary light, with tails leading off to the upper right.

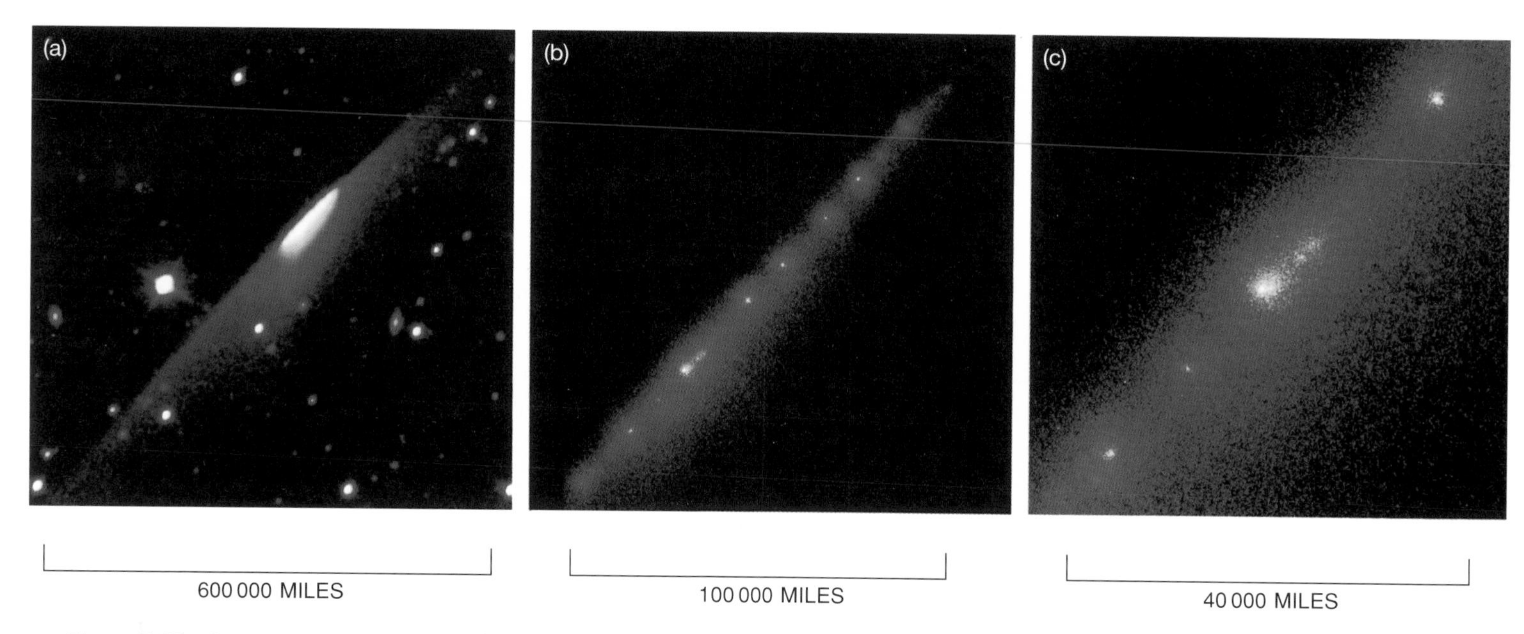


**Figure 3.26.** Several views of Periodic Comet Shoemaker-Levy 9, showing the 'String of Pearls'. (a) A ground-based image taken on March 30, 1993, with the Spacewatch Camera of the University of Arizona (J. V. Scotti, University of Arizona) (b) WF/PC-1 (in PC mode) image taken on July 1, 1993 (H. A. Weaver and T.E. Smith, STScI; NASA) (c) An enlargement of the WF/PC-1 image showing the region of the brightest nucleus.

There was something 1 arcminute long that looked like a cometary coma. Spreading out toward the upper right were little tails, maybe five or six of them. On either side of them was a pencil-thin 'wing' of light. It was the strangest thing they had ever seen. Their first thought was: 'How are we going to confirm this?'

Levy called his friend Jim Scotti at Kitt Peak and told him about the find and gave him the position. Then they sat back to wait while Scotti did his own observations. 'A couple of hours later I called him back,' said Levy. 'He said, "David, it's at least five to eight individual comets – each with a tail, and it's real all right."'

The trio reported their find and it was announced the following day as Comet Shoemaker-Levy, later renamed Periodic Comet Shoemaker-Levy 9. There were approximately 20 objects, moving along in formation, headed for a 6-day-long July 1994 collision with Jupiter. The largest 'pearl' in the first HST image measures only about 5 kilometers or less across. The pearls originally came from a larger comet that was torn apart by a very close approach to Jupiter in the summer of 1992.

As one might imagine, hurling 20 or so cometary objects into the Jovian atmosphere would probably stir up the clouds. Of major interest was how much energy would be released when the largest pieces hit. Astronomers spent the months before the impacts calculating energy release rates so that they could calibrate ground-based and orbital observatories for observations.

A special HST campaign was put together, with dozens of investigators from around the world participating in the observations. There was a great deal of speculation about whether the comet pieces (each piece named with a letter of the alphabet) would survive the tremendous gravitational field of Jupiter. It seemed that they might be torn apart into thousands more fragments. So, HST watched the fragments until 10 hours before the

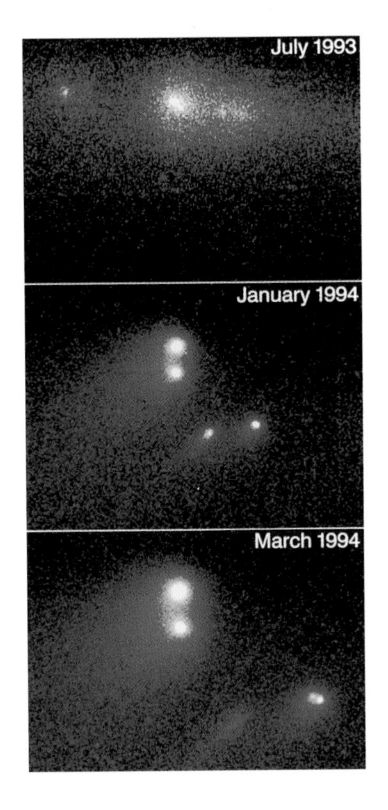


**Figure 3.27.** WF/PC-1 and WF/PC-2 images of Periodic Comet Shoemaker-Levy 9's brightest region. Some evolution has occurred, and one of the regions all but vanished between January and March 1994. (Hal Weaver and T. Ed Smith, STScI; NASA)

impacts to see if the largest pieces would break up. These observations indicated that each of the largest 'pearls' in Shoemaker-Levy 9 went into Jupiter whole, trailing a stream of dust as it went.

Unfortunately, the impacts occurred on the side of Jupiter facing away from the Earth. Nonetheless, Shoemaker-Levy's last hurrah was quite spectacular. Several impacts sent material high enough into the atmosphere to be seen from the Earth only a few minutes after impact. When the impact zones rotated into view, there were dark scars where the comet pieces had entered the Jovian atmosphere.

As each comet piece plowed into the clouds, it disintegrated. The heat of friction and disintegration powered huge explosions that sent material from Jupiter's lower clouds back out. The Faint Object Spectrograph detected the strong signatures of sulfur, carbon disulfide, hydrogen sulfide and ammonia. In addition, the FOS detected emissions of silicon, magnesium and iron that apparently came from the comet itself.

The cometary impacts also affected the Jovian magnetosphere. When the comet first entered the magnetosphere four days before impact, HST saw strong emissions of ionized magnesium at the comet. During the actual impact events, the so-called K fragment stirred up unusual auroral activity over Jupiter's poles. This occurred about 45 minutes after the K fragment entered the atmosphere. The auroral features were well outside the regions where aurorae are usually seen on Jupiter. The K fragment impact produced energetic, ionized particles that traveled along magnetic field lines into the upper atmosphere of the planet. When they came back down, these particles caused atmospheric gases to glow brightly in ultraviolet light.

For months after the impacts, scientists followed the evolution of the impact scars as Jupiter's winds stirred up the debris raised by the impacts. The fine dust is suspended in the Jovian upper atmosphere and will 'rain out' or diffuse itself into the lower clouds over the next few years. While the scars remained for many months after the impacts, it appears that Jupiter was not seriously damaged in any way by the events.

For planetary scientists like Gene Shoemaker, the chance to watch a series of impacts in almost 'real-time' was a once-in-a-lifetime experience. Using HST to do it added a high-resolution component to an already exciting scientific event. During a press conference in October 1993, he highly amused a roomful of reporters and scientists by

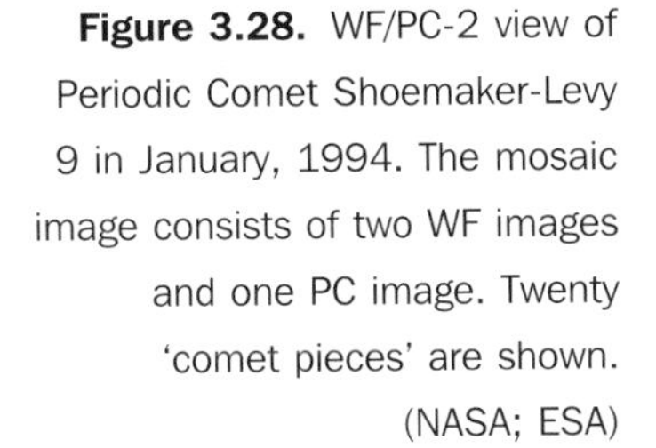

**Figure 3.28.** WF/PC-2 view of Periodic Comet Shoemaker-Levy 9 in January, 1994. The mosaic image consists of two WF images and one PC image. Twenty 'comet pieces' are shown. (NASA; ESA)

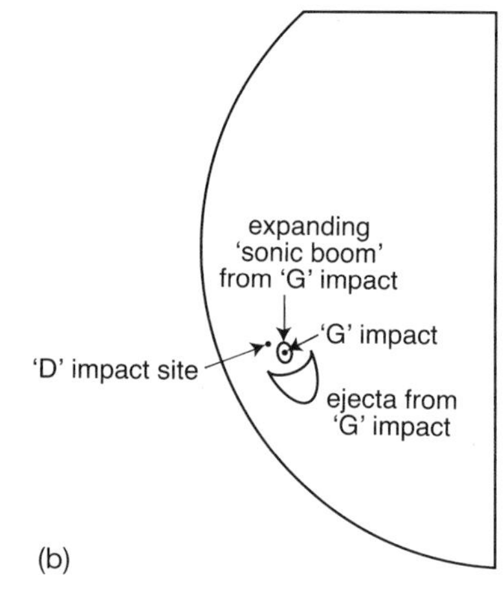


**Figure 3.29.** (a) The impact site of the 'G' fragment, the largest impact site from Periodic Comet Shoemaker-Levy 9, as seen by WF/PC-2 (in PC mode). The impact occurred on July 18, 1994, at 3:28 a.m. Eastern Daylight Time. The region also shows the impact of the smaller 'D' fragment, which occurred at 7:45 a.m. EDT, on July 17. (b) Schematic showing the impact sites and a possible explanation for the ring and the crescent features. (Hubble Space Telescope Comet Team; NASA)

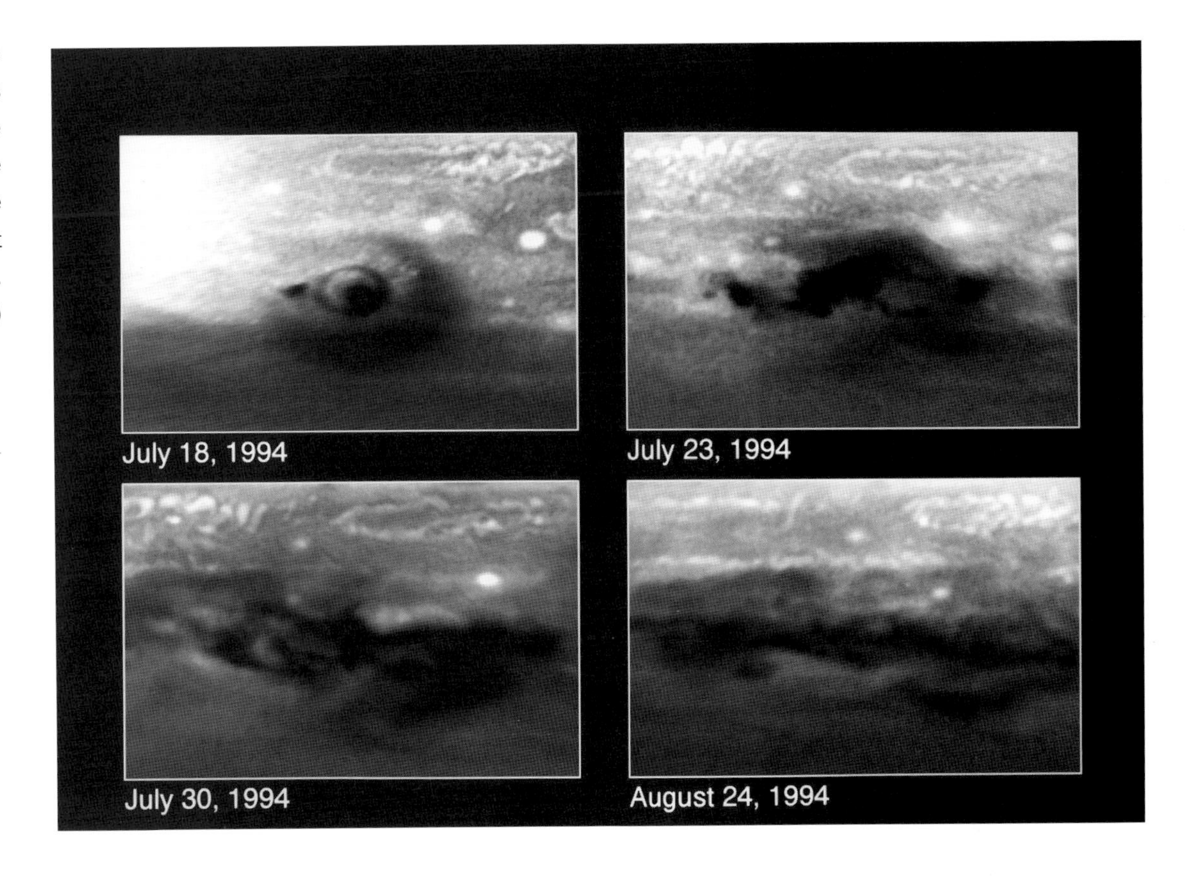


**Figure 3.30.** Evolution of the 'D/G' impact site on Jupiter as recorded by WF/PC-2. The predominant effect is the stretching and tearing of the features by strong east–west winds in the Jovian atmosphere. (H. Hammel, MIT; NASA)

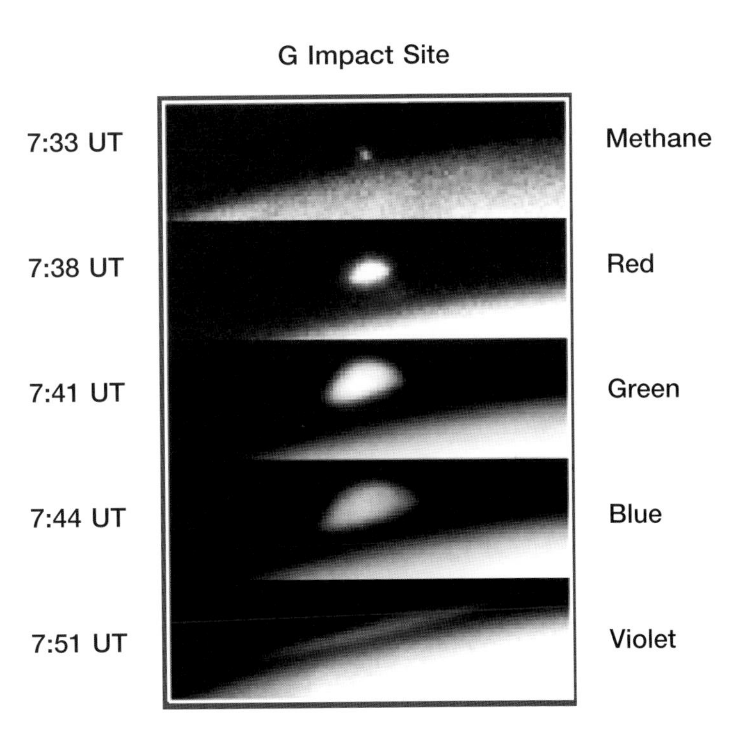


**Figure 3.31.** WF/PC-2 sequence of images taken on 16 July, 1994, showing a plume rising above the G impact site. (STScI; SL9 imaging team)

**Figure 3.32.** This mosaic of WF/PC-2 images shows the evolution of the G impact site on Jupiter. The images from lower left to upper right show: the impact plume at 07:38 universal time on July 18, 1994 (about 5 minutes after the impact); the fresh impact site at 09:19 UT (1.5 hours after impact); the impact site after evolution by the winds of Jupiter, along with the L impact site above, taken at 6:22 a.m. UT on July 21, 1994 (3 days after the G impact and 1.3 days after the L impact); further evolution of the G and L sites due to winds and the impact of the S fragment in the G vicinity, taken at 9:08 UT on July 23 (5 days after the G impact). (R. Evans, J.T. Trauger, H. Hammel and the HST Comet Science Team; NASA)

remarking, 'I've dreamt during most of my professional career that someday I'd get to witness an impact. Now, as a geologist, I sort of had in mind maybe something out in the middle of the Gibson desert in Australia and I'd rush out and map the crater. Little did I dream that the impact event I was going to get to witness was going to be at Jupiter. I sat back and thought about it, and wondered, what are the odds? And it turns out that Jupiter's a lot better bet! It happens there a lot more frequently, but we just don't actually get to see them very often.'

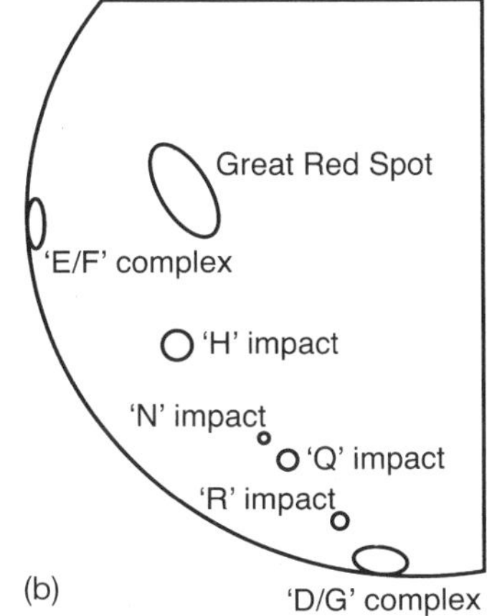

**Figure 3.33**. (a) WF/PC-2 (in PC mode) image of Jupiter showing multiple impact sites in July 1994. Several sites are visible in a row including the star-shaped 'H' site (below the Great Red Spot) and extending to the 'D/G' site near the limb (bottom). (b) Schematic showing the impact sites. (Hubble Space Telescope Comet Team; NASA)

Indeed, the Great Comet Crash of July 1994 has spurred planetary scientists to re-think their assumptions about impact events and collisions in the Solar System. Already a group of astronomers is mounting a search effort for Earth-crossing asteroids and small comets that could pose a similar threat to Earth. Many Voyager images of icy moons that exhibit crater chains are being re-examined as new evidence that other comets like Shoemaker-Levy may have existed in the past.

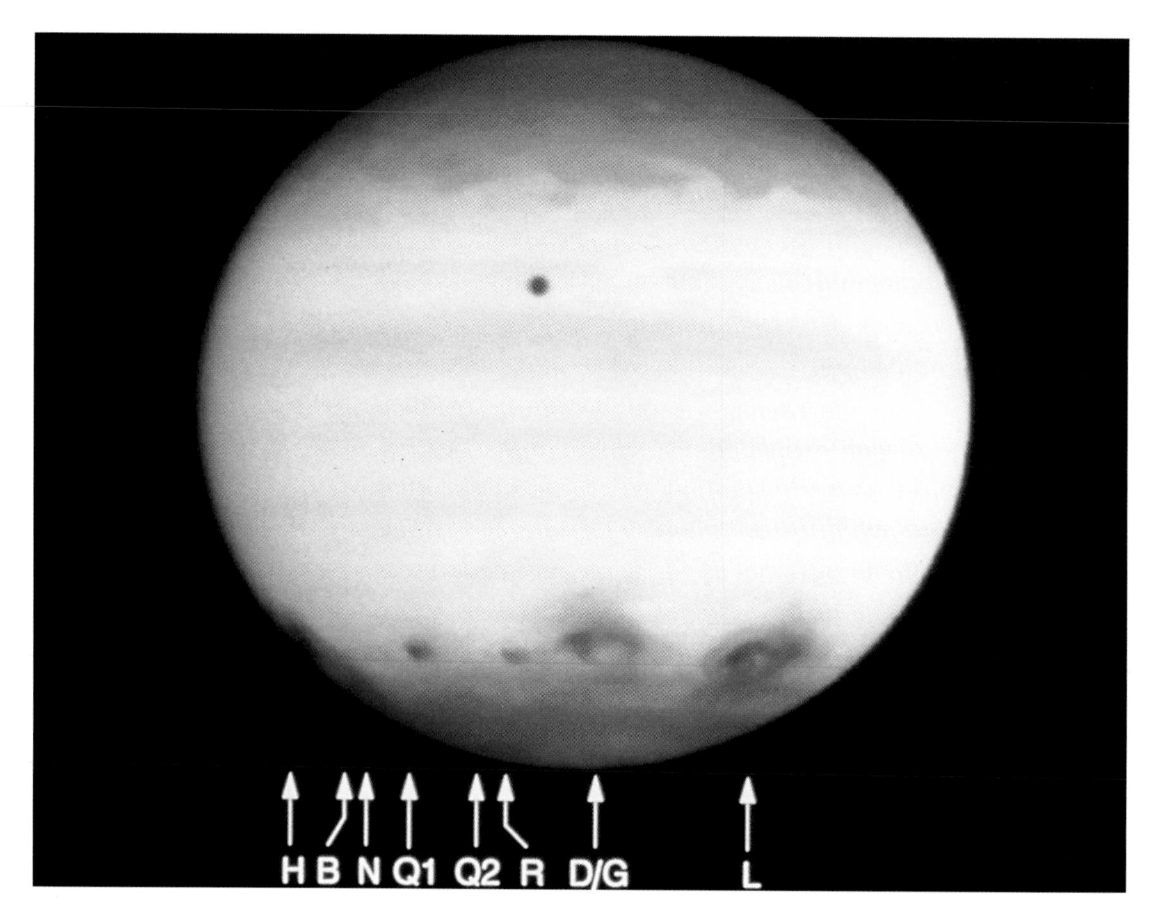


**Figure 3.34.** The impact sites from fragments H, Q1, R, D/G and L are seen, and small dark spots from the impacts of fragments B, N and Q2 are barely visible in this ultraviolet image made with WF/PC-2. (STScI)

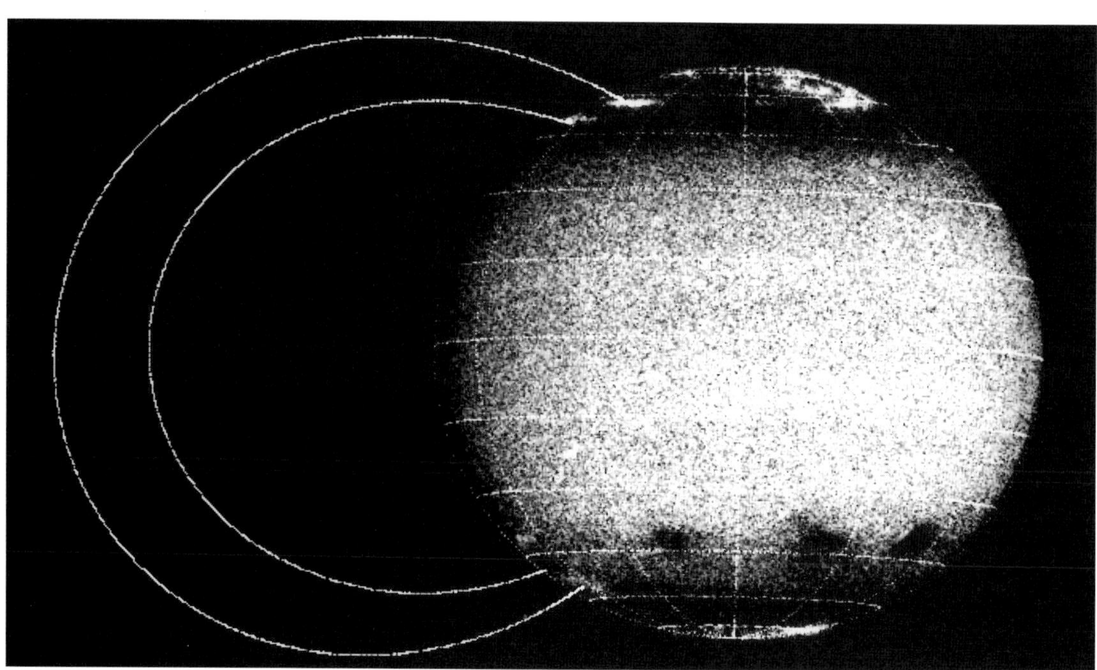

**Figure 3.35.** A WF/PC-2 image of Jupiter showing aurorae produced 45 minutes after the 'K' impact. The impact (site at bottom left) produced energetic particles which traveled along the Jovian magnetic field (marked by arcs at left) and slammed into Jupiter to produce the aurorae (bright regions at upper end of arcs). Jovian aurorae do not normally occur at these locations. The normal aurorae are visible at the north and south poles. (John T. Clarke, University of Michigan; NASA)

While the chances of other such spectacular events occurring during HST's 15-year lifetime are unknown, it is clear that the future of planetary science with HST is good. HST will continue to concentrate on Jovian cloud studies, for example, supplying valuable information to the Galileo team members as they study the planet with *their* spacecraft. Cometary nuclei will come under HST's high-resolution gaze, and of course, researchers will continue to watch for changes in the atmosphere of Mars. Planetary scientists faced with a dwindling number of fly-by missions may well rely on the HST for planetary observations for some time to come.

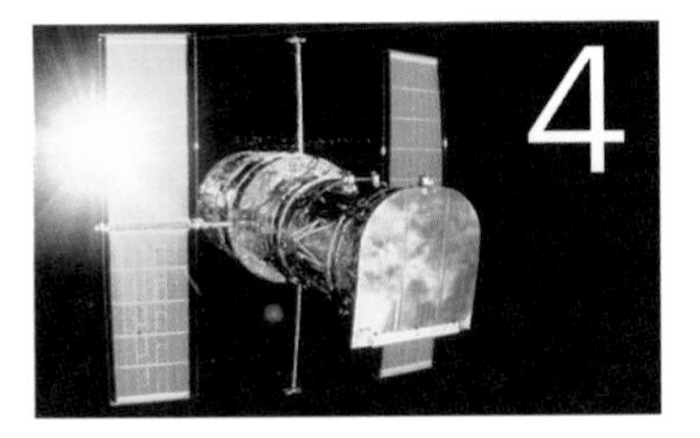

# 4 Stars and the interstellar medium

The next object I have observed is the essence or substance of the Milky Way. By the aid of a telescope anyone may behold this... the galaxy is nothing else but a mass of innumerable stars planted together in clusters.

*Galileo Galilei*

Stars are like animals in the wild. We may see the young, but never the actual birth, which is a veiled and secret event.

*Heinz R. Pagels*

Stars look serene, but they are incredibly violent furnaces that occasionally erupt in incredibly violent explosions.

*Isaac Asimov*

## The lives of stars

Stars. They fascinate us. They are born, live their lives, and die – on timescales that make the longest human lifespan seem like only a moment in time. The physics of stellar evolution is a theme that echoes endlessly across the universe. The stars are variations on that theme, each one adding a distinct voice to the cosmic symphony. There are newborn stars and supernovae, hot supergiants and white dwarf stars, cataclysmic stars and the ancient, slowly dying red dwarf stars. The understanding of all these variations forms an important recurring theme in the history of astronomy.

Stargazing prompts us to ask many questions. What are stars? How are they born? What happens when they die? Why are stars so different from each other? While we have begun to answer these questions in the past few decades, there is a lot of activity occurring in stars that we just do not understand yet. This may seem surprising, but remember that astronomers only recently developed the tools to interpret what stars are telling them.

Looking at a sky full of stars is like wandering through a forest and seeing trees of all different shapes, sizes, ages and states of health. There are tall trees next to short trees, and so one might decide that tall trees are older than short trees. That could be an erroneous conclusion since there are trees of different species in the forest, and some short trees might be older than their taller neighbors. Seedlings sprouting up here and there might lead one to conclude that they are weeds instead of baby trees. Dead logs and dried-out branches on the forest floor come from the surrounding trees, but an inexperienced observer might not be able to link specific branches to the surrounding trees, or determine why the trees and branches died. One might walk across piles of leaves and figure out that some leaves come from certain trees, but how to explain pine needles? Or pine cones? Or acorns? Or sap?

In short, to understand the rhythms of forest life, we have to know a lot about trees and their life cycles. It is the same way with stars, but, unfortunately, humans do not live as long as stars do, so the best thing we can do is to make repeated observations of different kinds of stars and come up with theories to describe what we see.

So, what exactly *are* stars, anyway? The simple definition is that they are self-luminous spheres of gas. To truly characterize stars takes a detailed study of the different 'interpretations' of that definition. Of course, the best-understood example of a star is our Sun, which makes it a good place to start when studying the general properties of stars. Astronomers usually refer to other stars in terms of *solar* brightness, luminosity, mass and radius. This gives a nice shorthand way to refer to any stellar characteristics: $L_s$ for luminosity; $M_s$ for mass, and $R_s$ to describe the radius. So, for example, if a star's mass is given as $10M_s$, that means it has ten times the mass of the Sun.

Stellar masses range from about $0.1M_s$ to roughly $100M_s$. At the lower end, we are really testing the definition of the term 'star' because a star needs enough mass to produce temperatures that will sustain the nuclear reactions at its center. At the upper end, say stars that are $100M_s$, their mass is harder to estimate, but there is no simple reason to assume an upper limit.

The radii of most stars range from $0.1R_s$ to $10R_s$. Exotic stars, such as white dwarf stars, neutron stars and black holes also test the definition of the term star. White dwarfs are about the size of the Earth and may be equal in mass to the Sun, while neutron stars are thought to be only about 30 kilometers across and yet be of several solar masses. Stellar black holes, on the other hand, could be very small, with their exact size very dependent on how much mass has been swallowed up. A black hole of $5M_s$, for example, would have a size (i.e. the radius of its event horizon) of about 15 kilometers.

Temperatures of stars are the only stellar measurement not given in solar units. Instead, they are given in kelvin (K), where 0 kelvin equals –273 °C. Stars are classified by their temperature and luminosity. Luminosity is simply a measure of the amount of energy emitted by the star each second. The luminosity of the Sun, for example, is $4 \times 10^{26}$ watts per second, and its temperature is 15 million kelvin at its core. Stellar surface temperatures range from 1000 kelvin for the coolest objects we call stars, to 5750 kelvin for the Sun, to super-hot stars measured at over 200 000 kelvin!

Astronomers plot the range of stars in a well-known chart called the *Hertzsprung–Russell diagram* (or H–R diagram for short). It is named after the two astronomers responsible for its introduction: Eijnar Hertzsprung and Henry Norris Russell, and its origin goes back many decades. It works simply by using stars as points on a plot, with the horizontal axis representing the effective temperature of a star and the vertical axis representing the star's luminosity. Temperatures can be estimated from the color of the star or by its spectral type – another form of classification that sorts stars by their chemical spectra.

Luminosity is estimated by using the star's absolute magnitude, the magnitude that the star would have at a standard distance of 10 parsecs. Since a star's apparent brightness

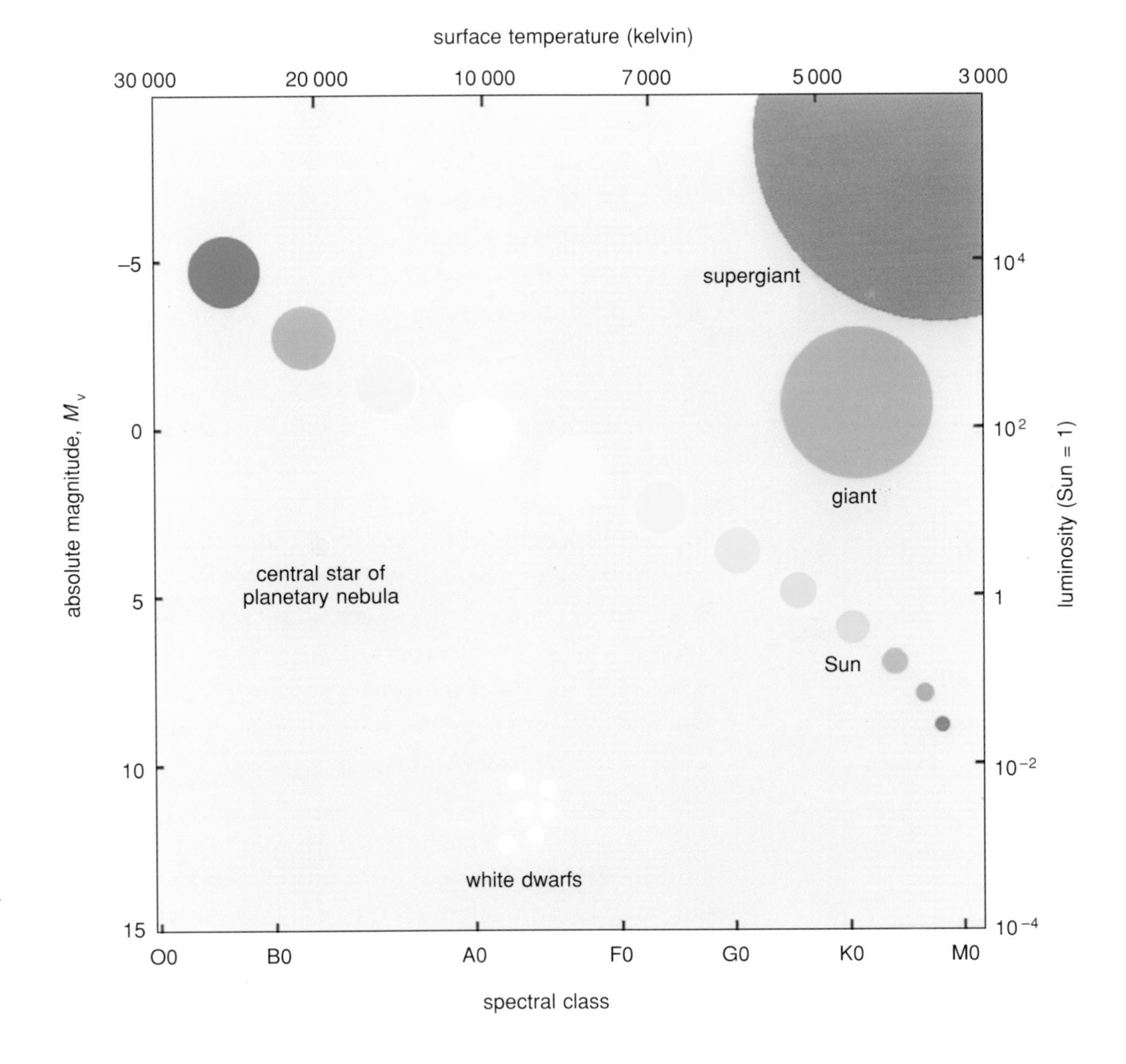


**Figure 4.1.** The Hertzsprung–Russell diagram is a graph of stellar brightnesses (labeled 'absolute magnitude') plotted against temperatures. The result is a classification of stars. Shown in unscaled sizes are the Sun and other stars of the main sequence (the band of stars that cuts across the graph from upper left to lower right), as well as a red giant, a supergiant, white dwarf stars and the central star of a planetary nebula. (T. Kuzniar, courtesy of Loch Ness Productions)

changes with distance (specifically as the inverse square of its distance), this has the effect of putting all stars on an equal footing.

What fuels a star? The energy source for stars was established in the 1930s (primarily by physicist Hans Bethe) as nuclear fusion. The fuel for stars is primarily hydrogen, and when nuclei of hydrogen fuse to form the next heaviest element – helium – energy is given off in the form of heat and light. There are three major nuclear cycles that can operate in stars. At lower temperatures, the *proton–proton chain* transforms hydrogen into helium, and releases energy. The *carbon cycle* takes place in slightly higher-temperature stars. Hydrogen-burning still occurs, but a carbon nucleus is needed as a catalyst. At the very highest temperatures, such as those found in red giant stars, the *triple-alpha process* transforms helium into carbon.

Study Figure 4.1 for a short while, and you can make some interesting deductions about how the stars are grouped. The belt of stars slanting down and to the right is called the *main sequence*. Most stars spend their lifetimes as main-sequence stars, but there are some stars that never make it onto the main sequence. Above the main sequence lie the giant and supergiant stars. Below the main sequence is the domain of the dwarf stars. Some have evolved to their present position from the main sequence; others formed as dwarfs or giants. These designations also describe the radius of the star, and show that the measured radii increase as we go up and to the right on the H–R diagram.

A star's life on the main sequence is relatively stable as it converts hydrogen into helium. Unfortunately for the star, the process of nuclear fusion changes the composition of the core where the conversion takes place. At first, the star is all hydrogen. Then, as nuclear fusion proceeds, a helium core grows. This stage of main-sequence life lasts until about 10% of the star's mass has been converted into helium. The Sun, for example, has been on the main sequence for about 5 billion years, and should go on for another 5 billion years peacefully converting hydrogen to helium.

When the helium core becomes so big that the star is no longer stable, the fun begins. The core contracts, and raises the temperature so that the helium burning can begin. This causes heavier and heavier elements to be created in the ongoing fusion process, and then these, too, are burned. In effect, ashes are being burned to squeeze out energy. A huge excess of energy in the core causes the star to swell into a 'giant' stage. At this point, the star has left the main sequence. The fusion process continues until iron is created in the core. To burn iron would require more energy than would be created, and therefore, everything comes to a standstill. In very massive stars, the core collapses and the outer part of the star blows away in an event called a *supernova explosion*.

When it comes to answering the questions of starbirth and stardeath, and all the variations in between, we have only to look at the sky. Everywhere we look, we can see the process of stellar evolution – places where stars are being born, where they are living and the aftermaths of their deaths. Sometimes in their death throes stars initiate the births of other stars. In other places, colliding galaxies excite star formation – leaving behind a twinkling trail of stellar newborns.

The manner in which stars die intrigues astronomers. Often, each new discovery spurs more questions than it answers. This is not to say that scientists have not collected a lot of data, however. Sites of stardeath, for example, appear everywhere in the universe, and they remain fertile areas of study. How a star dies is largely dependent on the mass it ends with, rather than its original mass. Stars like the Sun spend billions of years steadily converting hydrogen to helium. When the process ends, the star goes through the red giant phase. The end of nuclear burning signals the end of the giant phase. Then, the star shrinks, leaving behind a ghostly looking shell of gas surrounding it – called a planetary nebula. Finally, the star cools and contracts and becomes a white dwarf about the size of the Earth.

Stars slightly larger than the Sun – with final masses between $1.4M_s$ and $3M_s$ – also spend billions of years on the main sequence, converting their hydrogen to helium. When they contract, they can become *neutron stars*, so-called because the intense gravity at the core presses the protons and electrons together to form neutrons.

Stars with masses greater than $3M_s$ do not end up as neutron stars. They contract further to form a stellar black hole. The gravity is so strong that no light can escape the hole, although we do see radiation coming from hot material as it falls into the hole.

Stardeath in its many forms is one process that enriches the interstellar medium with chemical elements. Many stars lose mass throughout their lives, particularly in the form of stellar winds. Clearly, the most catastrophic mass loss they can experience comes when they die. As we will see later on in the chapter, supernovae eject essential elements such as iron into the universe, and these are found in the interstellar medium as concentrations of chemical elements. Eventually these elements turn up in other stars, planets and, ultimately, our bodies. How can this be? Was everything not created when the universe came into being?

The Big Bang and the resulting Primordial Fireball produced hydrogen, deuterium, helium and a little lithium. Ordinary stars are needed to produce carbon, oxygen, calcium and iron. Elements heavier than iron, such as lead and platinum, are produced in supernova explosions. The cycle of star formation out of interstellar gas and dust, and the subsequent re-depositing of stellar materials back into the interstellar medium, increases the amount of heavy elements in the universe over time. The material in the Sun and planets, for example, has been processed through at least one other star. We, having come to life here on Earth, also carry the atoms of long-dead stars in our bodies. Tracing the origins of these chemical elements back through the generations of stars that created them is one way of finding our own place in the universe. Ultimately, we will find the keys to understanding the evolution of the universe.

This brings us back to HST, because its primary stellar mission is to gaze at a variety of stars, catching them in very distinct, and sometimes unusual, stages of existence. Before we look at HST's studies of starbirth, stardeath and the denizens of its exotic stellar zoo, let us turn our gaze to a unique part of space that we rarely think about as we gaze out to the stars – the *interstellar medium*, the 'empty' space between stars. It plays such an important part in the lives of stars that several HST researchers have devoted large amounts of time to probe its mysteries.

## Chemical abundances in the interstellar medium

Imagine receiving an astronomy press release about stellar abundances in the mail. If you were a typical news editor, you probably would not read beyond the first sentence, unless the article happened to start off with the headline 'Hubble Gets The Lead Out – Telescope Finds Heavy Metals Between The Stars'. You might not expect to find such a humorous capsulization of chemical abundances between the stars, but according to University of Wisconsin-Madison astronomer Jason Cardelli – the man behind the headline – there are enough heavy elements in the interstellar medium to warrant using valuable HST time to find them.

So, what is the reason for the excitement over finding heavy elements out there? As we mentioned earlier, interstellar regions offer intriguing clues to the origins and amounts of chemical elements found in stars. If certain elements are especially 'abundant' in one area, then they should be plentiful in nearby stars. This, of course, assumes that the local stars formed in the same region and are not just 'passing through'. It is like walking through a forest and finding cabins built from the wood of surrounding trees. If there are more pine trees, the cabins will be built mostly of pine, while a scarcity of spruce trees means that fewer homes would have spruce wood.

Measuring chemical abundances between stars can also give astronomers a better idea of the life cycles of stars. As we have mentioned, older stars actually return much of their material back to the interstellar medium, and observations can be made to detect that sort of activity. Newborn stars take in that material as they form. By probing the elements shed by stars, astronomers can follow the processes of star formation and death.

When scientists discuss the amounts of elements in the universe, they usually compare what they have observed to something called a *cosmic abundance*. Because not all stable elements are found in a single object, an estimate of the cosmic abundance is cobbled together from a variety of sources, including the Sun, other stars, nebulae, meteorites and the Earth's crust. This cosmic distribution for some sample elements is shown in Figure 4.2. Only relative abundances are significant; the absolute scale is arbitrary. If astronomers find some amount of an element – say krypton – in a region of space, they can measure that amount against the overall abundances of krypton in the universe. Thus, the cosmic abundance of krypton becomes a sort of 'yardstick' against which the specific measurements of krypton in various regions of space may be compared.

To explore interstellar clouds of gas and dust in more detail, HST astronomers use the

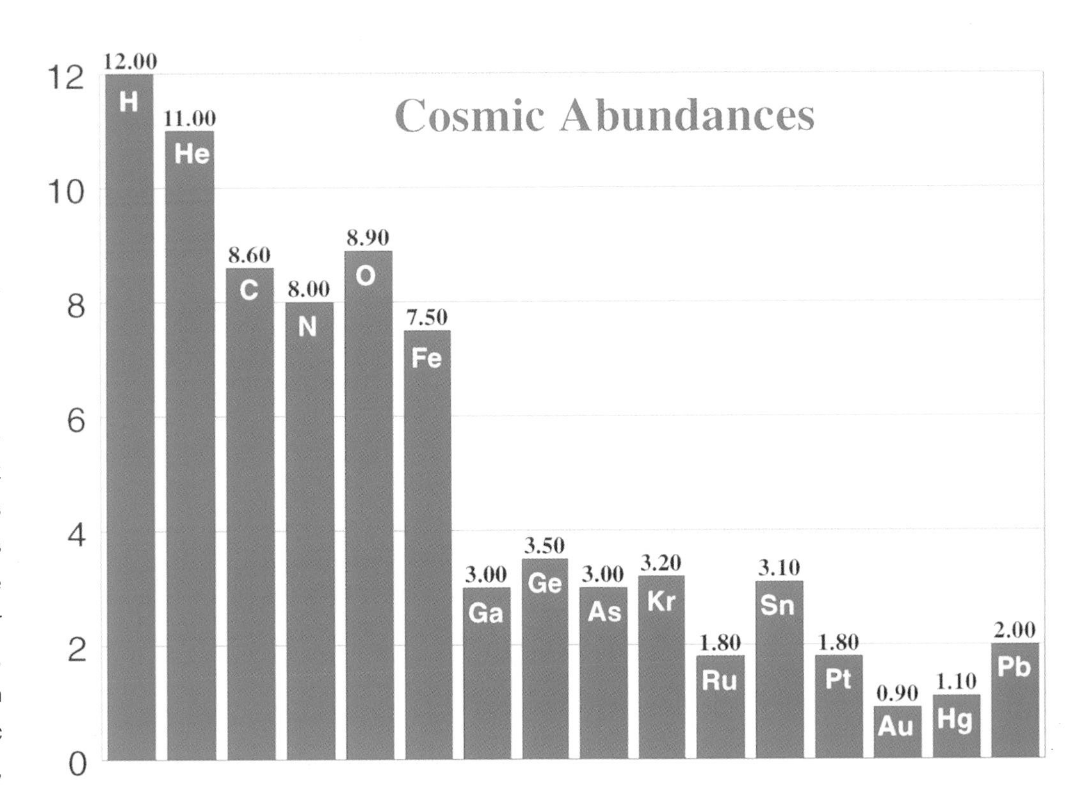


**Figure 4.2.** Some sample so-called 'cosmic abundances'. There are uncertainties in many of these abundances, but the trend is sound. The absolute abundance is arbitrary. The scale is logarithmic, which means that a change of one in the scale corresponds to a factor of ten in the abundance. Thus, helium (He) at 11.0 is ten times less abundant than hydrogen (H) at 12.0, and gallium (Ga) at 3.0 is $10^9$ (or 1 000 000 000) times less abundant than hydrogen. The other elements and their abbreviations are carbon (C), nitrogen (N), oxygen (O), iron (Fe), germanium (Ge), arsenic (As), krypton (Kr), ruthenium (Ru), tin (Sn), platinum (Pt), gold (Au), mercury (Hg) and lead (Pb).

telescope's spectrographs, which are built to examine ultraviolet light. Prior to HST's deployment, one way of performing useful ultraviolet spectroscopy was to use spacecraft designed for it – the Orbiting Astronomical Observatories and the International Ultraviolet Explorer (IUE), for example. IUE has done ground-breaking work since its launch in 1978, paving the way for HST's higher-resolution spectrographs to take much more detailed spectra of the interstellar medium. How does this work?

Cardelli and Ball Aerospace astronomer Dennis Ebbets used the Goddard High Resolution Spectrograph to look at several stars to determine the abundances of elements in the interstellar medium between Earth and a pair of stars. The stars they used as light sources to illuminate the intervening clouds of gas and dust were Xi Persei and Zeta Ophiuchus. Basically, the two scientists took advantage of the fact that certain elements absorb ultraviolet light. They examined the light dispersed with the GHRS and noted which lines indicated absorption. The heaviest element observed in the interstellar medium before the launch of HST was zinc. During HST's first three years on orbit, Cardelli and Ebbets took spectra that revealed the heavier elements lead, gallium, germanium, arsenic, krypton and tin. 'We have now added eight elements that are heavier than zinc,' Cardelli reported. 'These elements are produced by specific nuclear processes in certain types of stars and dumped into the interstellar medium.'

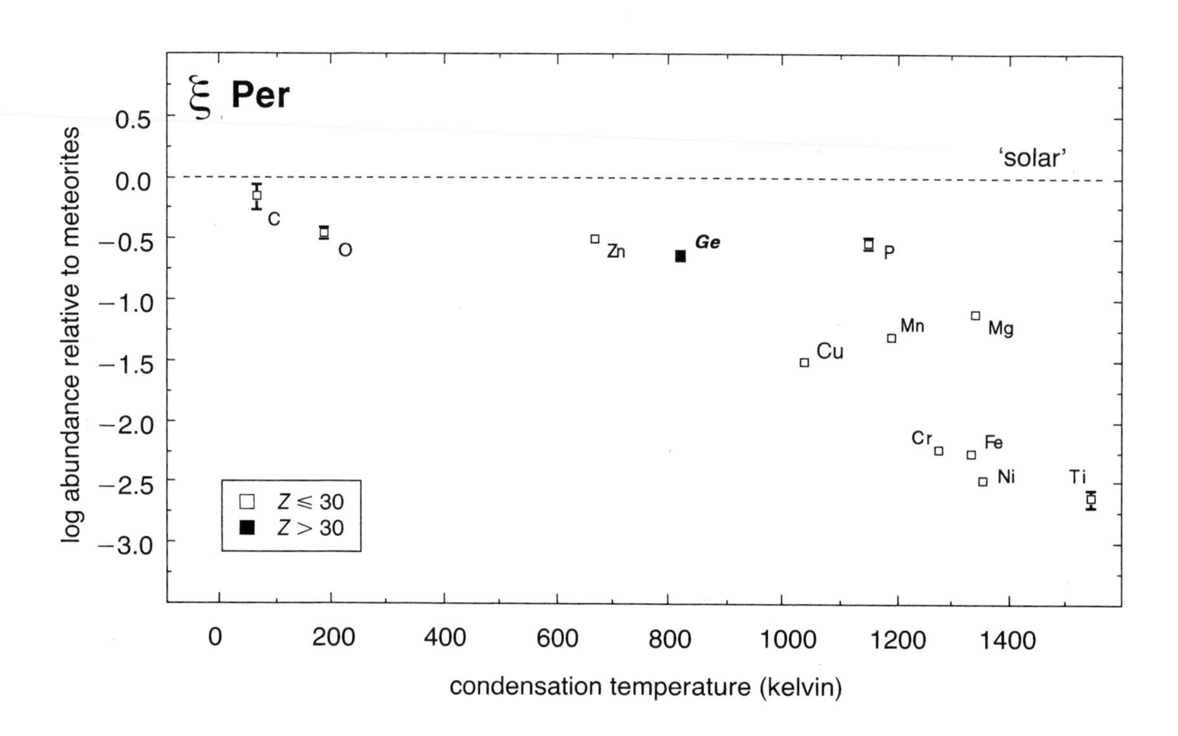


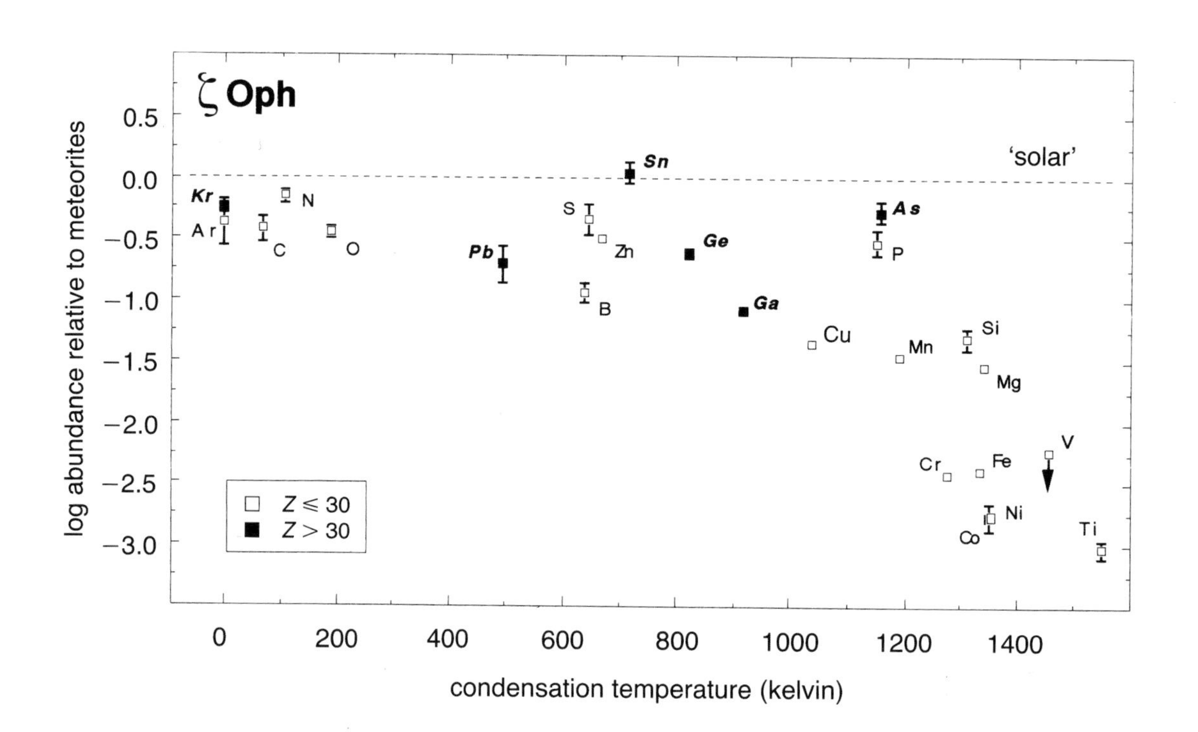


**Figure 4.3.** A plot of abundance differences (relative to cosmic abundances) toward the stars Xi Persei (top) and Zeta Ophiuchi (bottom) versus condensation temperature. The trend toward underabundances for most, but not all, elements is clear. A variety of elements are plotted, including (but not limited to) zinc (Zn), germanium (Ge), magnesium (Mg), iron (Fe), krypton (Kr), argon (Ar), and titanium (Ti). (Jason A. Cardelli, Villanova University)

Analyses of the spectra showed underabundances of several elements, when plotted against each element's condensation temperature – the temperature at which the element becomes a solid. In Figure 4.3, the pattern becomes clear. Generally, when astronomers study interstellar abundances, they find the amount of underabundance is greater for the heavier elements. The usual explanation is that the 'missing' elements *do* exist – but that they are 'hidden' as interstellar dust grains. Since spectroscopy samples elements only in gaseous form, these 'missing' elements would not show up. In addition, the heavier elements form grains at higher temperatures, so these elements condense first when stars form and therefore are much less abundant in the regions between the stars.

The exception to this is the noble gas krypton, which should, theoretically, show up all over the place. It does not form compounds with other elements very easily, so is unlikely to be incorporated into dust. In reality, astronomers only see about half the krypton they expect to find. This is puzzling. The most reasonable explanation for this underabundance is that there is an error in the previously adopted cosmic abundance of krypton, the measured abundance of krypton probably being the 'true' cosmic abundance. If this result is confirmed by other measurements of the interstellar medium with different stars, the cosmic abundance for krypton will have to be revised.

According to Cardelli and Ebbets, this result is important also because it helps astronomers get a more accurate feel for how much lead, krypton and other heavy elements exist in the space between stars. 'The confirmation of these elements represents the first time they have been detected in the interstellar medium,' said Cardelli. 'It's like the promising start of a new day. We've opened a door to new explorations of the interstellar medium and the stars that dump material into it. With our insights we can explore how a galaxy processes and then mixes interstellar material.'

Now that we have some idea of the elements that make up the interstellar medium, it is time to look at the processes of starbirth, starlife and stardeath. These are incredibly long processes, so the best we can do, even with HST, is to take snapshots of the universe, hoping to piece together a coherent understanding of just how stars do what they do.

## Newborns in space

Hubble Space Telescope's view of stellar evolution starts in the gas and dust clouds of the Orion Nebula – a stellar nursery about 1500 light years away from Earth. The nebula lies below the three distinctive stars of Orion's Belt and looks, to the naked eye, like a faint greenish haze. Larger telescopes reveal an irregularly shaped cloud scattered with glittering stars. This region of the sky has been so rewarding for HST researchers that it has been scheduled time and again for observation, including a so-called 'Early Release Observation' in the weeks just after the First Servicing Mission. HST's first look at the nebula came with WF/PC-1 in October, 1990. Understanding what is happening in

**Figure 4.4.** Nebula in Orion as photographed by the 4-meter Mayall Telescope of the Kitt Peak National Observatory. (National Optical Astronomy Observatories)

this image requires us to take a brief look at the physics of gas clouds and hot young stars.

When we look at the Orion Nebula, we are seeing a diffuse nebula called an HII region (HII is ionized hydrogen.) Most HII regions are heated by nearby stars, and they produce bright emission-line spectra. Of course, we see visible light when we gaze at the Orion Nebula through binoculars and small telescopes, whereas HST sees radiation over a wider range of wavelengths. Each gas in the nebula radiates at a slightly different wavelength, so if you know which region of the spectrum is specific to each gas, you can construct a 'color map' of the nebula based on the wavelengths you see. In Figure 4.4, light from neutral hydrogen is coded green, red light is the code for once-ionized nitrogen, and the blue light indicates twice-ionized oxygen.

The Orion Nebula is a good example of what astronomers call a 'blister' nebula – an illuminated region on the edge of a dark cloud of material. The blister is the area where the ultraviolet light from young stars is lighting up the surface of the cloud. The stars in the cloud are around 300 000 years old – which makes them mere babies in the cosmic scheme of things.

The search for hot, young stars in the Orion Nebula has led astronomers, such as Robert O'Dell, to study the region in great detail. Figure 4.5 shows O'Dell's 'big picture' taken in December, 1993. The image covers a small part of the northeastern edge of the nebula, where more than 100 young stars have been surveyed. At least one has formed a great plume of gas, blowing matter away from itself, sending shock waves through the clouds. The shocks 'eat away' large scalloped regions of the nebula.

**Figure 4.5.** WF/PC-2 image of part of the Orion Nebula taken on December 29, 1993. The nebula is 1500 light years away, and the length of the image (diagonally) is 1.6 light years. Many young stars with disks of material around them are visible. Refer to the text for color coding. (C.R. O'Dell, Rice University; NASA)

How can something as diffuse and tenuous as an interstellar cloud of gas and dust give rise to hot young stars? Starbirth is inevitable in interstellar hydrogen clouds. It may take millions of years before a star is born, but it will happen. Molecules of gas and dust grains move around and bump into each other in a process called 'mixing'. Give these cold molecular clouds enough time, and maybe a little push from the outside, and they will start to form a star. That little push may be in the form of shock waves from a nearby supernova or the action of a passing star as it plows through the cloud. A region of higher density starts to form a disk-like shape and material falls toward the center. The cloud spins faster, just as when an ice skater spins faster and faster during a pirouette. At some point, the temperature in the center becomes high enough for a star to 'turn on'.

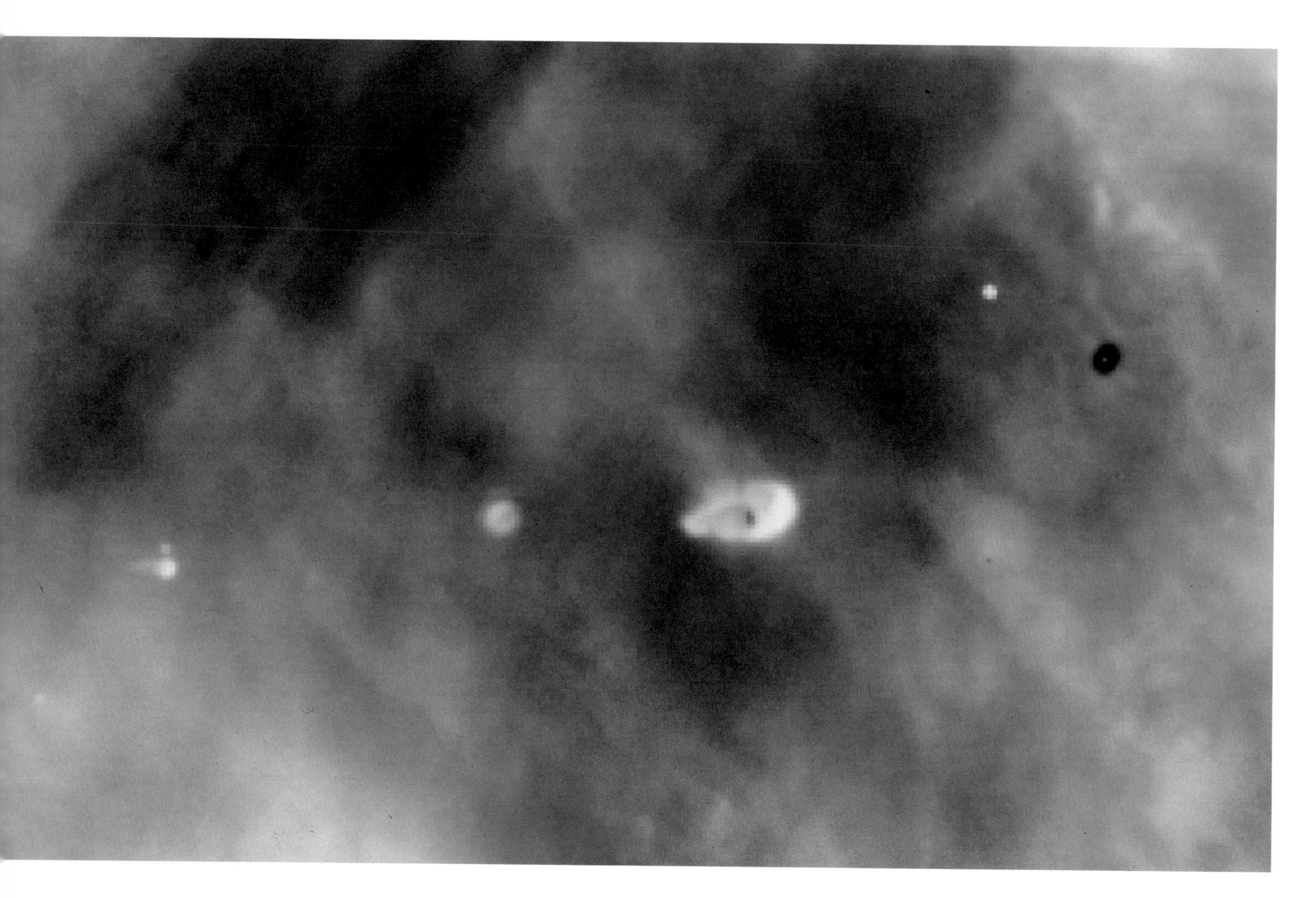

**Figure 4.6.** A close-up view of the central part of Figure 4.5, clearly showing the disks, the proplyds, of material around the young stars. The field of view is only 0.14 light years across. (Color coding is as in Figure 4.5.) (C.R. O'Dell, Rice University; NASA)

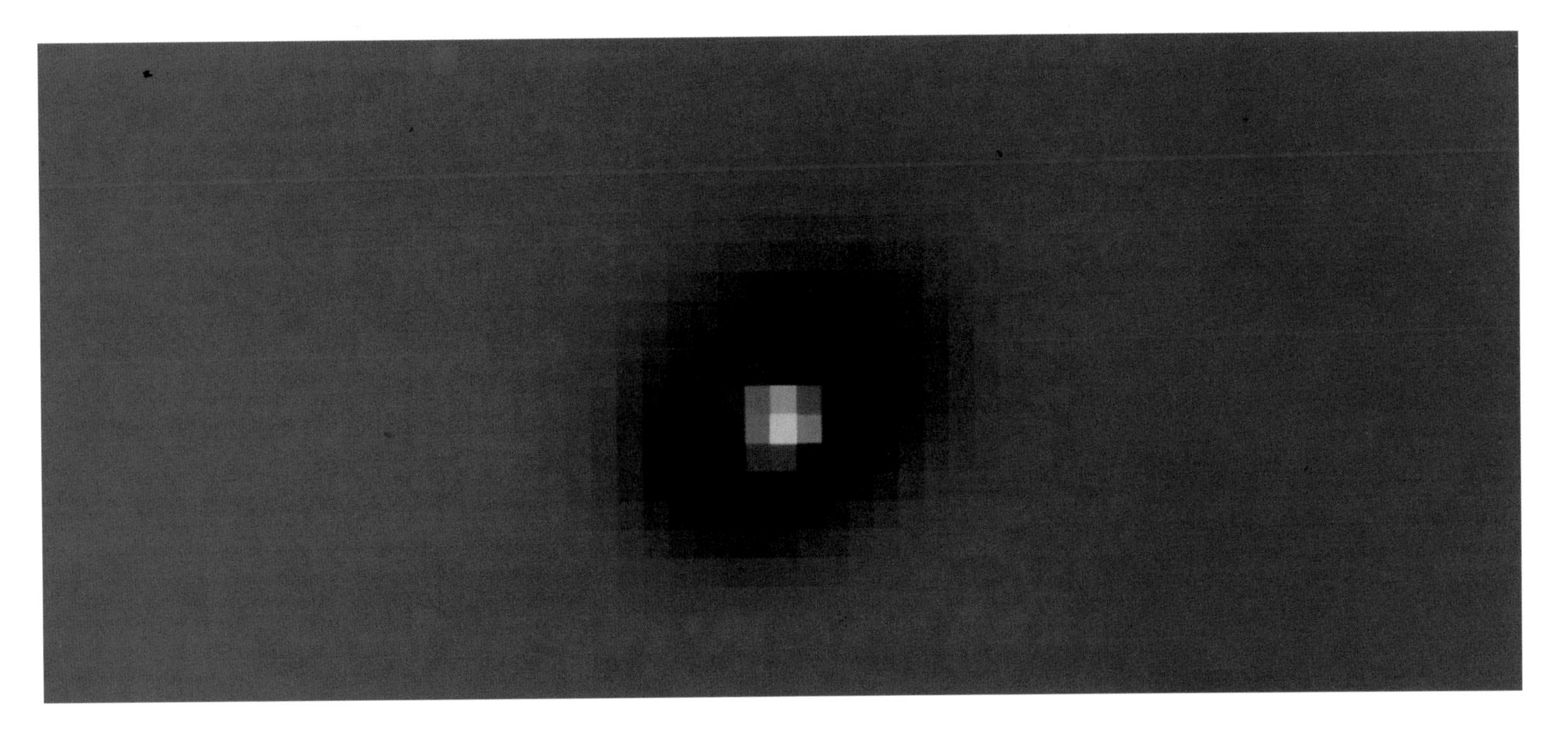

**Figure 4.7.** A closer view of one of the proplyds. The young star is surrounded by a dark disk, which is seen in silhouette against the bright background of the Orion Nebula. The disk is 90 billion kilometers across or about 7.5 times the diameter of the Solar System. (C.R. O'Dell, Rice University; NASA)

Once the star is born, a great deal of material in the disk is still present, spinning around the star and reflecting starlight. In some cases, material falls onto the star and is ejected out in a plume. If there remains enough gas and dust, eventually *protoplanets* will form in regions of higher density within this *protoplanetary disk*. The protoplanets will continue to accrete material, and over time something approximating our Solar System will be formed. Anything remaining from that process might inhabit the farthest reaches of the new system, as the Kuiper Belt and Oort Cloud of comets do for our own Solar System.

Now, with all this starbirth going on in the universe, one might think that astronomers would see disks around every newborn star. They do not, but this does not mean the disks are not there. These accretion disks are very hard to see because at some stages they are composed mainly of cold gas and dust particles, which are easily outshone by the nearby star. These disks are usually more detectable in the infrared than in visual or ultraviolet wavelengths. O'Dell says he was lucky to find them in the Orion Nebula for a couple of reasons – they are near the edge of the nebula, and they are being lit by hot young stars that have only recently 'turned on'. Because they are so hot, these stars are great emitters of a lot of ultraviolet light, which lights up the surrounding cloud of gas, and, in turn, the gas emits light.

If the stars are lighting up gas clouds in their region of the Orion Nebula, it stands to reason that they will also illuminate clouds of denser material that are the birthing grounds for planets. Figure 4.6 shows three protoplanetary disks – or 'proplyds' as O'Dell has dubbed them – in the Orion Nebula, and Figure 4.7 shows an expanded look at one of them. Proplyds were a serendipitous discovery. According to O'Dell, what he was really studying was the dynamics of star formation in the Nebula. In the process, he also found large flattened disks of material clumped around them. 'Those disks are a good example of something we should have expected to see,' he said. 'Once we understood what they were, it's obvious that we should have been looking for them all along.' Figure 4.6 shows a region of the nebula about 0.14 light years across, with five young stars. Four of them are surrounded by disks of material.

O'Dell's large-scale image captures an area of the Nebula about 12 light days across. The generic model given for the proplyds in Figure 4.8 shows a core region about 50 astronomical units wide, illuminated by a hot star (represented by a black dot in the center). The dark ring-like region around the star is the part of the system swept clear of the gases that make up the star – and the region where rocky planets are most likely to form. The outer ring of material is being blown away from the star, and formation of gas giant planets would take place there.

Since starting their exploration of the Orion Nebula, members of O'Dell's team have

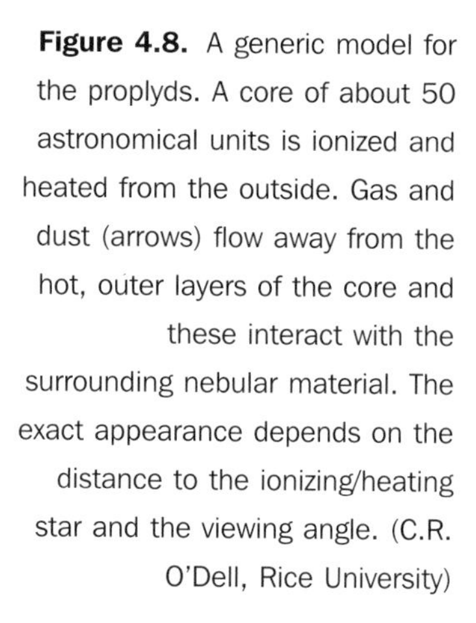

**Figure 4.8.** A generic model for the proplyds. A core of about 50 astronomical units is ionized and heated from the outside. Gas and dust (arrows) flow away from the hot, outer layers of the core and these interact with the surrounding nebular material. The exact appearance depends on the distance to the ionizing/heating star and the viewing angle. (C.R. O'Dell, Rice University)

found proplyds around 56 of 100 stars in the region. No one expects to see actual planets embedded in these pancake-shaped disks of dust and gas, however – they would be far too small and faint to be detectable in images.

## Beta Pictoris

As we have mentioned above, protoplanetary disks around stars are not unheard of in astronomy, and have been seen in ground-based images and radio telescope surveys. The best-known protoplanetary disk was found around the star Beta Pictoris using a telescope at Hawaii's Mauna Kea observatory, and it was subsequently studied by the Infrared Astronomy Satellite. Beta Pictoris looks like a star with a cloudy ring of material around it. HST zeroed in on the star – which lies 54 light years away in the constellation Pictor.

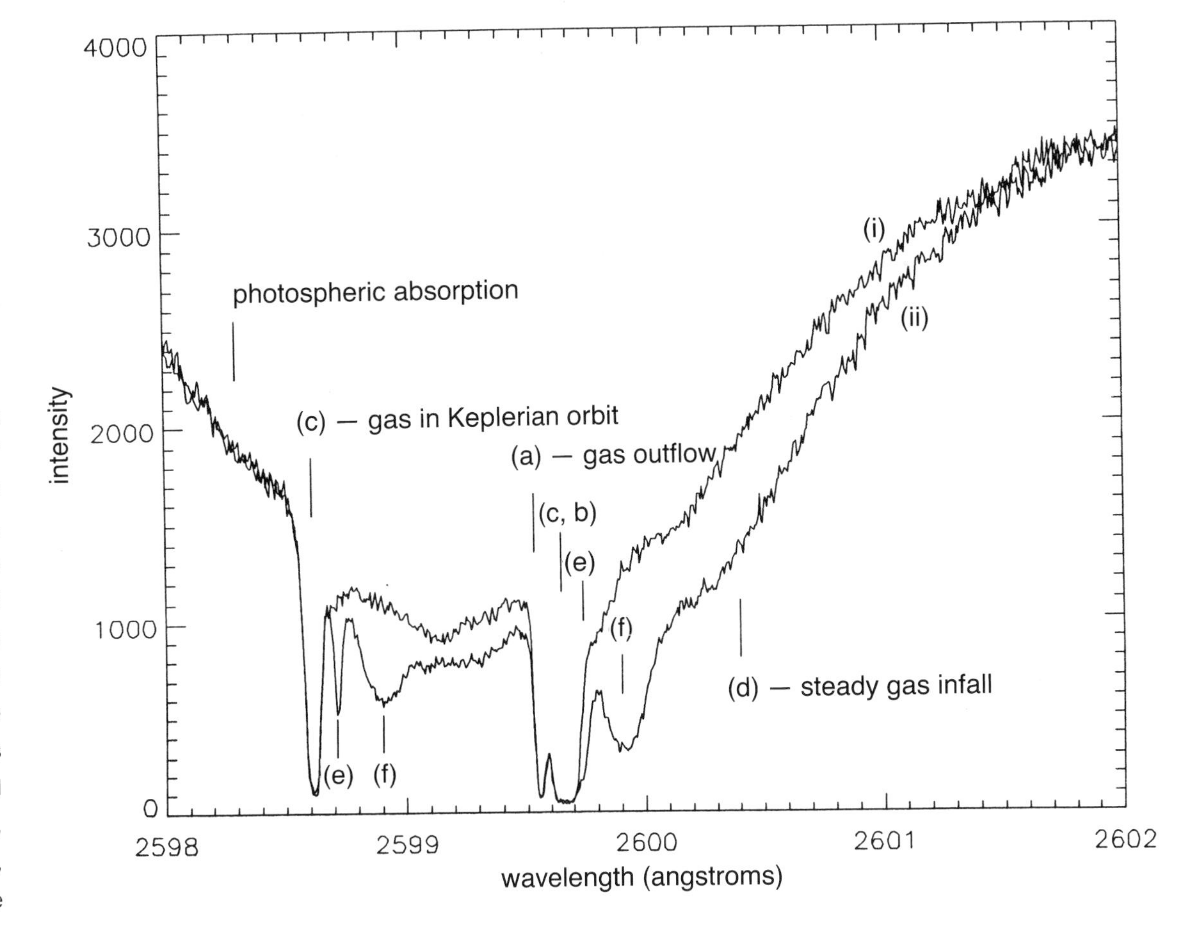


**Figure 4.9.** GHRS spectra of Beta Pictoris, taken on January 12, 1991 (i), and February 4, 1991 (ii), covering the spectral region of singly ionized iron (Fe II). The spectra are the same at either end (representing the star's spectrum), but the central part shows changes due to material between the Earth and the star, material presumed to be mostly in the disk surrounding Beta Pictoris. Parts of the spectra are keyed to the schematic diagram in Figure 4.10. The lowering of most parts of the spectrum between January 12 and February 4 means that much more material is between the star and the observer. By noting the wavelength, whether the gas is flowing outward or falling inward can be determined. (Dara Norman, Computer Sciences Corporation, University of Washington and the GHRS Science Team; Allen Home, Catholic University of America and the GHRS Science Team)

The Goddard High Resolution Spectrograph was programmed to complete a spectroscopic study of Beta Pictoris. The resulting spectra tell us that there are clumps of gas falling into the central region. In fact, both the GHRS data and earlier IUE data suggest that 100 to 150 clumps of material fall into the star each year. Figure 4.10 is a schematic attempt to describe this newborn system. With the Orion proplyds and the Beta Pictoris disk, astronomers may be seeing different stages of the same process of star system formation. Perhaps after a few more observations with HST, they will be able to piece together a model of starbirth that incorporates these different stages.

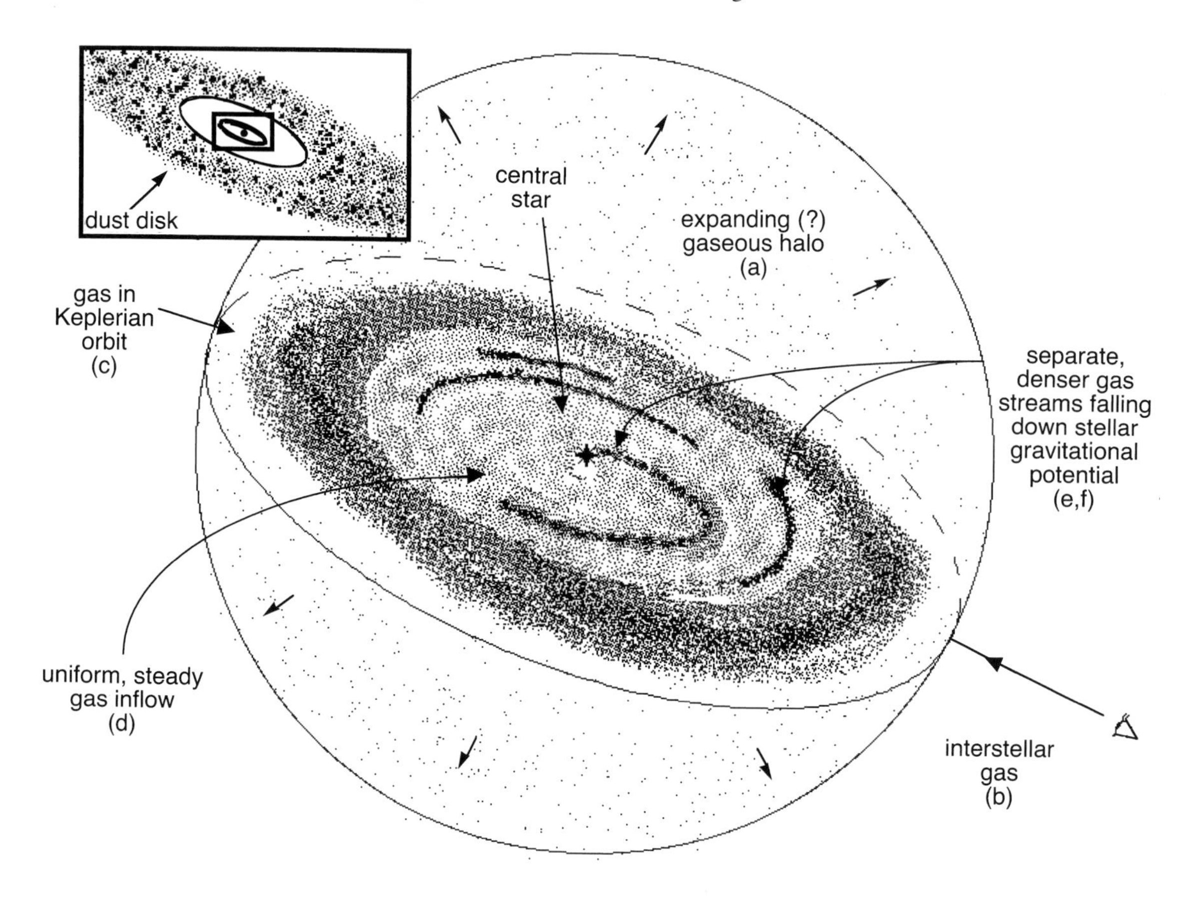


**Figure 4.10.** Schematic diagram keyed to the spectra in Figure 4.9. Possible identifications are: (a) an expanding halo of gas; (b) the interstellar gas around Beta Pictoris; (c) gas in (Keplerian) orbit around the star; (d) uniform, steady gas inflow; (e) and (f) denser, separate gas streams falling down (the stellar gravitational potential) toward the star. (Dara Norman, Computer Sciences Corporation, University of Washington and the GHRS Science Team; Allen Home, Catholic University of America and the GHRS Science Team)

## HST and the stellar zoo

When it comes right down to it, the stars HST looks at fall into three categories: those that are *getting ready* to do something; those that are doing something *right now*; and those that have *just finished* doing something. The telescope is rarely programmed to look at stars which make up a fourth category: those that live their lives quietly, doing nothing spectacular for millions of years. The reason is obvious – it is the same reason that spectacular events grab headlines in the media. No one is really interested in seeing stories that proclaim, 'Entire City Goes About Its Business'. However, if one morning the newspaper headline read 'Entire City Disappears Overnight', it would be sure to catch our attention.

With stars, the headline-grabbing stories cover such exotic things as supermassive stars, violently exploding stars, interacting double stars and stars that are not like anything seen before. These stories describe stellar activities that would command our attention if we did not live in a blissfully happy state of existence around a medium-sized yellow star in a fairly quiet stellar neighborhood. We are safe in our relatively uneventful part of the galaxy, but all around us are neighborhoods of stars whose lives make the Sun look quite bland indeed.

## Supergiants, stellar winds and recycled stars

The Large and Small Magellanic Clouds are a familiar sight to southern hemisphere stargazers. They are actually companion galaxies to the Milky Way, and lie about 170 000 light years away from us. The Large Magellanic Cloud is home to a collection of hot, massive young stars called blue supergiants. Astronomer Sara Heap has long been fascinated with the characteristics of these stars. Of particular interest is why these stars form in such massive proportions. Heap and her colleagues analyze the ultraviolet light streaming from those stars, using spectra taken with the GHRS.

*Melnick 42* Figure 4.11 is the spectrum of one such star called Melnick 42, a roughly $100M_s$ star about 2.5 million times brighter than the Sun. It takes a little work to interpret the spectrum of Melnick 42, but it is well worth the effort. Notice the shape of the spectral line marked 'ionized carbon', which comes from carbon that has lost three electrons. This is the typical spectral signature of a star that is losing mass, so we know that Melnick 42 is blowing away large amounts of its gaseous outer envelope. This characteristic profile is also called a *P Cygni profile* – after a star in the constellation Cygnus that has this readily recognizable type of stellar wind. While all stars blow material away from themselves, supergiants blow away proportionally more than smaller stars, and at a higher velocity. How fast does this wind blow out from Melnick 42? To find out, Heap and her colleagues compared the profile of Melnick 42 to a detailed computer model for a stellar wind. It appears that the wind is blowing away at about 2900 kilometers per second. This means that the star is losing mass at a rate of $4 \times 10^{-6}M_s$ per year. In other words, if the star continues with this rate of loss, Melnick 42 would lose the equivalent of four Suns every million years. By comparison, this is about 100 million times stronger than the mass loss in the wind streaming out from our Sun.

Using other detailed models and HST observations, Heap came up with an independent check on the mass of Melnick 42. The WF/PC-1 image of the area shows no evidence

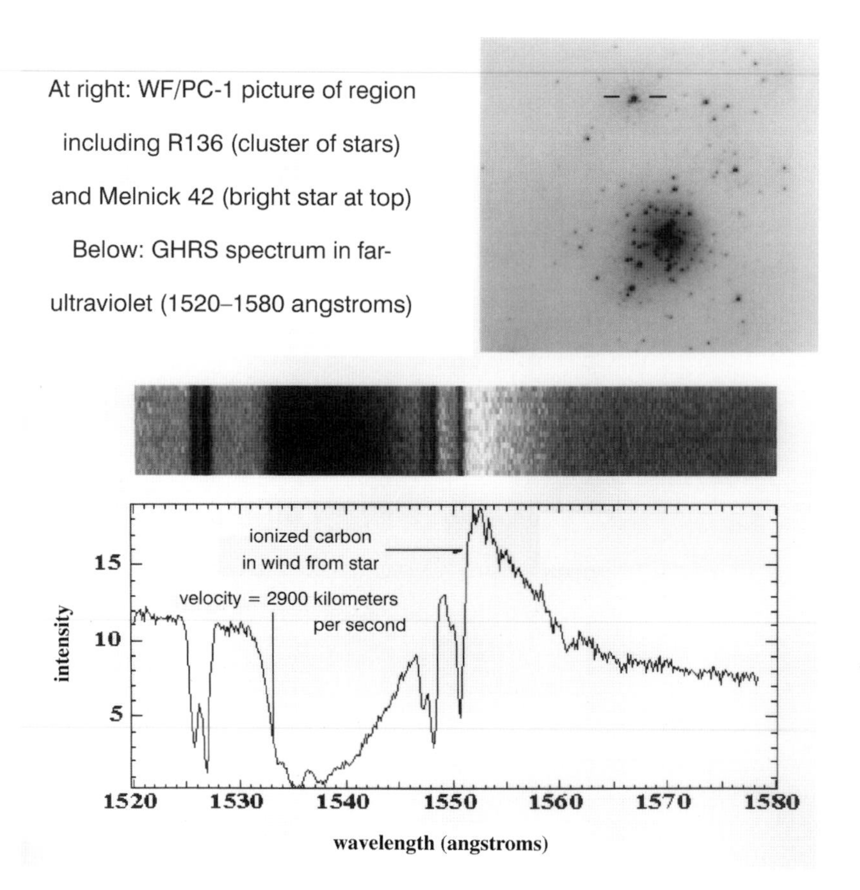


**Figure 4.11.** WF/PC-1 and GHRS observations of Melnick 42. The characteristic shape of the line marked 'ionized carbon' – with emission (brighter) toward the longer wavelengths and absorption (darker) toward the shorter wavelengths – comes from gas flowing away from the star in all directions being illuminated by light from the star. See the text for a discussion of the analysis. The presentation is also a reminder of the correspondence between bright and dark parts of a spectrum (above) and the graphical representation (below) used by astronomers. (S.R. Heap, NASA–Goddard Space Flight Center)

for more than one star (it would confuse things if there were several stars there and you did not know it). Once Heap catalogued the characteristics of this hot young supergiant star, she turned her attention to others in the same region and took spectra of another star in 30 Doradus, the cluster where Melnick 42 is found. Since then, Heap has taken more pictures with the Wide Field and Planetary Camera and has found that the whole region is a cluster of stars. If you could take all the stars in the region and lump them together, you would have a conglomeration of 100 000 solar masses.

What Heap thinks she has really found here is another stellar nursery of hot young supergiants. 'This is fantastic to find in another galaxy,' she said. 'It's kind of your "prototype starburst" – and it is close enough so that you can actually resolve stars. For 30 Doradus we now know that the age for the stars is about 30 million years, which is young.'

Interestingly, the same cluster may have given birth to the progenitor star for Supernova 1987a, and Heap thinks it could be a site for a cluster of supernovae over the next million years, as these short-lived supergiants reach the end of their lives and die, spreading their elements out to be reused in future stars.

## Binary stars

Stars are often grouped in associations, in pairs or clusters, though some, like the Sun, go their way alone through the galaxy. Binary star systems are probably the most common grouping of stars, and they seem to be found everywhere. These are pairs of stars in mutual orbit around each other, and they have come in for a fair amount of study by the HST's Astrometry Team. Studying how binary stars orbit each other is useful for determining masses of the two stars, and if astronomers can see them eclipse each other, they can confirm the radius of each star. This is one of the goals of astrometry.

HST is looking at many binary stars in an effort to determine their mutual interactions. Otto Franz of Lowell Observatory is part of the HST Astrometry Team making observations of binary stars. Although he and others have calculated binary orbits using earth-bound instruments, one drawback with these types of observations is that it is nearly impossible to resolve the two stars when they are closest to each other. The Astrometry Team uses HST to 'fill in' those parts of the orbit that cannot be seen with ground-based instruments.

This was a difficult project in the first years of HST's existence because of the pre-servicing mission problems with spherical aberration and jitter. Even before the repair, Franz said the data were pretty good. 'We had data with an accuracy of 1 milli-arcsecond per observation. That is pretty spectacular.'

One milli-arcsecond is a very small area of space. If you had that sort of resolving power with your eyes, for example, you would be able to spot a letter on this page from a distance of 200 kilometers.

*NGC 6624* Sometimes binary stars are so close together that the members interact gravitationally and, in some cases, dump material on each other. HST has observed at least three incidents of this sort of 'stellar dumping'.

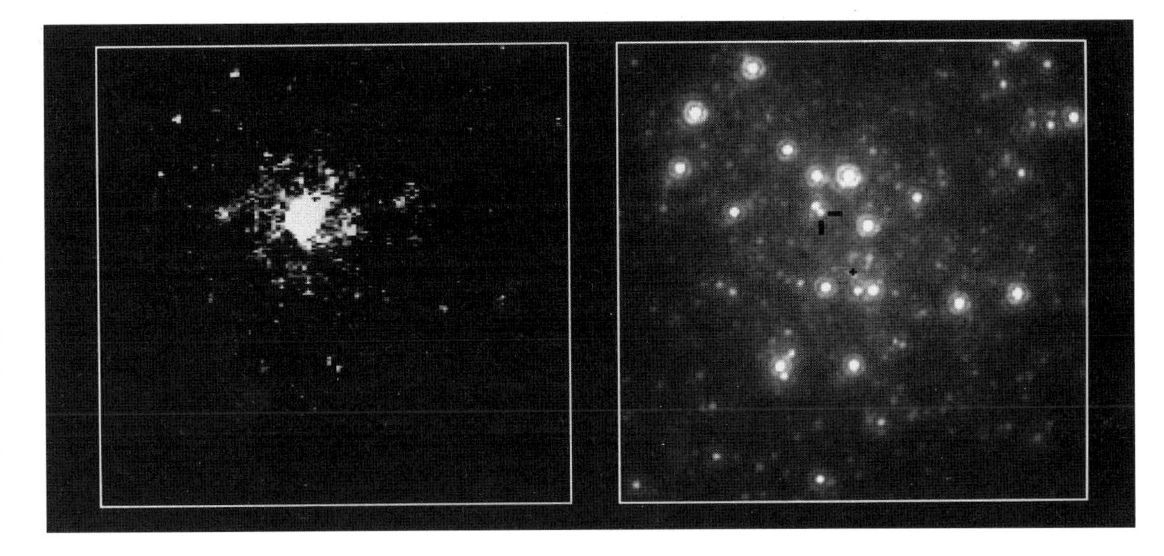

**Figure 4.12.** Ultraviolet (left) and blue (right) FOC images of the core of the globular cluster NGC 6624. The ultraviolet image is dominated by a single source. In the blue image, this source is inconspicuous (small black blocks) and is displaced from the center of the cluster (+). This source is a very hot star and X-ray source. (Ivan King, University of California at Berkeley; NASA; ESA)

NGC 6624 is a globular cluster approximately 28 000 light years away from Earth in the direction of the constellation Sagittarius. Astronomers have known for years that this region is a strong source of X-ray bursts. Figure 4.12 shows ultraviolet and visible light images of the core of NGC 6624. The HST's Faint Object Camera isolated the X-ray source by searching for the ultraviolet light given off by a hot disk of gas surrounding a binary pair. The system is denoted by black blocks in the visible light image (right); the cross marks the exact center of the globular cluster, which lies nearly one-tenth of a light year away from the position of the X-ray source.

The physical system is fascinating. One star is a white dwarf and the other is a neutron star. Figure 4.13 shows a schematic of the orbiting pair. The white dwarf (left, below) is distorted by the intense pull of the neutron star (center of the disk at right) into a pear shape. The two stars are only about 160 000 kilometers apart (less than one-half of the distance between the Earth and the Moon), and they orbit around each other every 11 minutes. The gravitational pull constantly strips helium from the surface of the white dwarf, producing a steady fall of material onto the neutron star, in turn producing a steady

**Figure 4.13.** An artist's model for the X-ray source in NGC 6624. A neutron star (right) is surrounded by material in an accretion disk, and this material comes through an accretion bridge (narrow connecting feature) from the white dwarf star (left). See the text for additional discussion. (Dana Berry, STScI)

stream of X-rays. When large amounts of helium accumulate, nuclear fusion occurs to produce the very intense X-ray emissions that earned the name 'X-ray burster' for this pair.

*Vela X-1* The presence of X-rays tipped off astronomers to violent activity in another binary system. Dick McCray of the University of Colorado's Joint Institute for Laboratory Astrophysics has long been interested in the X-ray binary system called Vela X-1. His HST observations of the system revealed a new phenomenon for these peculiar stellar twins. With most X-ray binary systems, a neutron star is rapidly rotating, sending out an X-ray beam each time the star swings around. The rotating X-ray beam shines on the gas, and then heats it. The gas responds with ultraviolet emission and absorption lines. McCray saw changes in the spectrum of Vela X-1 on time scales as short as a few minutes.

To follow the movement of gas in these systems, McCray looked for pulsations in the spectra that indicated gas flowing between regions. 'It's a new area of science and nobody's had any technique that's enabled them to map these gas flows,' he explained. 'It's all been theory up to now.'

The first of McCray's observations was done simultaneously with HST and the Japanese X-ray satellite GINGA. He also planned a simultaneous observation with HST and the German ROSAT satellite. McCray and other astronomers have identified about 30 X-ray bursters in the Milky Way Galaxy alone.

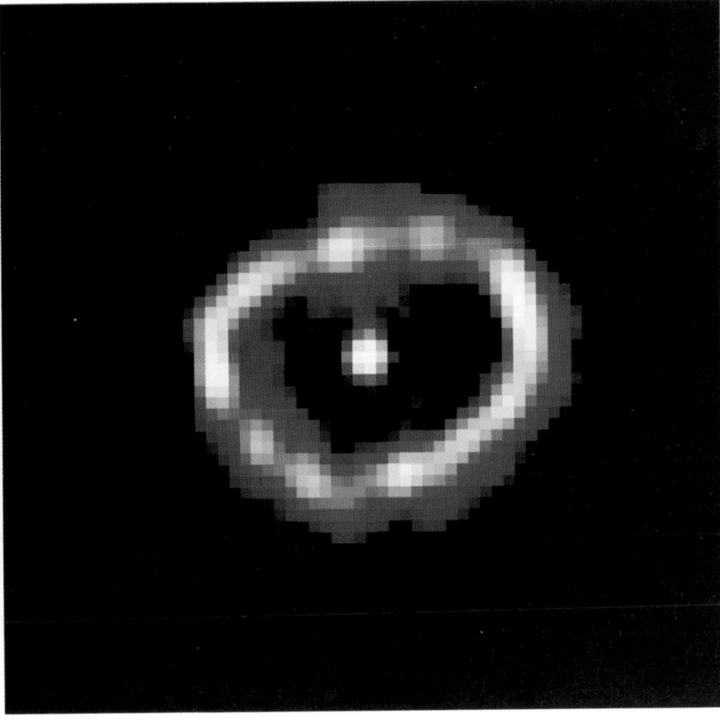

**Figure 4.14.** FOC image of Nova Cygni 1992, before COSTAR (May 31, 1993) and after COSTAR (January 1994, at right). The images show the expansion of the ring of ejected material and the improved resolution of the FOC with COSTAR. See the text for further discussion. (F. Paresce and R. Jedrzejewski, STScI; NASA; ESA)

*Nova Cygni 1992* One example of binary star interactions is the sharing of material that produces what astronomers call a 'nova'. Novae occur when one star in a binary pair dumps material onto the surface of the white dwarf: what we see is basically a thermonuclear explosion. Nova Cygni 1992 was discovered from the ground, and was imaged by HST soon after its latest eruption. The FOC imaged the ring of light formed on the rapidly ballooning bubble of gas coming from the star in 1993, and again right after the First Servicing Mission. The ring had expanded 35 billion kilometers in the intervening seven months, giving scientists a good idea of how fast the gases are moving out from the site of the explosion. As the cloud of material expands from the star, it grows thinner and more tenuous, and we may see the circular ring grow more egg-shaped as the gas spreads out. If the Sun had a shell of gas expanding at the same rate, the edges of the shell would leave the Sun (traveling at almost 1700 kilometers per second) in January and would reach the orbit of Pluto in early February! In seven months, the edge of the shell would be 1% of the distance between the Sun and its nearest stellar neighbor, Proxima Centauri.

As we can see, binaries exhibit a rich set of phenomena. We should remember that most stars are in binary systems, so studying different types of binary systems and the circumstances of their interactions provides a unique window to understanding many of the stars in the galaxy.

## Star clusters

Stars can also be bound together gravitationally into associations called clusters. There are open clusters, which are generally thought of as loose-knit associations of stars that are found almost exclusively in the plane of the galaxy. Generally, they contain up to a few hundred stars, and, over time, the members of the cluster can escape the weak gravitational field that binds the cluster together. The Pleiades and Hyades in the constellation of Taurus are good examples of open clusters.

Sometimes when you scan the sky at night with binoculars, you might view tightly packed associations of stars called globular clusters. There are more than 146 of them associated with the Milky Way. Globular clusters typically contain tens or hundreds of thousands of very old stars, bound together in a tight gravitational ball from which there is no escape. The night sky from anywhere inside a globular cluster would blaze with the light of a hundred thousand suns!

*M15* One of the chief difficulties with observing globular clusters is in distinguishing individual stars, particularly when they are tightly packed together in the core of the cluster. Despite the density of the starfields in globulars, HST has had good success resolving single stars in very crowded fields. Figure 4.15 shows the globular cluster M15 located in the constellation Pegasus, some 30 000 light years away from Earth. To an observer using a pair of binoculars, it is visible as a hazy-looking patch of light about one-third of

**Figure 4.15.** M15 (NGC 7078), the globular star cluster in Pegasus, as photographed by the 4-meter Mayall Telescope of the Kitt Peak National Observatory. (National Optical Astronomy Observatories)

the Moon's diameter. Through a small telescope it begins to look more like a shimmering globe of stars. This cluster has long been known to be home to a large number of variable stars. HST imaged the central region of M15, and Figure 4.16 shows HST's view – a region about 2 light years across. In the central core, 15 very hot, blue stars are clearly visible. These stars pose a mystery, since hot blue stars are generally thought to be young stars, and globular clusters are not usually known for their young star populations. Indeed, the historical view of globulars is that they consist of stars that may be as much as 13 to 15 billion years old! Why would a globular cluster contain what appears to be hot young stars? One good explanation for these blue stars is that they are 'naked cores' of old stars – that is, the central portions of stars that have been stripped of their outer envelopes of gases by collisions with each other. This type of interaction can only take place where the stars are so crowded together that the chance for very close encounter (which gravitationally pulls material from both stars) is exceptionally good.

According to astronomer Guido De Marchi, the find was amazing. 'These objects represent a totally new population of very blue stars. When we started wondering what they

**Figure 4.16.** FOC image of faint, blue stars at the core of M15. Fifteen very hot, faint stars are shown in this ultraviolet image, which is about 2 light years across. Most of them are concentrated within 0.5 light years of the center. A likely explanation is that the crowded conditions at the center of M15 produced close encounters which resulted in the outer layers of the stars being stripped off leaving the 'naked cores'. Also see Figure 4.17. (G. De Marchi, STScI and the University of Florence, Italy; F. Paresce, STScI; NASA; ESA)

could be, we realized that they may be among the first observed stars to have been stripped of their atmospheres.'

This result lends support to the idea that the evolution of stars can be changed drastically by close encounters with other stars.

Another globular cluster under study – 47 Tucanae – also has a good-sized population of massive blue stars. These were first imaged in late 1990 by the FOC. Scientists readily spotted a collection of 'blue straggler' stars in the crowded confines of the cluster. These are massive stars that have passed into old age, but have somehow managed to rejuvenate themselves back to a hotter and brighter state, most likely by colliding with each other and merging. Because these stars appear to be so massive, it is also possible tht they may actually be double star systems. If so, their action may influence the motions of other stars within the cluster.

**Figure 4.17.** An artist's model of a close encounter at the center of M15 as viewed from a hypothetical planet. Here, the average distance between stars is a fraction of a light year. The outer layers are stripped away into the wisps, leaving the cores of the stars exposed. (Dana Berry, STScI)

## Chemically peculiar stars

Stars do not have to flaunt their naked cores to catch an astronomer's interest. Sometimes their chemical compositions send out a coy 'come look at me' call. Recall chemical abundances from our discussion of the interstellar medium. Many stars have abundances of chemical elements that are close enough to the standard cosmic abundances that astronomers call them 'normal'. Sometimes, however, astronomers find a star where the observed abundances are so abnormal – so far from the so-called normal abundances – that they call out for special study. When these kinds of stars are studied in depth, the following questions come to mind. Do we really understand the nucleosynthesis of the elements, or is it possible that we are observing a special case? Is the chemical makeup of the region on the star that we are observing typical of the entire star? This is an important question, because one of the keys to understanding the star lies in determining its chemical makeup. If elements and their isotopes are stratified on the star – that is, if they only show up in particular layers of the star – it would be a mistake to assume that those elements exist throughout the star.

Astronomers call stars with unusual chemical abundances 'chemically peculiar' stars. GHRS Co-Investigator Dave Leckrone made observations of Chi Lupi, a chemically

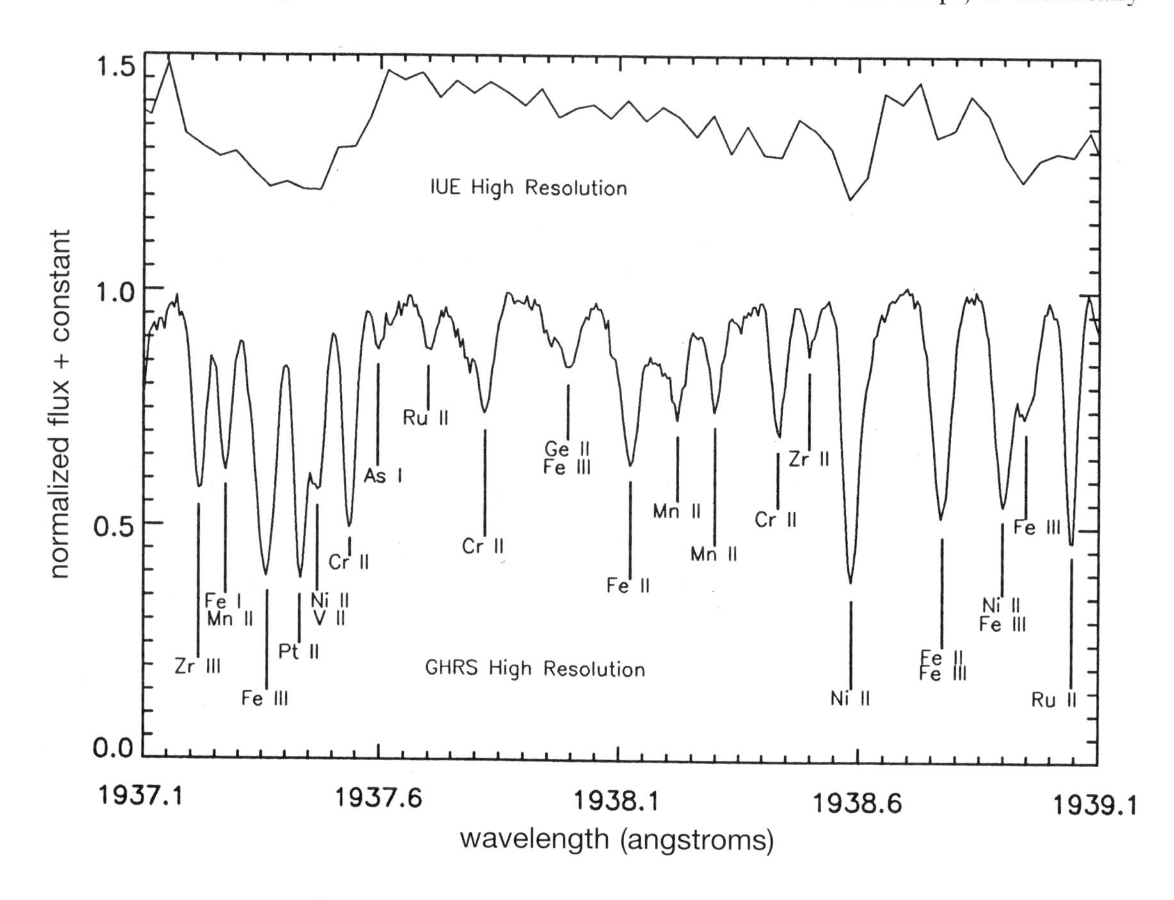


**Figure 4.18.** A 2 angstrom segment of the Chi Lupi spectrum taken with the highest resolution mode of the GHRS. The lines in the spectrum are cleanly resolved, whereas they were not in the IUE spectrum (above). The elements identified are: arsenic (As); chromium (Cr); iron (Fe); germanium (Ge); manganese (Mn); nickle (Ni); platinum (Pt); ruthenium (Ru); and zirconium (Z). (David Leckrone, NASA–Goddard Space Flight Center; Glenn Wahlgren, Computer Sciences Corporation and GHRS Science Team)

peculiar star with a surface temperature of about 10 000 kelvin and a small binary companion. A section of Chi Lupi's spectrum is shown in Figure 4.18. Leckrone described these observations as the most detailed ultraviolet spectrum of a star other than the Sun ever obtained. 'What we're trying to do with Chi Lupi is to define a pathfinder study,' he said. 'Obviously, we would like to observe many different stars with many different kinds of peculiarities, covering a range of temperature, gravity, magnetic and non-magnetic characteristics, and so on. We want to look at the basic "zoology" of these stars and see what determines some of these peculiarities we see.'

For chemically peculiar stars, the biggest challenge is to figure out why they are peculiar. Leckrone thinks the peculiarity is something confined to the surface layers of the stars. The element mercury (Hg) beautifully illustrates some of the bizarre results. Figure 4.19 shows the observations of a mercury line and the best-fit theoretical profile.

Two results come out of this analysis. First, the mercury seems to be in the form of mercury 204. If you buy a mercury thermometer at a hardware store, the mercury is a mixture of the mercury isotopes 198, 199, 200, 201, 202, (no 203), and 204. The fit shown is for 100% mercury 204, the heaviest stable mercury isotope.

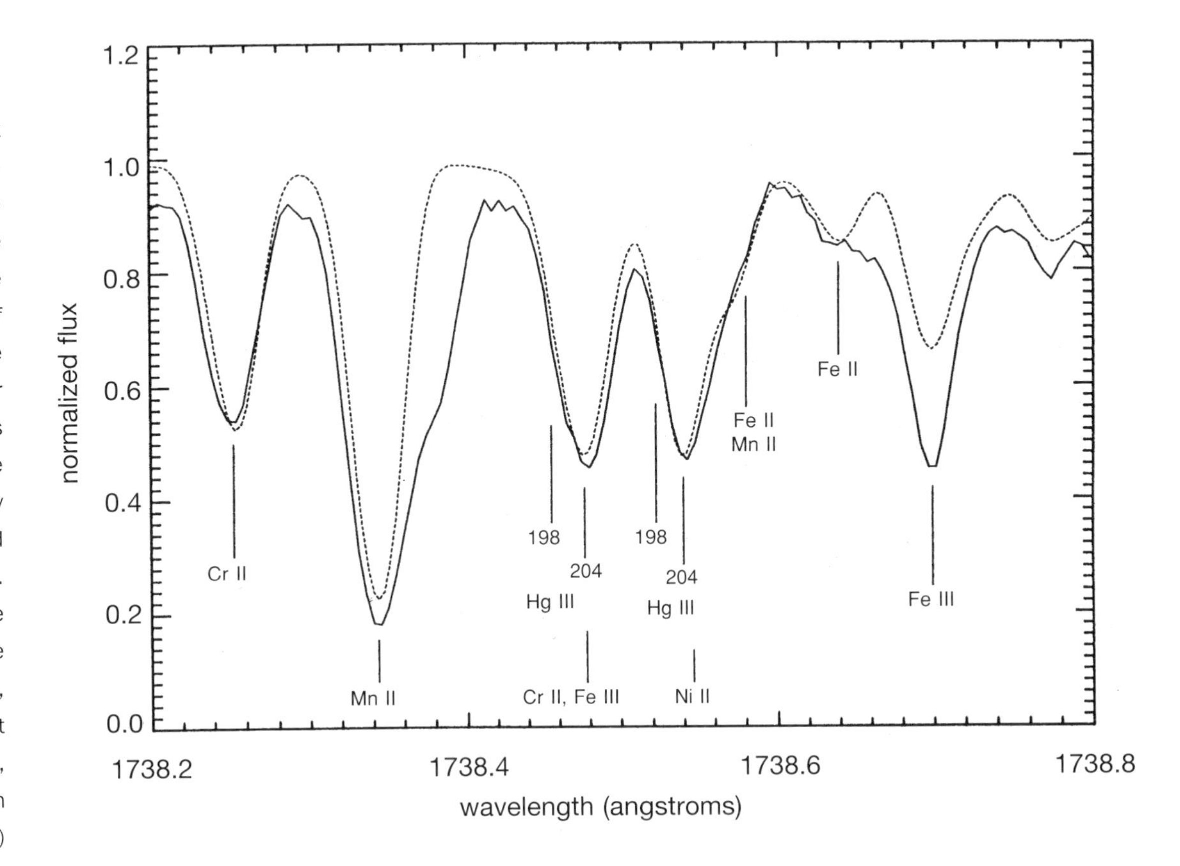


**Figure 4.19.** A 0.6 angstrom segment of the Chi Lupi spectrum (solid line) taken with the highest resolution mode of the GHRS. In addition to lines from some elements already identified in the caption to Figure 4.18, the segment shows two lines of mercury (Hg). The position of the lines changes with the particular isotope present, and the positions for the isotopes 198 and 204 are marked. The spectrum clearly matches isotope 204 (dotted line) and is not close for 198. The analysis indicates that the mercury is at least 99% isotope 204. (David Leckrone, NASA–Goddard Space Flight Center; Glenn Wahlgren, Computer Sciences Corporation and GHRS Science Team)

Secondly, the abundance value is very high, some 100 000 times the value in the Solar System. Similar very high values are found for platinum (Pt) and gold (Au). Figure 4.20 shows some sample abundances for Chi Lupi.

It is very likely that Chi Lupi shows such a high amount of these heavy elements in its spectrum because we are only seeing the top layers of the star. Chi Lupi's photosphere may itself be layered, which means that specific elements have coalesced in specific places and we just happen to be examining light emitted from those places. If that has not happened, then we must consider other explanations, such as some unusual nuclear processes running amok deep within the star.

The more likely possibility is that the atmosphere of the star is unusually stable. According to Leckrone, 'Chi Lupi basically has no rotational velocity in space. It's extraordinarily slow and the question you have to ask is: what happened to it? Was that just a pure accident of its birth, or is it because of interactions with its binary companion?'

It is also possible that small differences in radiation pressure – the pressure exerted by light streaming away from the core of the star – could concentrate isotopes and elements in the region under observation. All of these possibilities shape Leckrone's study of Chi Lupi, which remains very much a work in progress.

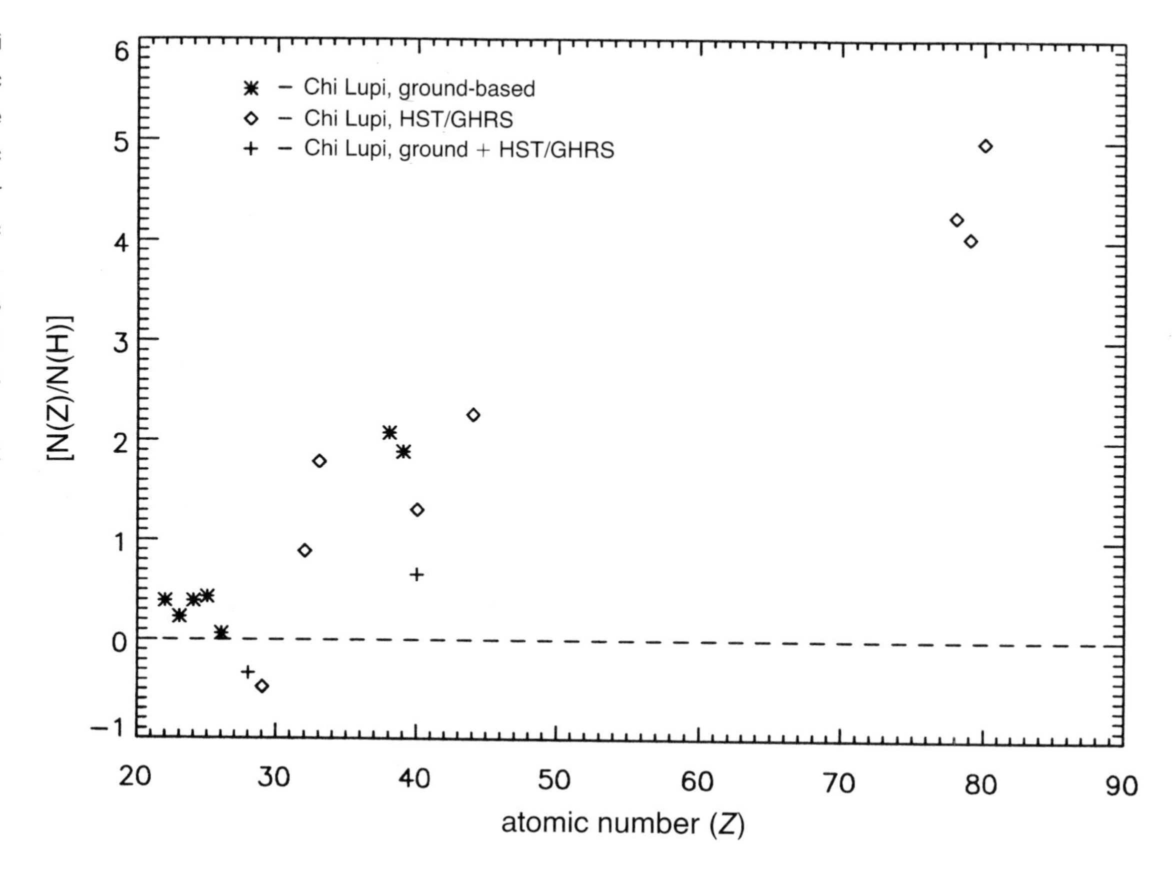


**Figure 4.20.** Abundances in Chi Lupi plotted against the atomic number of the element. The trend, compared to cosmic abundances, is for higher abundances for higher atomic number. The elements platinum, gold and mercury are plotted as the diamonds at upper right and are 10 000 to 100 000 times overabundant. (David Leckrone, NASA–Goddard Space Flight Center; Glenn Wahlgren, Computer Sciences Corporation and GHRS Science Team)

One interesting side effect of Leckrone's Chi Lupi work is the impact it has had on atomic physics. Understanding the spectrum in Figure 4.18 requires atomic physics data of higher quality than are sometimes available to fit the strength and location of the observed lines. This work has led to a unique collaboration between atomic physicists – who try to pinpoint elemental species in the laboratory setting – and astrophysicists, who observe these species in space. In the case of Chi Lupi – as with the data from observations of the interstellar medium – Leckrone and Swedish astrophysicist Sveneric Johanssen of the University of Lund are 'filling in the table of the elements' using these observations of naturally occurring elements in space. It is important to remember that these advances could not have taken place before an operational HST and GHRS, and Figure 4.18 shows a comparison of the observations of Chi Lupi made by HST and IUE. Recognizing the unique value of this work, the Swedish Academy of Sciences awarded Johanssen a prize for his achievements, citing his 'research in the border field between atomic physics and astronomy, particularly high-resolution spectroscopy using the Hubble Space Telescope'.

## Planetary nebulae

When stars about the size of Earth reach their final stage of evolution, they become planetary nebulae (so named because they appear in telescope views as stars with surrounding rings of gas and dust). What is happening inside the star is quite interesting. Nuclear fusion in the core has ceased, and the core is a degenerate carbon–oxygen melange about half the mass of the Sun. Surrounding the core are hydrogen and helium-burning shells. The whole thing is wrapped in a hydrogen envelope, and stretching away from the star is a huge shell of gas blown off during mass loss. The star continues to produce energy, and burns off the outer layers. This process exposes the hot core, and the star's surface temperature increases. Eventually it reaches a temperature of about 25 000 kelvin, by which time it is hot enough to produce ultraviolet radiation, which ionizes the shell and lights it up, creating a thing of ghostly beauty. The glowing shell is like a fossil record of the end stages of the dying star within.

*Henize 1357* HST focused its gaze on several planetary nebulae, including the highly complex Henize 1357. Pre-HST spectroscopic observations established the region as a young planetary nebula, and it appears that the central star began expelling its cloud of gas only a few thousand years ago. Only recently has there been enough radiation from the star to make the gas glow. At the equator, wind blowing out from the star has compressed the inner part of the gas to form the ring shape as shown in Figure 4.21. This is one of many examples of stars returning material to their environment as they cool and slide toward inevitable extinction.

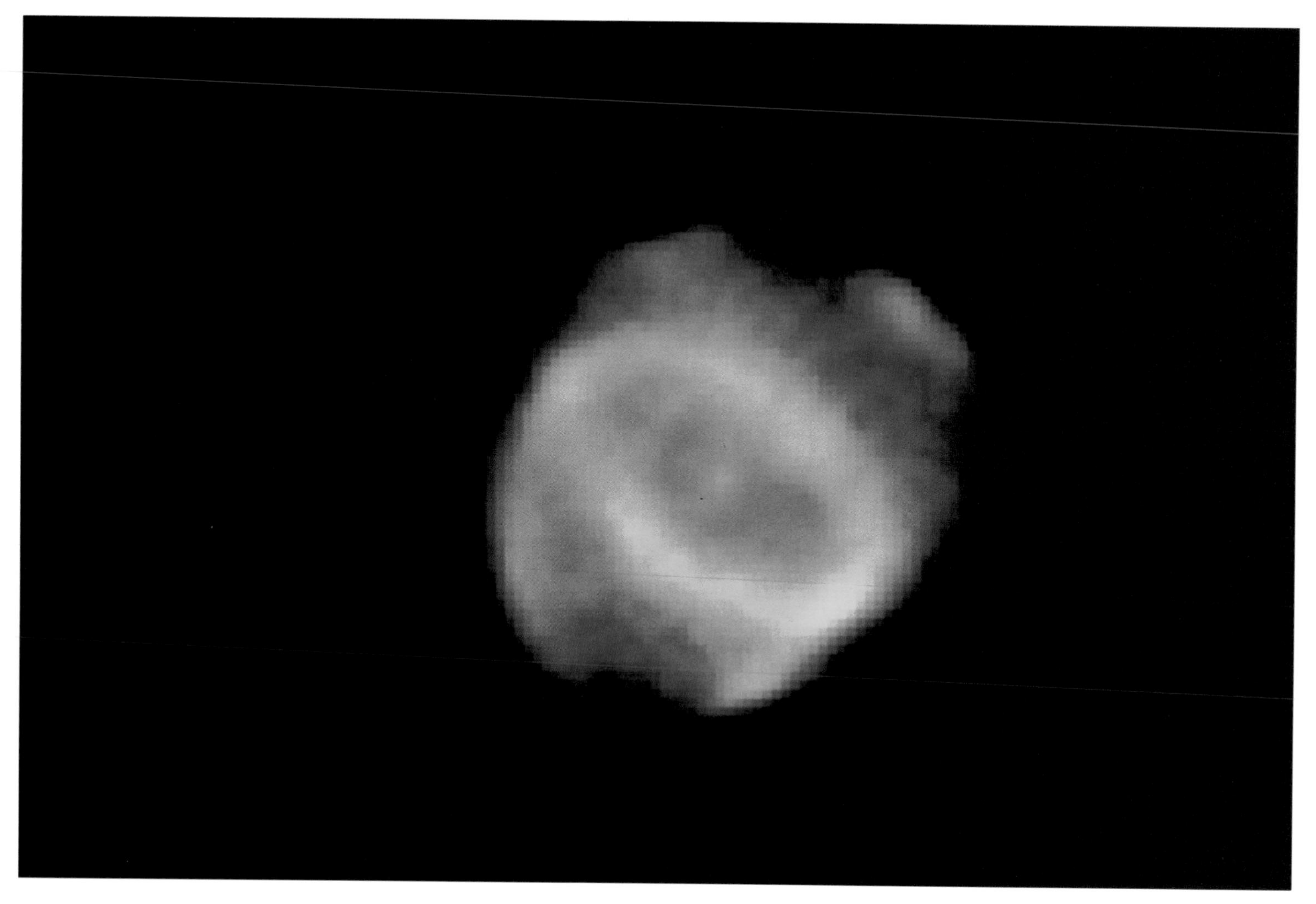

**Figure 4.21.** The young planetary nebula, Henize 1357, as imaged by WF/PC-1 (in PC mode). This image reveals detail and structure not visible from the ground. See text for discussion. (Matt Bobrowsky, CTA Inc.; NASA)

*NGC 2440* While some stars pique astronomers' interest because of their chemical makeup, others grab attention with their temperatures. One question that astronomers often hear (and ask) is, 'just how hot can stars get?' The WF/PC-1 image of NGC 2440 shown in Figure 4.22 shows a good example of an extremely hot star. The star is so hot and bright that its surrounding nebula is also extremely luminous. The temperature of the star is at least 200 000 kelvin, making it one of the hottest stars known. (By comparison, the temperature of the Sun is only about 5750 kelvin.) In earlier, ground-based images of the nebula, the central star and the nebula were blurred together in one brilliant core, and it was very hard to tell the two apart.

NGC 2440 is an example of a massive star in the center of a planetary nebula. Such stars can become very hot, and, if they continue to evolve, they become hotter and hotter. Eventually they get so hot that they become dimmer in the visual part of the spectrum and

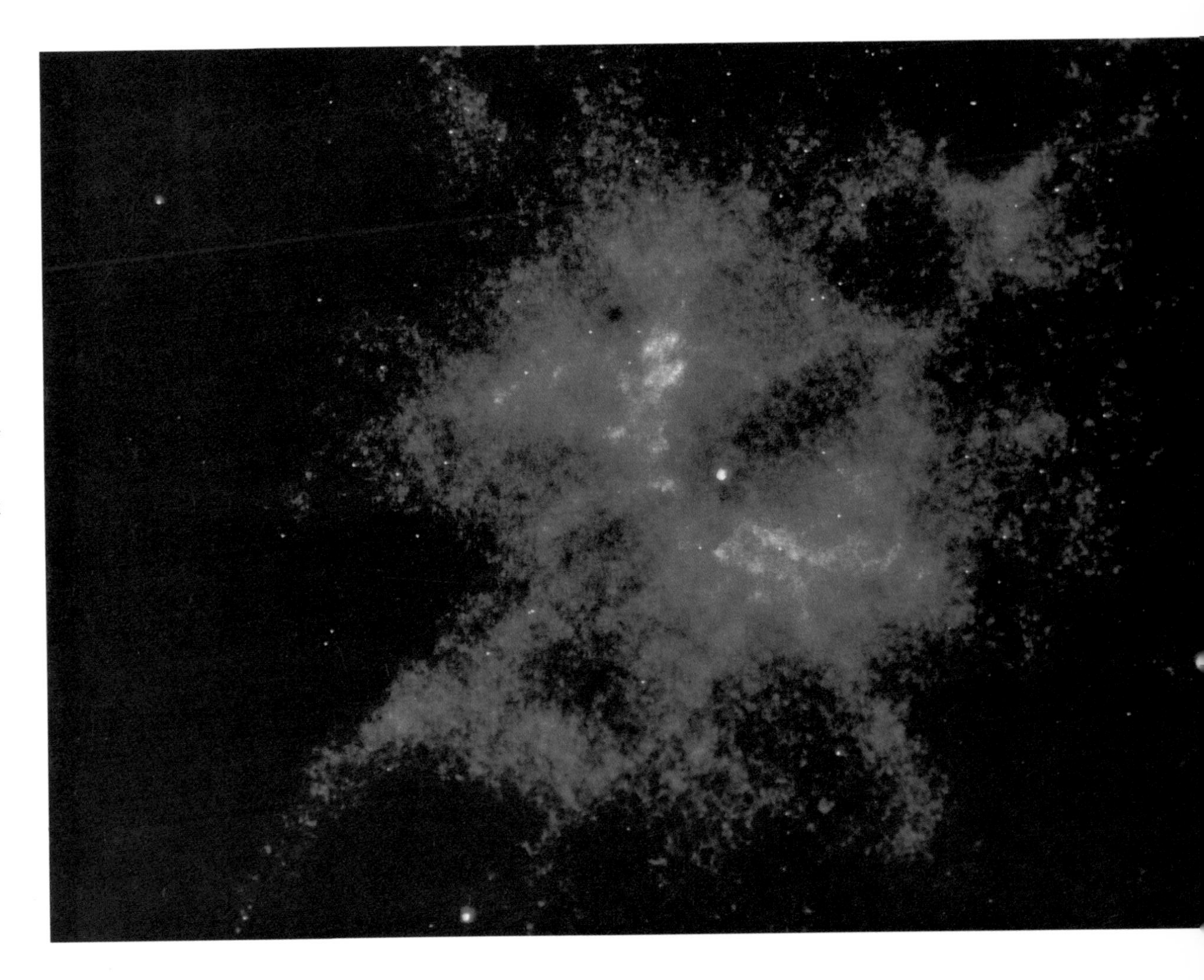

**Figure 4.22.** WF/PC-1 (in PC mode) image of the central region of NGC 2440; the image has been sharpened by computer image restoration. Because the central star is clearly separated from the surrounding nebula, its temperature can be estimated and is at least 200 000 kelvin, one of the hottest known stars. The image also shows the incredibly complex structure of the nebula. (S.R. Heap, NASA–Goddard Space Flight Center)

appear brighter in ultraviolet wavelengths. Therefore, observers must employ higher spatial resolution to see the star against the background nebula. The HST image, sharpened by computerized image restoration, clearly separates the star from the nebular glow, and it shows great detail in the structure of the star's nebula. Notice the filaments, blobs and streamers.

*Cat's-Eye Nebula* One of the most complex planetary nebulae observed, NGC 6543, lies in the constellation Draco. Nicknamed the 'Cat's-Eye' Nebula, it is estimated to be at least 1000 years old. Seen through WF/PC-2 in September, 1994, the structure of the nebula shows jets of high-speed gas, concentric shells and shock-induced knots of gas. At least one theory suggests that the star at the center might be a binary system. These dying binaries probably spent their last millennia losing mass through stellar winds. Because of their motion around each other, the shells of gas were twisted into complicated shapes. A fast stellar wind blowing off the central stars might have created the curious ellipsoidal shape of the inner gas shell. Surrounding it are two larger lobes of gas blown away from the star. The bright arcs and curly-shaped structures might be formed by jets of gas. These jets seem to point in different directions from some of the features they have highlighted, suggesting they are wobbling as the stars orbit each other, possibly turning on and off like a beacon. To help understand the dynamics of the system, astronomers mapped the glowing gas clouds at wavelengths specific to particular colors. In Figure 4.23(a) red is hydrogen, blue denotes neutral oxygen and green is ionized nitrogen.

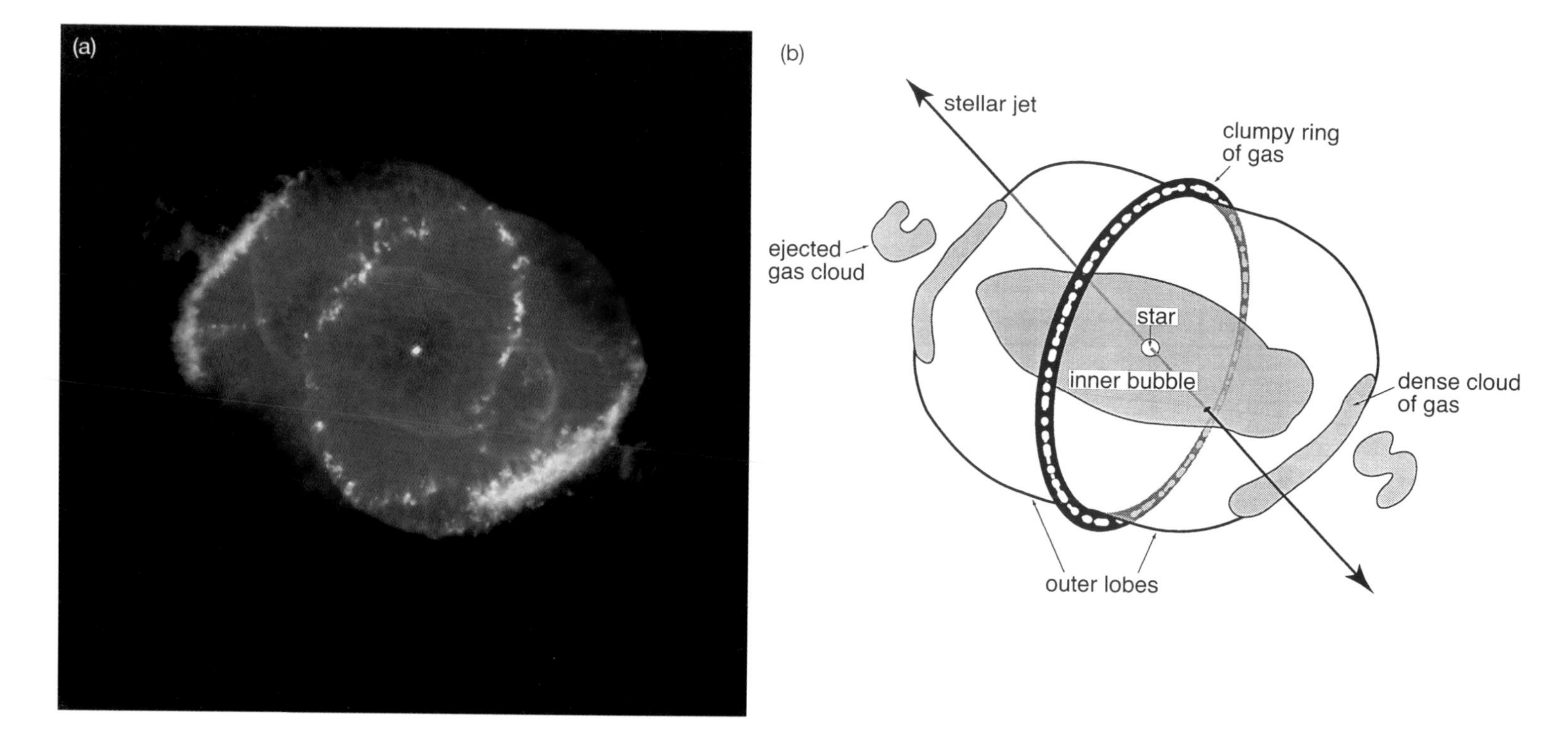


**Figure 4.23.** (a) WF/PC-2 image of NGC 6543, the Cat's-Eye Nebula. This planetary nebula lies 3000 light years away from Earth. (b) A schematic for the Cat's-Eye Nebula. (STScI; P. Harrington, University of Maryland; NASA)

## Stellar death throes

Beyond giants and supergiants, planetary nebulae and chemically peculiar stars, astronomers reach into the realm of the outlandish stars – weird beasts like white dwarfs, neutron stars and the ever-popular black holes. These intriguing members of the 'stellar zoo' are extreme examples of stars that are 'doing something'. What they are doing is dying, and that activity can range from the quiet fading away of a white dwarf to the violent explosion of a supergiant star as a supernova.

## White dwarf stars

White dwarfs are faint old stars with masses less than $1.4M_s$. Most of them are about the same size as the Earth, but they are 300 000 times its mass. Some may have started life as stars a few times the size of the Sun, but blew off much of their material into surrounding space in a gentle stellar wind. Others had masses below $1.4M_s$. At the white dwarf stage of their lives, the nuclear furnaces inside these stars have stopped burning. They are dimming down, and at some point they will stop contracting. This happens when the electrons in the atoms of gas that remain in their cores push against the contraction and the star enters a state of equilibrium – no more contraction, no more energy source. What is left is a stellar mass so dense that a small spoonful of the star's material would weigh 10 tons! As an important end-point of stellar evolution, white dwarfs are a natural target for HST scientists, especially if they want to take spectra of those stars as they die. Because the gravity for white dwarfs is very high, the spectral lines are usually broad, and so the Faint Object Spectrograph, with its high sensitivity and coverage in ultraviolet wavelengths, is the instrument of choice.

If you study a white dwarf star, you should not expect to see much wind (if any) blowing away from it. If you have a spectrograph able to resolve the light from such a dim, distant object, you might be able to form an idea of which elements still exist in the leftover atmosphere of the star.

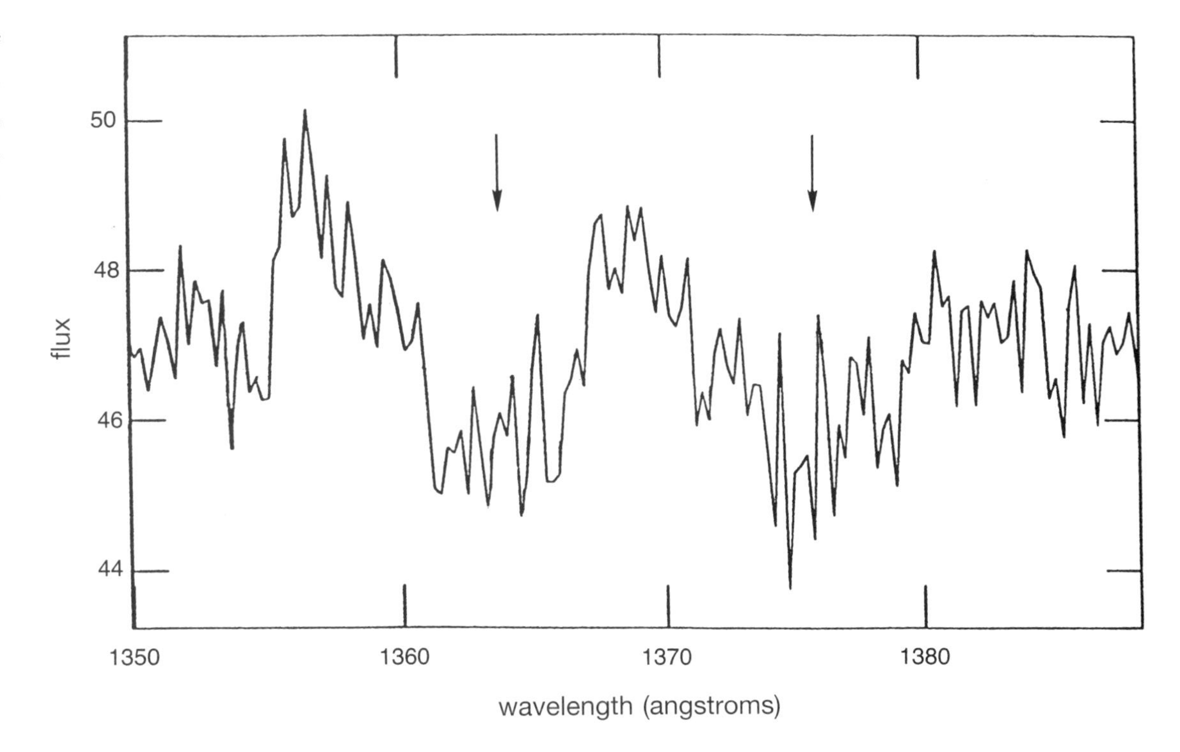


**Figure 4.24.** FOS spectrum of the white dwarf G191.B2B. Two of the strongest iron (Fe V) lines are marked by the arrows. See the text for discussion. (Edward M. Sion, Villanova University)

HST observations of the hot (effective temperature of about 62 000 kelvin) white dwarf G191.B2B have revealed the presence of highly ionized metals in the star. Ionized metals betray themselves through their spectra. In Figure 4.24, two broad absorption features at 1364 angstroms and 1376 angstroms are produced by iron ionized four times (Fe V). While Fe V lines are the most numerous, absorption lines of other elements show up too.

Why should hot iron be so important here? Unless there is some opposing mechanism, heavy elements like iron should sink into the white dwarf and be invisible in a spectrum. If we still see these elements in the spectrum, then something is happening to levitate or even expel these elements from the star. It is possible that high radiation pressure from within the star is causing this levitation effect. Maybe there are dense shells of material surrounding the star, and these shells are showing off their ionized iron content. Perhaps mass loss from white dwarfs actually occurs. If so, it challenges the conventional understanding of these slowly dying hot stars.

## Supernova 1987a

Nothing catches the attention of astronomers quite like the final, cataclysmic explosion of a supergiant star called a supernova. These temporary but spectacular outbursts can be unbelievably bright – sometimes rivaling the luminosity of entire galaxies. For that reason,

**Figure 4.25.** Supernova 1987a (the bright star) in the Large Magellanic Cloud on March 2, 1987, as photographed with the CTIO Curtis Schmidt telescope by M. Bass. (National Optical Astronomy Observatories)

supernovae are often used as standard candles to determine the distances to the galaxies they inhabit.

Because of their pivotal role in stellar evolution – from enrichment of the interstellar medium in heavy elements, creation of elements heavier than iron, production of neutron stars and black holes, and their probable role in the conception of new stars – supernovae afford astronomers who use HST a welcome research opportunity. Supernova 1987a (SN 1987a) in the Large Magellanic Cloud presented a spectacular chance to watch the aftermath of just such a stellar explosion. Since it was first observed by HST in August, 1990, studies have focused on the cloud of debris rushing out from the supernova. Now astronomers are searching for the pulsar expected to be at the heart of the event.

The Faint Object Camera image of SN 1987a and two other stars in the field of view is shown in Figure 4.26. It was taken through a narrow band filter designed to isolate light (lines) from doubly ionized oxygen, a common emission from gaseous nebulae in galaxies. The image shows an elliptical, luminescent ring surrounding SN 1987a. This ring was seen glowing very faintly in previous ground-based and space observations, but the FOC image supplies a much improved view. The angular separation between the ring and the supernova is just 0.8 arcsecond. The ring is too far away from the supernova to consist solely of material that has been blown off during the event. Rather, the material was already in the neighborhood, probably as the result of massive, ongoing stellar wind activity by the

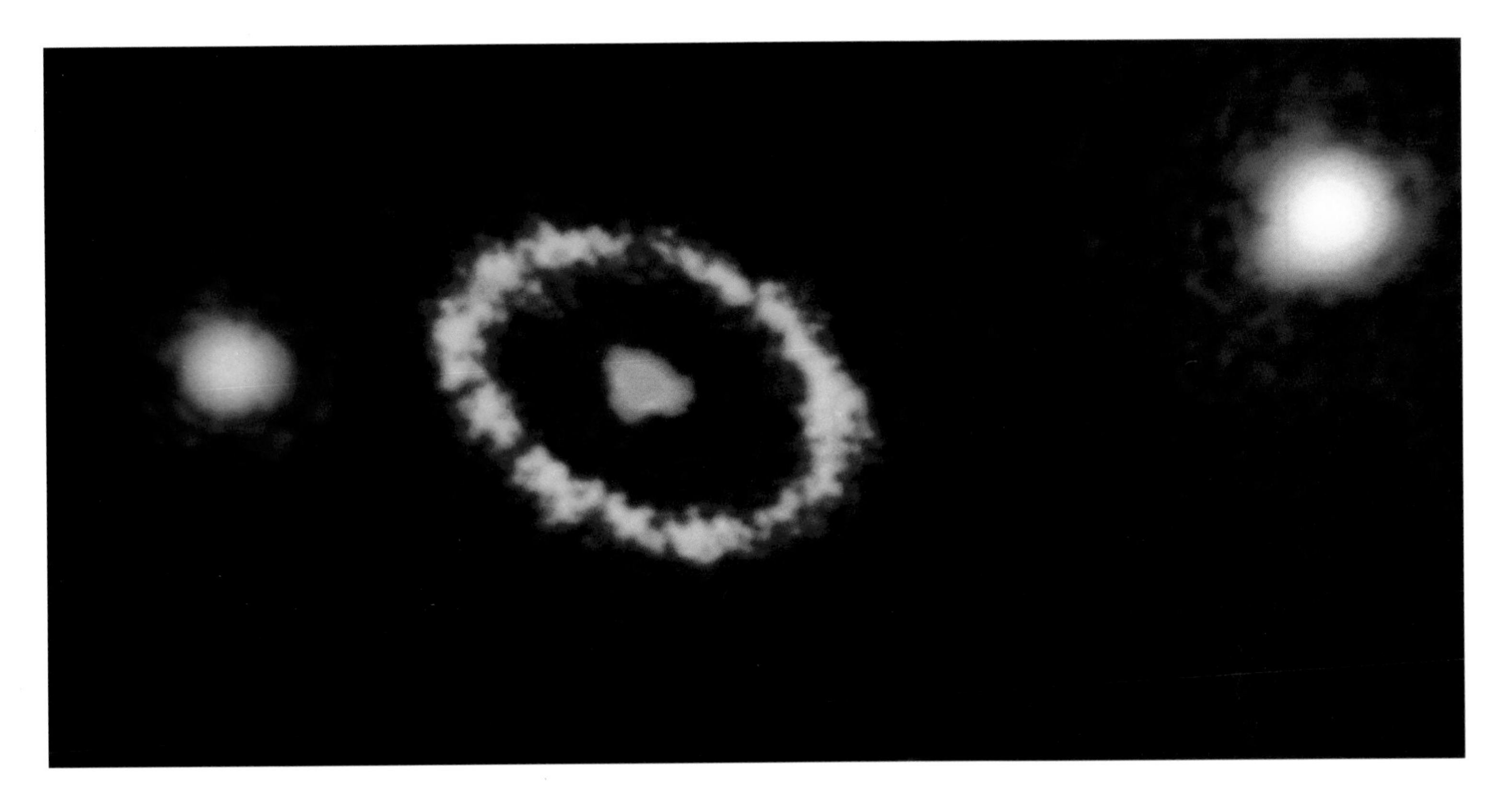

**Figure 4.26.** FOC image (pre-COSTAR) of Supernova 1987a. This is a raw image with no image restoration attempted. The original image was made in yellow light (for doubly ionized oxygen), and the blue and red were added to reflect the object's true color. The supernova is red, the ring around it is yellow, and the two blue stars are unrelated to the supernova. (NASA; ESA)

progenitor blue supergiant star. The material was brightened by the light flash from the supernova, which reached the ring 240 days after the explosion.

What will happen next with SN 1987a? A cloud of stellar debris is rushing out from the center of the explosion, and, according to University of Colorado astronomer Dick McCray, calculating the exact time at which the expanding debris will contact the outer ring of material is an interesting problem. He is heavily involved in theoretical modeling of the scenario unfolding at SN 1987a. 'The bottom line is that it's going to impact the ring,' he says. 'Anywhere from 1996 to 2002 is when it's going to hit. We'll learn what that ring is. It's going to sparkle, it's going to have line profiles, it's going to be seen in X-rays as well as with HST.'

McCray believes that the ring is a unique phenomenon. In some ways it is still a mystery, but he has his own theory about how it formed: 'Everybody's been saying that the ring was ejected by the progenitor star, the supernova – but this new idea is that the ring was not ejected, but was a protostellar disk that was formed at the same time as the supernova progenitor was formed. When the supernova progenitor became a blue star, it eroded away this disk – ate a big hole in the middle, and the outer part of the disk is still there and the ring is the inner rim. When the supernova went off, it made a flash and the flash ionized the inner rim of this disk and made it fluoresce. Space Telescope will make it absolutely clear whether it was a protostellar disk or was ejected.'

Supernova 1987a has not yet finished astounding astronomers. A spectacular image taken in early 1994 shows the same ring at the center of the supernova site, plus a double set of rings flanking the supernova, a distinct change in appearance that has had McCray and other astronomers stumped for a workable explanation. Chris Burrows, at the Space Telescope Science Institute, who was one proponent of the compressed gas shells theory for the central 'celestial hula hoop' around the supernova site, has proposed that these faint outer rings might have been the result of two phenomena. First, there could be jets of material coming from an unseen neutron star or black hole very close to the supernova. These jets compressed part of the shell of gas around the blast site into a circular shape. Secondly, when radiation from the exploded star hit the rings, they lit up to form the ghostly images we see in Figure 4.27. McCray disagrees with that interpretation, suggesting that Burrows' theory violates the 'tooth fairy rule' (which allows a credible theory to invoke only one mysterious, unknown and unseen agent to explain a rare phenomenon). However, both scientists agree that there is not yet a better explanation for these gracefully beautiful rings.

According to current theories, there should be a pulsar showing up at the heart of SN 1987a, but observations by the High Speed Photometer showed no evidence for a pulsar, the rapidly spinning remains of a supernova's stellar core. As the core rotates, sometimes many thousands of times per second, a pulse of light flashes out like a beacon shining from a lighthouse. There should be a pulsar at SN1987a, so why are not astronomers seeing it? There could be several reasons: (i) the pulsar is there, but is obscured by the expanding cloud of debris; (ii) the remnant of the star may be too massive (i.e. greater than about

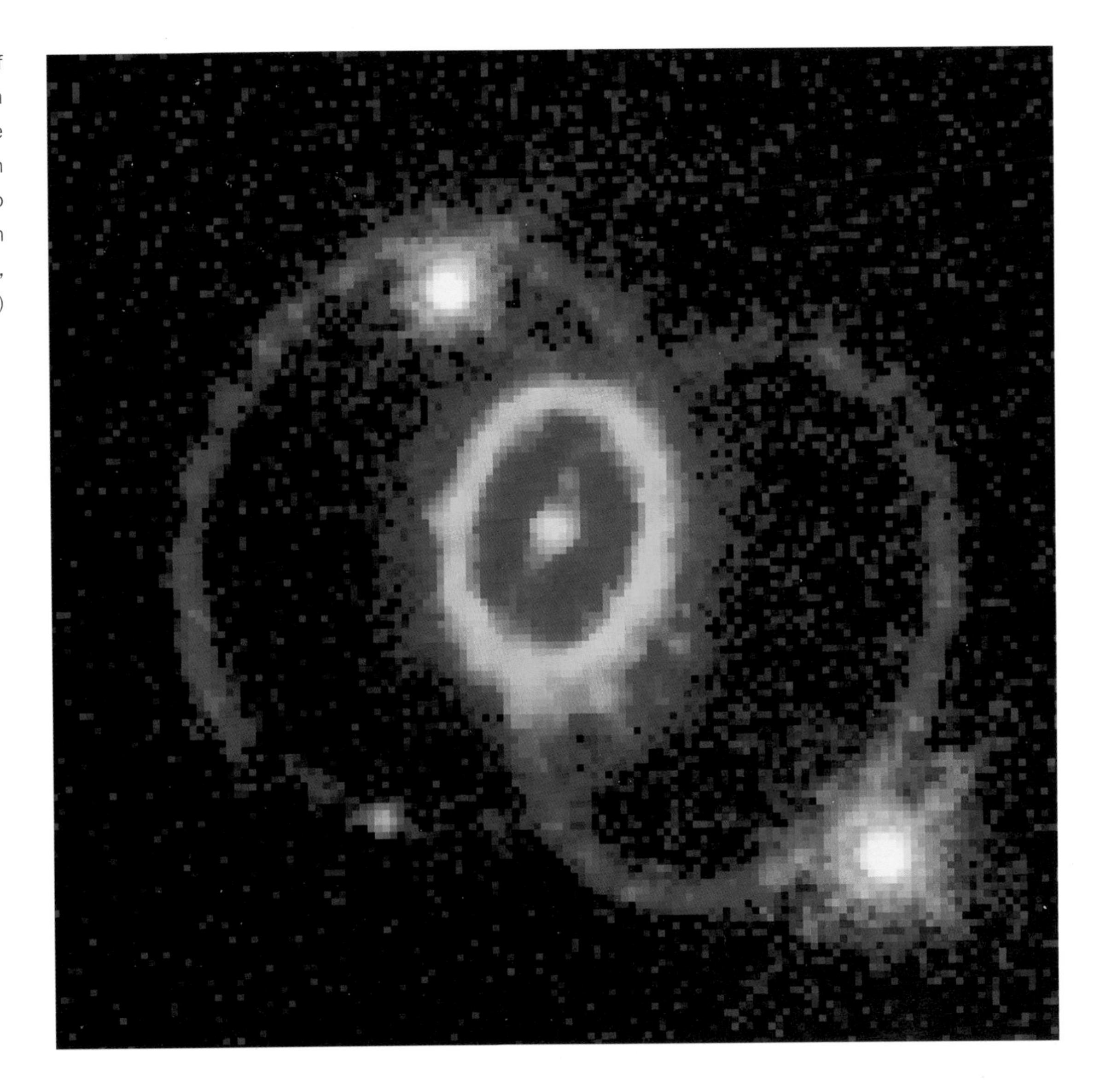

**Figure 4.27.** WF/PC-2 image of Supernova 1987a taken in February, 1994. In addition to the central ring (yellow) already seen by the FOC (Figure 4.26), two fainter, outer rings have been discovered. (Christopher Burrows, ESA and STScI; NASA)

three solar masses) to become a neutron star and has collapsed to a black hole; or (iii) the Earth may simply be out of the line of sight of the pulsar beam and we are just missing seeing it. Joe Dolan, a Goddard astronomer on the High Speed Photometry Team, has speculated that the brightness for the leftover star may be very low indeed. He estimated a visual magnitude of no more than 27 for the pulsed radiation from the supernova remnant. Any possible pulsar is certainly very faint, if it indeed exists.

## The Crab Nebula pulsar

The supernova that produced the Crab Nebula did form a pulsar – a rapidly spinning neutron star. The pulses are believed to be generated by electrons being whirled around in the

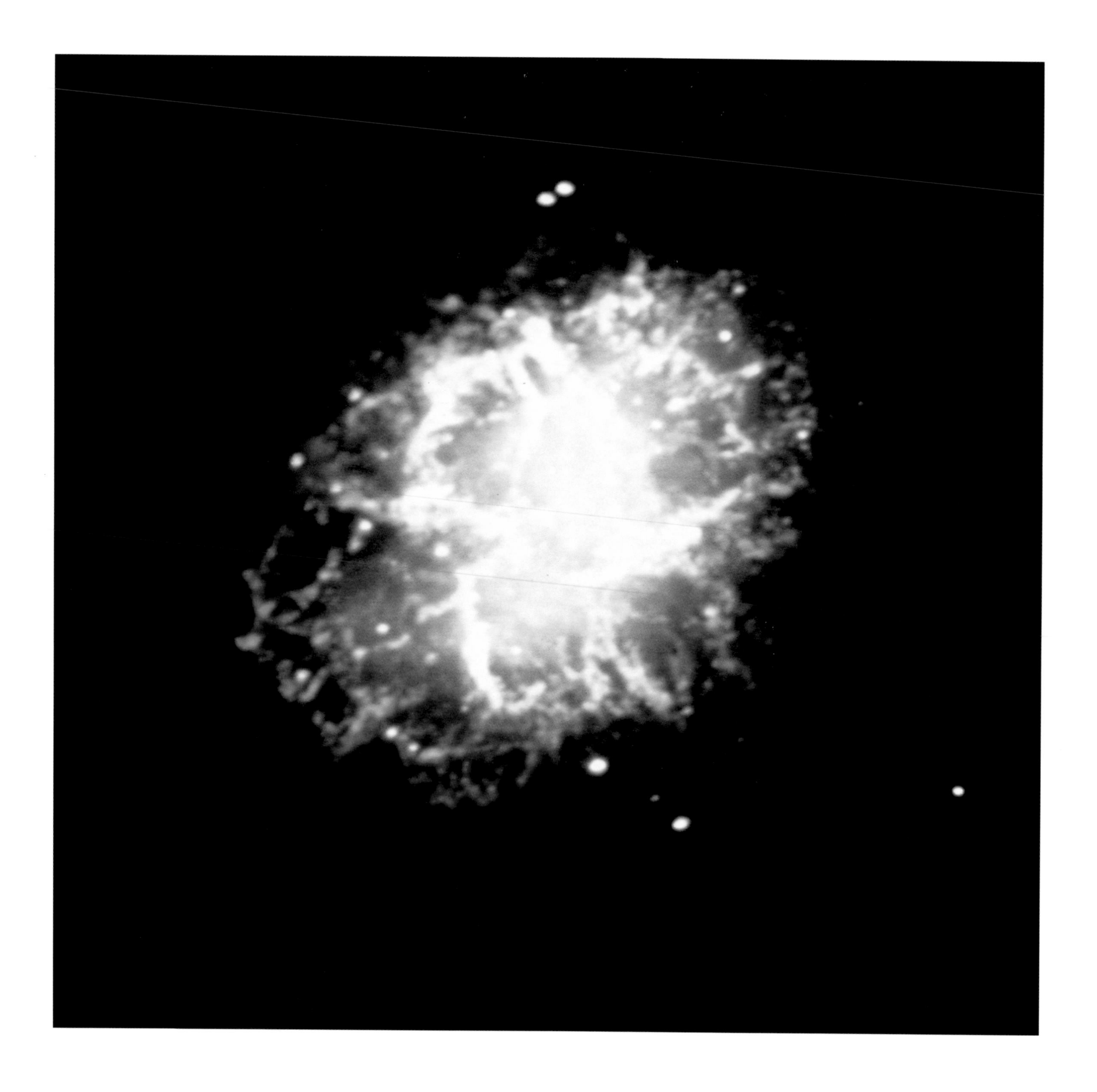

**Figure 4.28.** (facing page) The Crab Nebula (M1 or NGC 1952) in the constellation Taurus as photographed by the 4-meter Mayall Telescope of the Kitt Peak National Observatory. This color composite consists of red (hydrogen) and blue (sulfur) emissions. The Crab Nebula is approximately 6 light years in diameter. (National Optical Astronomy Observatories)

strong magnetic field of the star. As these particles are hurled into the surrounding gases by the spinning of the magnetic field, they radiate energy, producing synchrotron radiation. This radiation is what causes the Crab to glow.

What appears to be a torus of gas surrounds the pulsar itself, and one of the X-ray jets is forming a ghostly looking halo of synchrotron radiation. In the direction of the opposite jet lies a shocked gas region where stellar winds collide with clouds of gas. Clearly, jets from the star are interacting with the clouds of gas and dust ejected when it exploded more than 900 years ago. University of Arizona scientists Paul Scowen and Jeff Hester have determined that the cooler regions of the nebula coincide with the thickest clouds of gas and dust. The hotter areas contain mostly thin, tenuous gas clouds.

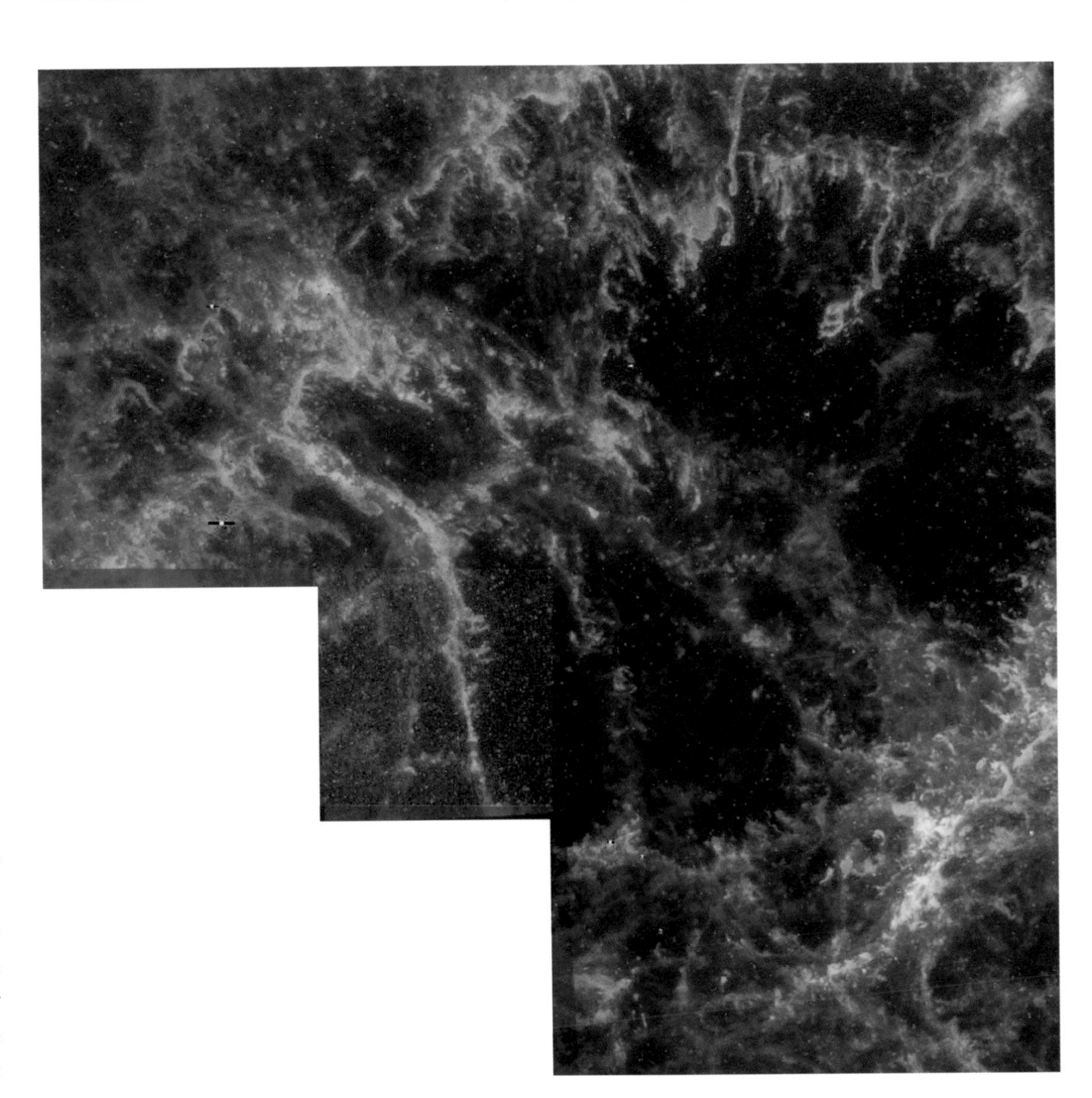

**Figure 4.29.** The filamentary remains of the star that created the Crab Nebula more than 900 years ago. This image was taken with the WF/PC-2. (Jeff Hester and Paul Scowen, Arizona State University)

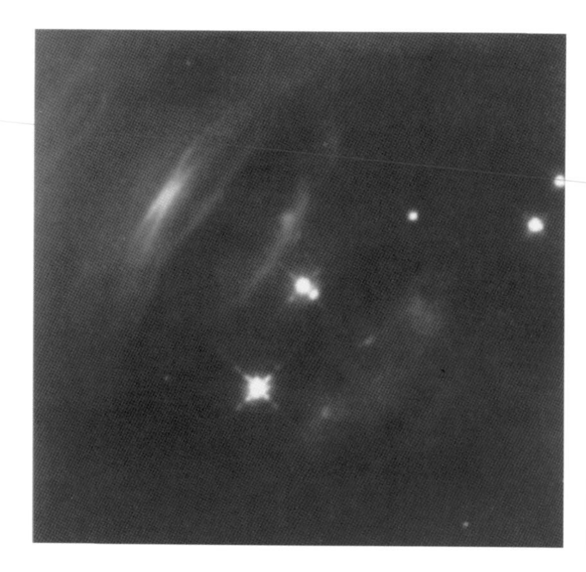

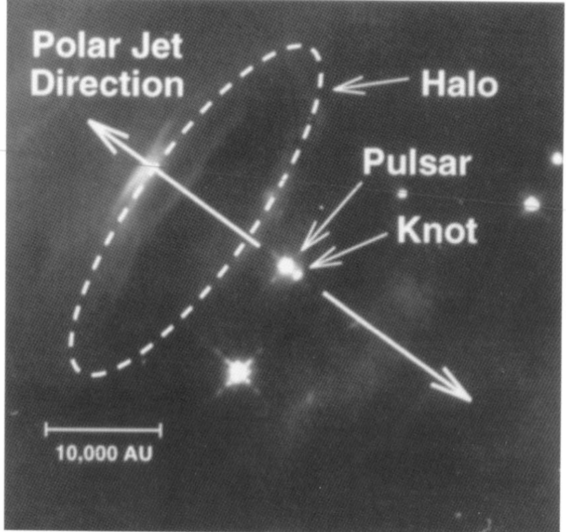


**Figure 4.30.** This WF/PC-2 image of the center of the Crab Nebula was taken in the light of synchrotron emission from the pulsar. This image showcases two discoveries. The first is that there is a small knot of bright emission located about 15 astronomical units from the pulsar. This knot went undetected by ground-based equipment because it was lost in the glow of the adjacent pulsar. The knot and the pulsar line up along the direction of a jet of X-ray emission. The knot may be a 'shock' in the jet – a region where the wind streaming away from the pole of the pulsar piles up. The second discovery is that in the direction opposite the knot, the Crab pulsar is capped by a ring-like 'halo' of emission tipped at about 20 degrees to our line of sight. The ring may mark the boundary between the polar wind and jet, and an equatorial wind that powers a larger ring of emission surrounding the pulsar. (Jeff Hester and Paul Scowen, Arizona State University)

The sensitivity (resulting in good temporal resolution) and wavelength coverage of the HST's High Speed Photometer allowed a test of how the pulses of radiation travel through the surrounding shells. The HSP team wanted to see if the arrival time of the main pulse would coincide with the arrival times of all wavelengths of light – from gamma rays to infrared radiation.

The HSP obtained observations of the Crab Nebula pulsar in visible light (which is in the 4000 to 7000 angstrom range) and in ultraviolet light (in the 1700 to 2900 angstrom range). What HST 'saw' in the middle of the Crab Nebula was a pulsar ticking away 30 times a second, and both wavelength ranges showed the same arrival time for the light.

This is good news because it is a confirmation of the standard model for pulsed light emission. The positive check supports the position that we understand the production of pulsed radiation from rapidly rotating neutron stars.

## Supernova 1006

As we have seen with Supernova 1987a and the Crab Nebula, supernova remnants – the rapidly expanding shells of gas and dust formed when a dying star spreads its rich mixture of heavy elements out into the interstellar medium – are some of the most beautiful objects in astronomy. The remnant from the supernova of AD 1006 is an astronomical curiosity. A small-radius hot star is located behind the remnant, and the line-of-sight passes within about 2.5 arcminutes of the center of the remnant. This arrangement permits a study of the remnant through absorption lines produced when light from the background star passes through the remaining gas and dust.

The Faint Object Spectrograph spectrum showing two broad absorption lines produced by once-ionized iron is given in Figure 4.33. From these lines we can make the following

**Figure 4.31.** A colorized version of the central region of the Crab Nebula. (Jeff Hester and Paul Scowen, Arizona State University)

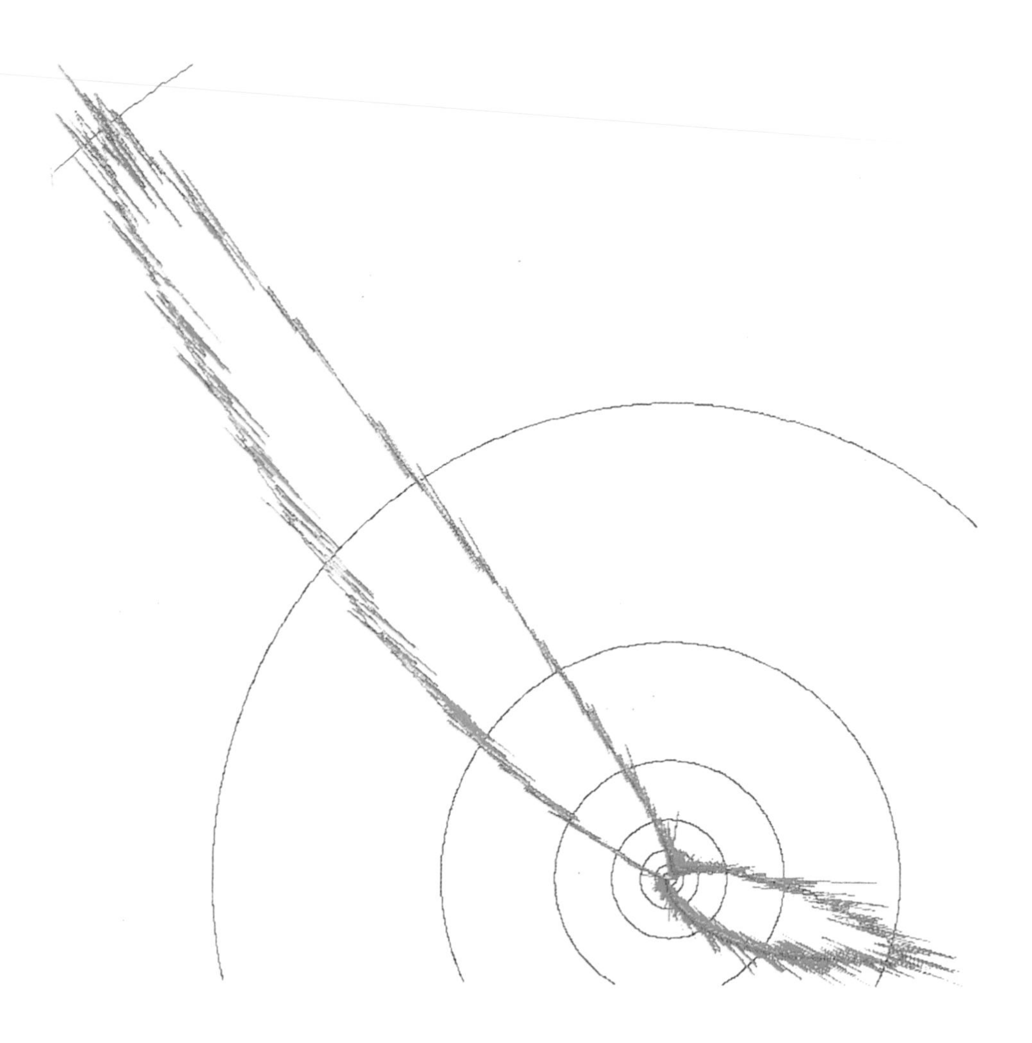

**Figure 4.32.** HSP measurements of the Crab Nebula pulsar plotted in a format where a complete revolution (360 degrees) is one period, and where the length is proportional to intensity. Visible light is plotted in red and ultraviolet light is plotted in green. The curves in both wavelengths are remarkably similar. (Robert Bless, University of Wisconsin and the HSP Team)

deductions about the material rushing out from the star: (i) the lines are symmetrical around their center, which implies that the original explosion was symmetrical – that is, the star exploded out equally in all directions, and (ii) the width of the line is produced by material receding from us at about 8000 kilometers per second and by material approaching us at the same speed. Using spectra to characterize the size and shape of this supernova explosion, as well as its speed and the chemical mix of the material rushing out from the star, has allowed astronomers to gain more insight into the complex dynamics of supernova explosions.

## The Cygnus Loop

The last supernova remnant that we will study here is a beautiful arc of light called the Cygnus Loop. It stretches about 3 degrees across the sky in the constellation of Cygnus,

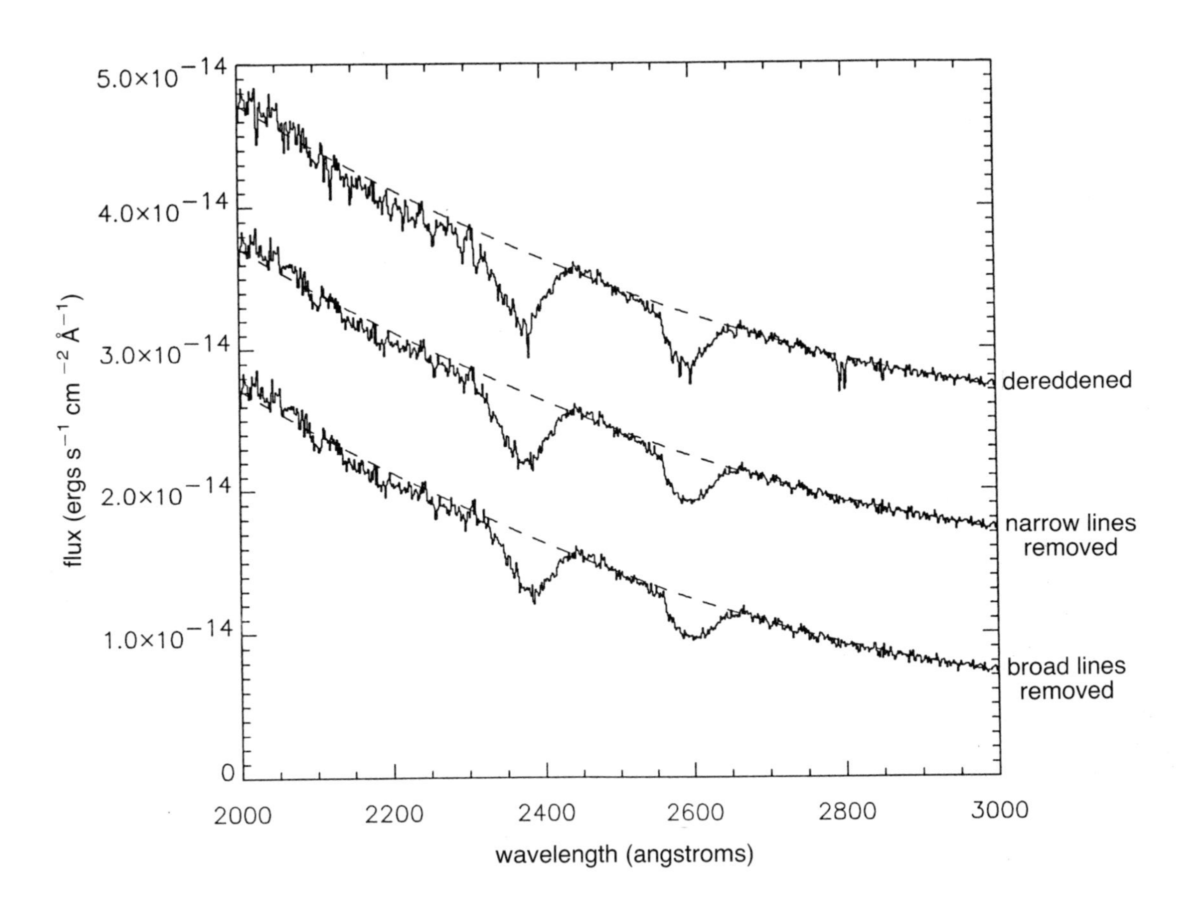


**Figure 4.33.** FOS spectrum of a hot star located behind the remnant of Supernova 1006. The different curves show the kinds of adjustments made to obtain the true profile of the iron (Fe II) lines. The bottom curve is the final result; the upper two are displaced for clarity. See the text for the interpretation of these absorption lines. (Chi-Chao Wu, Computer Sciences Corporation)

and lies in the plane of the Milky Way some 2600 light years away. The Loop is a blast wave from a supernova explosion estimated to have occurred about 15 000 years ago. Figure 4.34 shows part of the Loop known as the Veil Nebula.

The WF/PC-1 image in Figure 4.35 shows a small part of the Loop at better resolution, resolving details that are about the size of our Solar System. The blast wave from the supernova is slamming into clouds of interstellar gas. The shock process heats the gas and causes it to glow. Ejecta from the supernova envelope may be catching up with the blast wave, which has been slowed by interactions with the ambient gas. This gas is the bluish ribbon of light stretching across the image.

## Eta Carinae, the cataclysmic variable star

The Milky Way Galaxy has a number of stars called cataclysmic variables, part of a larger class of variable or unstable stars, and we bring our exploration of stars to an end with one that has not quite died yet. This is the luminous blue star Eta Carinae, and one look at it would be enough to convince anyone that this star is in its death throes.

Eta Carinae is a southern hemisphere object, and has been quite familiar to observers throughout the centuries. Edmond Halley watched it brighten up to 4th magnitude in 1677.

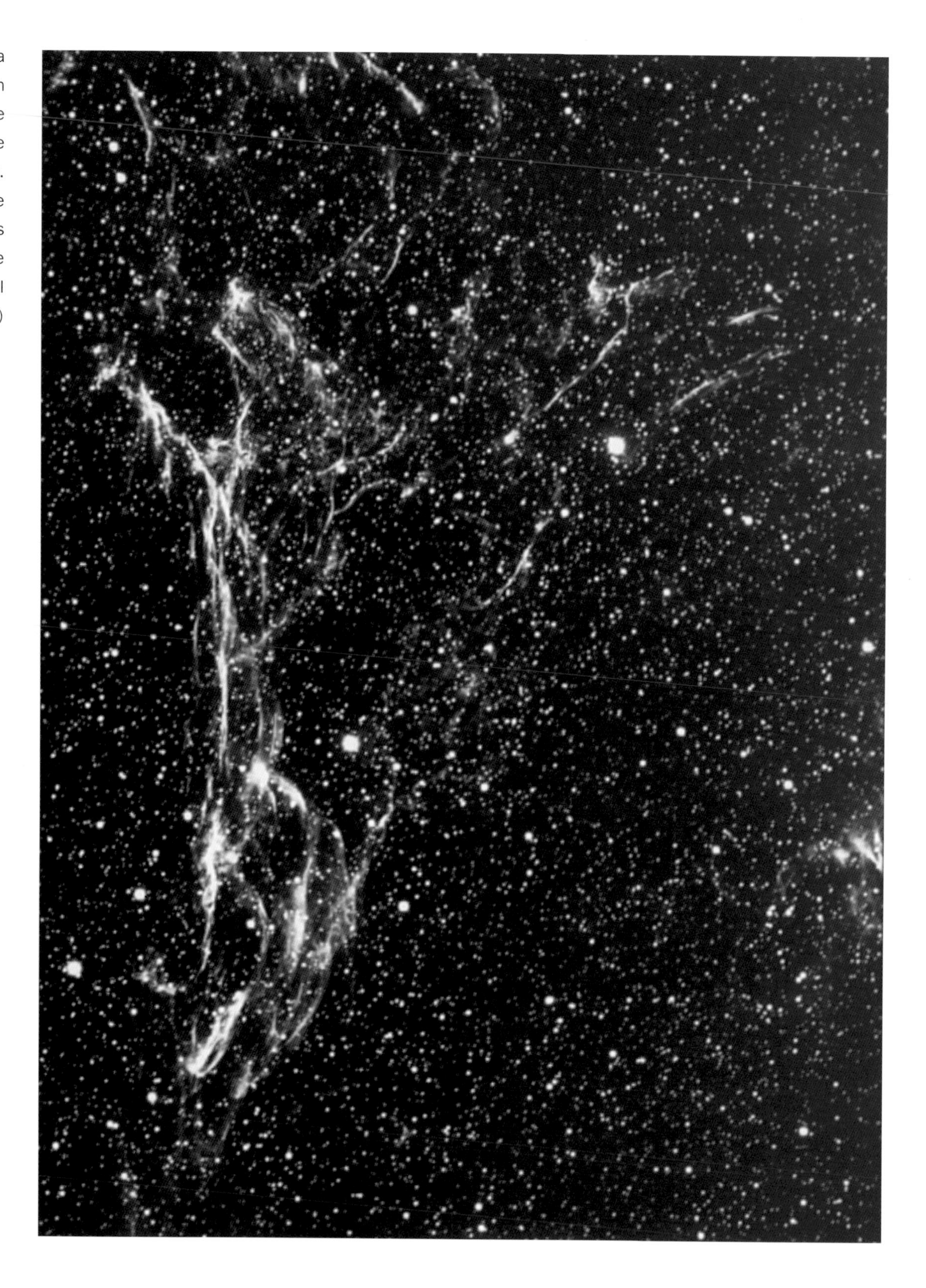

**Figure 4.34.** The Veil Nebula (NGC 6979) in the constellation Cygnus as photographed by the 4-meter Mayall Telescope of the Kitt Peak National Observatory. The nebula is also known as the Cygnus Loop. The image shows the north-central portion of the nebula. (National Optical Astronomy Observatories)

**Figure 4.35.** WF/PC-1 image of a part of the Cygnus Loop supernova remnant taken on April 24, 1991. The image is a color composite composed of blue light (oxygen atoms), green light (hydrogen atoms) and red light (sulfur atoms); the different atoms arise from gas at different temperatures. The image shows remarkable detail in the nebula. (Jeff Hester, Arizona State University; NASA)

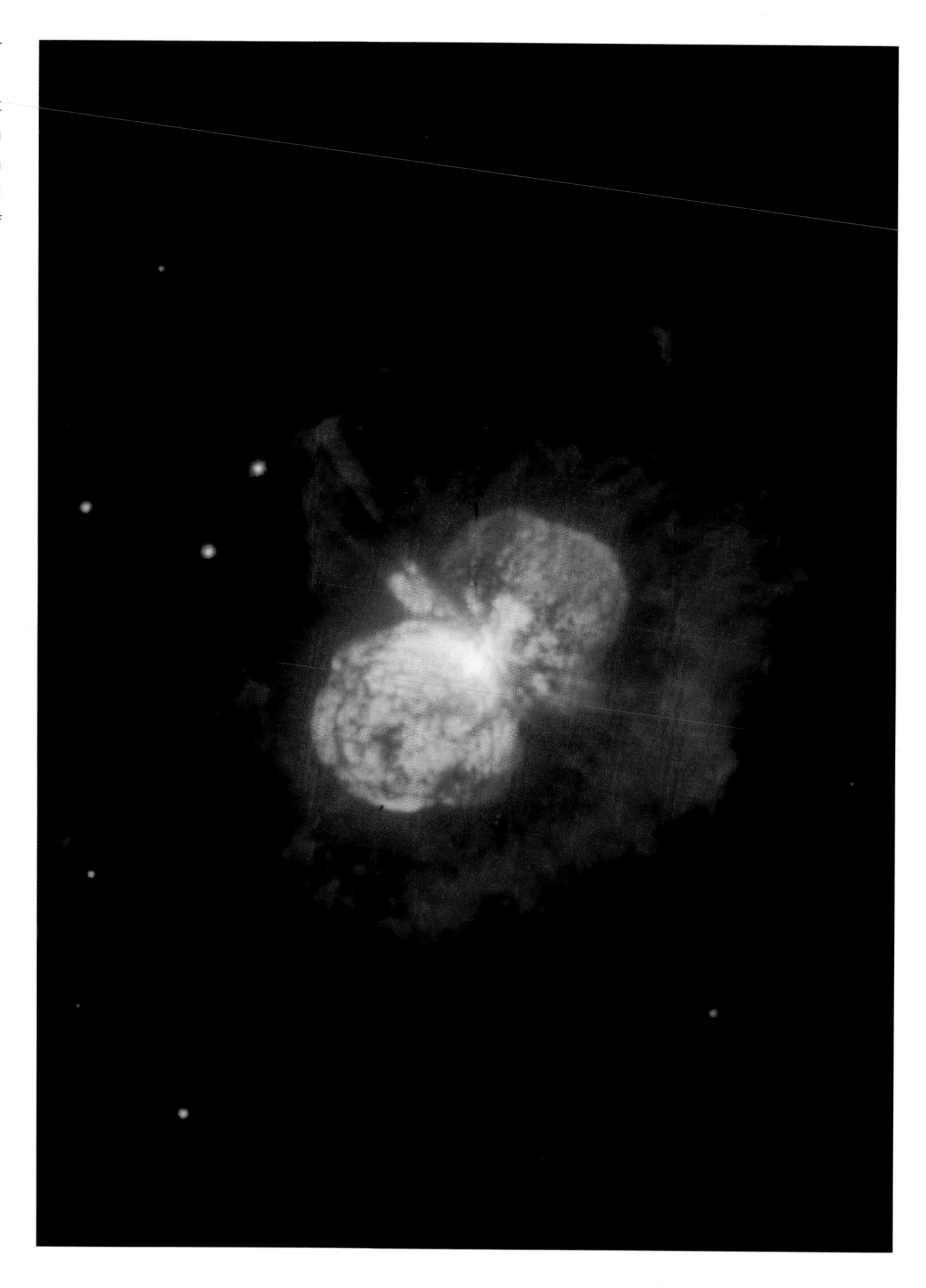

**Figure 4.36.** WF/PC-2 color composite image of Eta Carinae, a highly unstable star. The last outburst occurred in the 19th century when, in 1843, Eta Carinae became the second brightest star in the sky. (Jeff Hester, Arizona State University; NASA)

Its last outburst was in the 1840s, when it flared up to become the second brightest star in the sky. A.M. Clerk wrote in 1905 about the apparition:

> ... after a partial decline and several preliminary 'flutterings' it reached a final maximum in April 1843 when Sirius alone among the fixed stars slightly outshone it.

What we see today is a rapidly expanding shell of material, radiating out in 'lobe-like' structures from the central star. The reddish glow is actually light from fast-moving nitrogen and other gases ejected from the interior of the star at more than $3 \times 10^6$ kilometers per hour. The bright white material is very dusty and reflects starlight back to us. For scale, note that the two lobes cover an area about the size of our Solar System.

Astronomers are still trying to explain the dynamics of Eta Carinae's explosion. It certainly looks like a supernova, but so far it exhibits few other characteristics that would label it as such. The violence of the outburst of this cataclysmic star leads one to wonder what it will look like when it really does blow up!

From the first brilliance of a newborn star to the last gasp of a dying supergiant, HST follows the threads of creation and destruction across the galaxy. The 'zoo' of stars we have visited in this chapter are just a very few of the more unusual denizens of the Milky Way Galaxy, and represent some of the most interesting examples of the lives of stars. Yet, understanding the differences and similarities between them – these variations on a stellar theme – prepares us for the wider task of studying the galaxies. HST observations which are revealing the secrets of stars, are also starting to show astronomers the dynamics of large-scale stellar systems – and, as we shall see in the next chapter, the theme of creation and destruction is played out in grand scale across the face of the universe.

# 5 Galaxies

> In a sense the galaxy hardest for us to see is our own. For one thing, we are imprisoned within it, while the other galaxies can be viewed as a whole from outside... furthermore, we are far out from the center, and to make matters worse, we lie in a spiral arm clogged with dust. In other words, we are on a low roof on the outskirts of the city on a foggy day.
>
> *Isaac Asimov*

> Immense as the Milky Way appears, it does not travel the universe alone. Other distant islands of light wheel through the cosmos. We find no two exactly alike – galactic snowflakes, drifting through the universe.
>
> *Carolyn Collins Petersen*

Images of galaxies are among the most beautiful and evocative pieces of cosmic artwork we can imagine. Open any coffee-table book of astronomical photography and you will marvel at the splendor of these collections of stars, the stellar cities of the universe.

In Chapter 4, we pointed out that studies of stars are limited necessarily by our own short-term perspective on them. We do not live for millions of years, and thus rarely see stars change drastically over time. The rare occasions when we do afford us the exquisite sight of an expanding supernova cloud or the eerie sight of a planetary nebula. It stands to reason that we do not see much change in galaxies with their multi-*billion* year lifetimes. What we see of galaxies are – at best – snapshots of stellar conglomerations, frozen in time. The time we see is in the past, for the farther out from Earth we look, the further back in time we are seeing.

The realization that stars are organized into galaxies is a comparatively recent one, and it parallels the gradual discovery of our own place in the cosmos. Of course, the starry swath that the Milky Way traces out across the sky has evoked awe and wonder in the human mind since prehistoric times, but it was not until the early seventeenth century, when Galileo turned his telescope to it, that astronomers saw that the Milky Way was not a smooth pathway of light but a myriad of stars. The idea that we were looking edgewise through 'our own galaxy' was – and perhaps still is – completely astounding to some.

**Figure 5.1.** The Andromeda Galaxy, M31, in the constellation Andromeda. This Sb spiral galaxy is the nearest and most easily visible spiral galaxy. This is a computer enhanced image of a plate taken at the Kitt Peak National Observatory. (National Optical Astronomy Observatories)

Even the words used to describe galaxies through the years illustrate the confusion about just what they really were. The term 'nebula' was originally applied to any fuzzy cloud-like structure in the night-time sky that obviously was not a star. Compounding the problem was that some fuzzy things in the sky actually moved over short periods of time. As it turns out, these were actually comets. (In fact, Charles Messier compiled his famous Messier catalog of 'fuzzy-looking things' (which is where the 'M' numbers you see applied to some astronomical objects originate) so that comet seekers would not be confused by the permanent nebulae in the sky.)

Everywhere you look there are galaxies. Some are quite distant and hard to see with-

**Figure 5.2.** NGC 4565, an Sb spiral in the constellation Coma Berenices seen edge-on. (National Optical Astronomy Observatories)

out at least a fairly good-sized telescope, but others are easier to find. They come in a veritable rogue's gallery of shapes, sizes and masses. And they evolve.

In the northern hemisphere sky, stargazers can find the Andromeda Galaxy, also known as M31, not far from the W-shaped constellation Cassiopeia. Figure 5.1 shows a ground-based image of the galaxy, which lies 2.2 billion light years from us and comprises more than 300 billion stars.

*Spiral* galaxies are typified by M31, and by galaxies like the one seen in Figure 5.2. We see this galaxy in Coma Berenices 'edge-on', but we know it is like the Milky Way in many important respects. Spiral galaxies rotate, taking hundreds of millions of years to make a complete revolution. We do not see this stately motion visually, but if we take spectra we can measure a redshift on the side of the galaxy moving away from us, while the side of the galaxy moving toward us will be blueshifted.

The arms of spiral galaxies are of special interest to us for a couple of reasons:

**Figure 5.3.** M81 (NGC 3031), an Sb spiral galaxy in the constellation Ursa Major showing prominent spiral arms. (National Optical Astronomy Observatories)

**Figure 5.4.** M104 (NGC 4594), an Sa spiral galaxy with a large central bulge. The dark band across the galaxy's central region is composed of dust and gas. This is a 4-meter Mayall Telescope photograph from the Kitt Peak National Observatory. (National Optical Astronomy Observatories)

**Figure 5.5.** NGC 5383, a type SBb, barred spiral galaxy in the constellation Canes Venatici. This type of galaxy has a bar of stars passing through the center. This is a 4-meter Mayall Telescope photograph from the Kitt Peak National Observatory. (National Optical Astronomy Observatories)

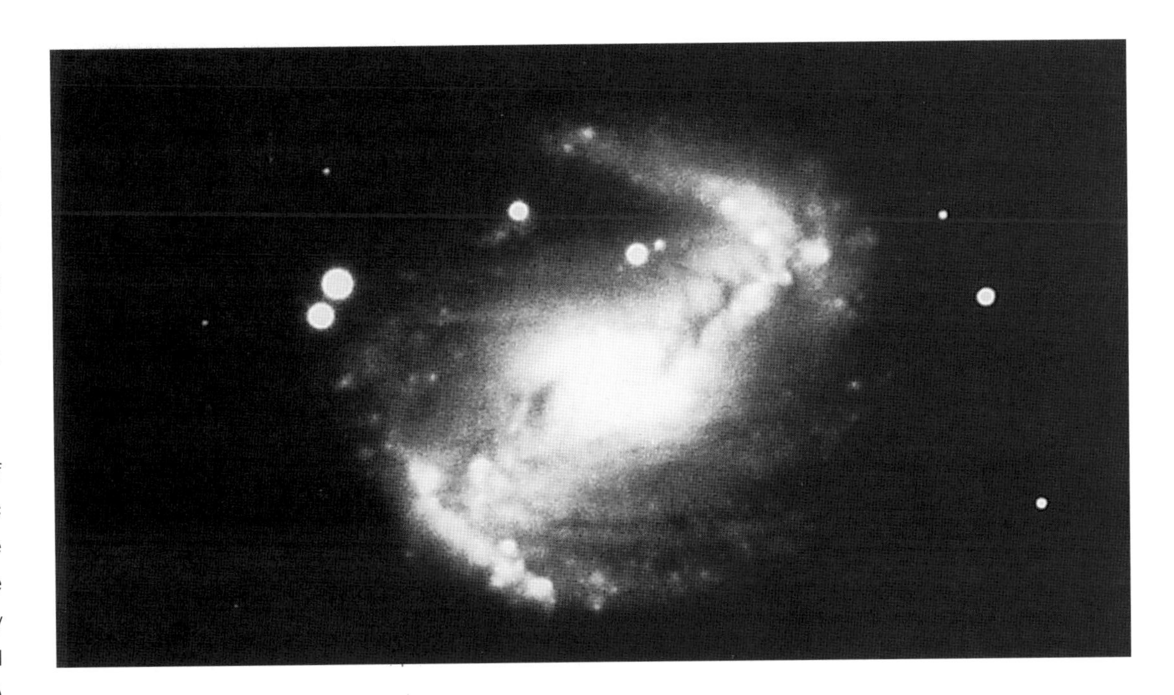

**Figure 5.6.** Wide-angle view of the Large and Small Magellanic Clouds, which are visible from the southern hemisphere. These clouds are satellites to the Milky Way Galaxy. (National Optical Astronomy Observatories)

(i) because our Solar System lies in a spiral arm of the Milky Way; and (ii) because spiral arms are prime sites for the birth of stars (for example the stellar nurseries in the Orion Nebula). Figure 5.3 shows the galaxy M81, its spiral arms shining with the light of young and middle-aged stars. As a general rule, galaxies with more open arms tend to have more gas and dust, and show a greater tendency for brisk episodes of star formation. The central bulges of spiral galaxies assume different sizes and shapes, and generally have older type stars. Figure 5.4 shows galaxy NGC 4594; its central bulge is among the largest known.

*Barred spiral* galaxies are a special case. They have long bars of stars passing through their central regions; for an example, see Figure 5.5, which shows the galaxy NGC 5383 in the constellation of Canes Venatici.

**Figure 5.7.** The irregular and active galaxy, M82. The composite approximates the true appearance of the galaxy and was taken by George Jacoby at the Kitt Peak National Observatory. (National Optical Astronomy Observatories)

**Figure 5.8.** An E4 elliptical galaxy, NGC 4435, in the constellation Virgo. A peculiar spiral galaxy is also shown. (National Optical Astronomy Observatories)

The Large and Small Magellanic Clouds (shown in Figure 5.6) are a pair of companion galaxies to the Milky Way. They are a pair of *irregular* galaxies that lie in the constellations of Doradus and Tucana, respectively. M82 in Ursa Major (Figure 5.7) is another irregular galaxy. We may think of them as chaotic systems of stars and interstellar material, with very little apparent structure. Irregulars undergo sporadic waves of starbirth, which can be triggered by supernovae and other events.

Lastly, we have the *elliptical* galaxies, named for their ellipsoidal appearance (see Figure 5.8). An elliptical galaxy shows a smooth outline, and its shape depends on its structure and orientation in space. Ellipticals are generally not very active galaxies in that they boast no regions of star birth, and indeed have very few young stars, or little of the gas and dust needed for star formation.

## The evolution of galaxies

In Chapter 4 we looked at a variety of stars to further our understanding of stellar evolution, but to truly understand why galaxies appear the way they do, and hence, galactic evolution, we need to study not only the constituent stars, but also clusters of galaxies.

Alan Dressler, of the Observatories of the Carnegie Institution of Washington, and a group of colleagues used both Wide Field and Planetary Cameras to make observations of a remote cluster of galaxies, CL 0939+4713. According to Dressler's data, the age of this cluster is about two-thirds that of the universe's present age, or, put another way, about one-third of the way back to the Big Bang. Depending on how old the universe is, this cluster could be around 3–4 billion years old.

Figure 5.9 shows the WF/PC-1 image of the cluster. The resolution is good enough to distinguish between different types of spiral and elliptical galaxies (see below) and to record galaxies in collision. This image illustrates two important points. First, there are many more spiral galaxies in early clusters than are found in more recent clusters. This confirms previous studies, which implied that star-forming galaxies were more prevalent in the early universe. Because we know that spiral galaxies are prime sites of star formation, by studying them we can look back to a time when clusters of young galaxies went through massive

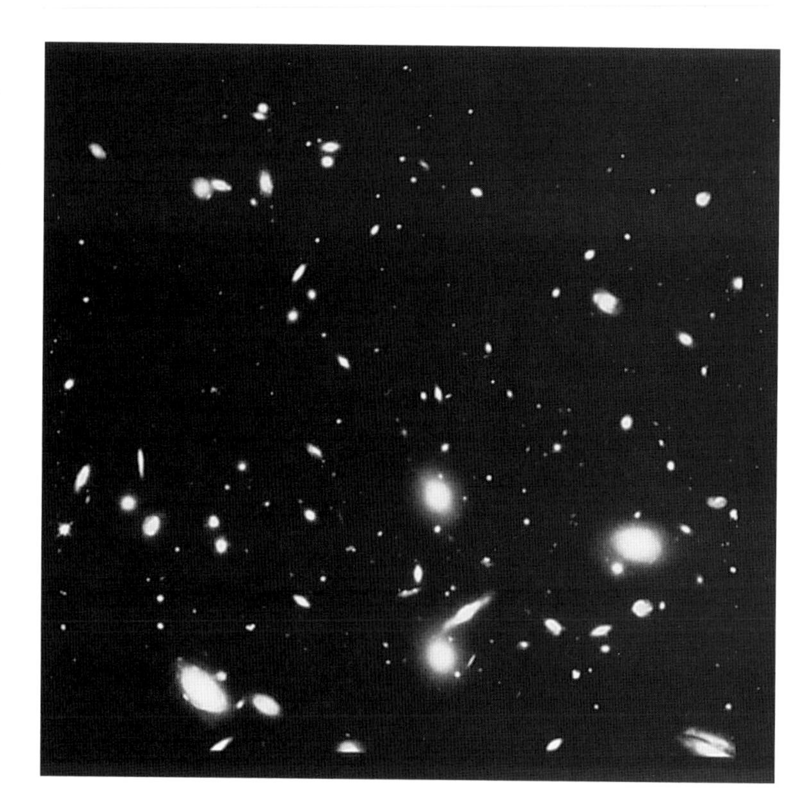

**Figure 5.9.** The central portion of the remote cluster of galaxies CL 0939+4713 as it looked when the universe was two-thirds of its present age. This image is sharp enough for us to distinguish between various forms of spiral galaxies. Most have odd features, suggesting that they were distorted within the cluster environment. Fragments of galaxies seem to be interspersed about the cluster. (Alan Dressler, Carnegie Institution of Washington; NASA)

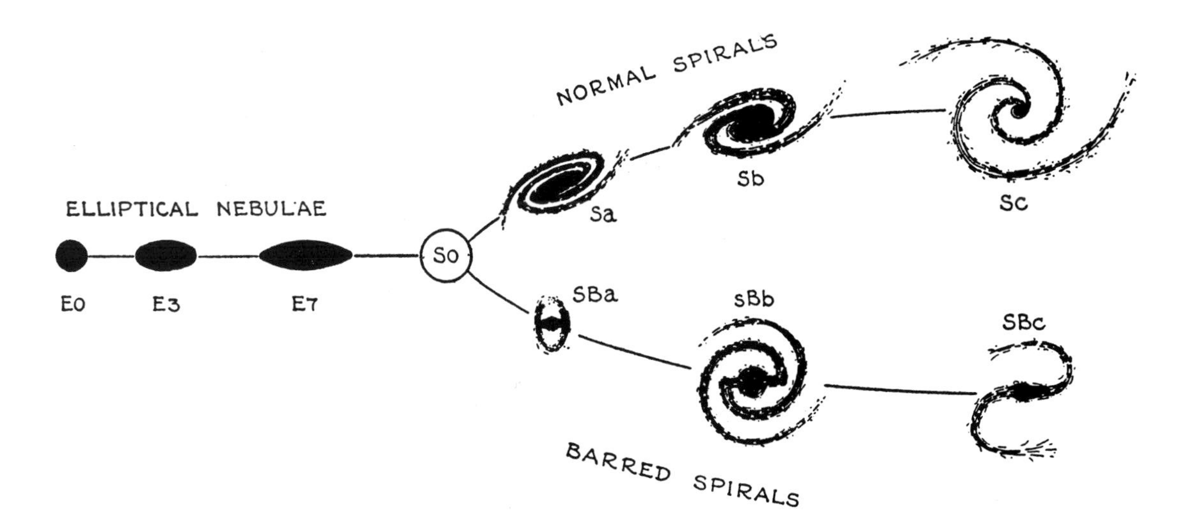


**Figure 5.10.** Hubble's famous 'tuning fork' diagram showing his classification scheme for galaxies. (Yerkes Observatory, University of Chicago)

amounts of starburst activity – the formation of new stars. Galactic starbursts take place over large scales and often involve collisions of galaxies.

Secondly, the number of interacting galaxies is large, demonstrating that the processes of disruption and merger are very important in the normal evolution of galaxies, as are the changes caused by stellar evolution.

These points are more significant if you look at a galaxy classification scheme proposed by Edwin Hubble in the 1920s: the famous 'tuning fork' diagram of galaxy shapes shown in Figure 5.10. The ellipticals are labeled according to their apparent ellipticity. The normal spirals vary from types labeled Sa, for those with strong central bulges and tightly wound arms, to types called Sc, for those with little central bulge and open arms. A similar sequence exists for the barred spiral galaxies. The class S0 is a hypothetical transition type. Some of these types were referred to as 'early' or 'late', designations with evolutionary connotations. Today we think the tuning fork diagram does not represent an evolutionary sequence. Hubble included other galaxies, such as Irr (for irregular), and the scheme has been updated from time to time; for example, astronomers have added a type Sd.

The galaxies in CL 0939+4713 have been arranged according to the scheme described above (see Figure 5.11), and from that work comes a startling insight into the nature of galactic evolution. Clusters of spiral galaxies, which were more numerous in the early universe, seem to undergo great changes as they mature. If we compare the galaxies in this distant cluster to what we see in more contemporary clusters, we can see these changes.

The top three rows show elliptical galaxies, with various ellipticities, and possibly some S0 galaxies. These types of galaxies are common in nearby clusters at the present time.

Rows 4–7 show spiral galaxies with differing degrees of openness of the spiral patterns from Sa to Sd. The Sds in row 7 are shaped atypically – that is, they do not look like normal spirals. These are common in CL 0939+4713, but they are not common in nearby clusters today.

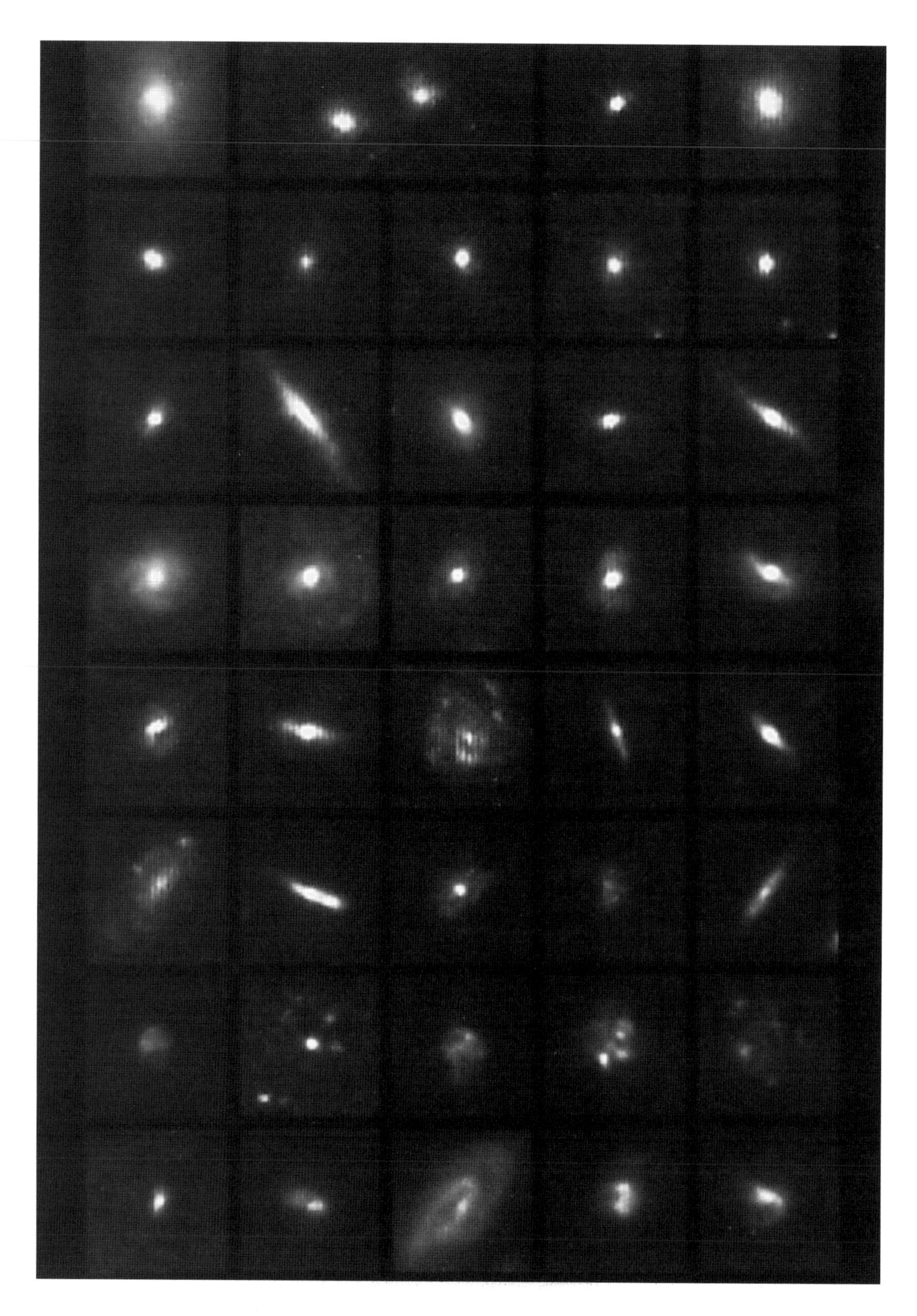

**Figure 5.11.** A Hubble classification of galaxies from the cluster CL 0939+4713. The original image was obtained with WF/PC-1. See the text for a discussion. (Alan Dressler, Carnegie Institution; NASA Co-Investigators: Augustus Oemler, Yale University, James E. Gunn, Princeton University, and Harvey Butcher, Netherlands Foundation for Research in Astronomy)

Row 8 gives another clue as to why the spirals are less common today. These are galaxies apparently merging into single galaxies. The interaction between galaxies – causing merging and disruption – must be considered along with a phenomenon called fading due to the effects of stellar evolution (stars are brighter when they are younger and dimmer as they grow old) when explaining why star-forming galaxies – the spirals – were more prevalent in young clusters than in the older contemporary clusters.

Alan Dressler sees the galactic changes happening over a short time period – only about 4 billion years: 'It seems that as soon as nature builds spiral galaxies in clusters, it begins tearing them apart.'

Why would the universe take so much time to build up spirals in clusters, only to destroy them? The reasons are still unknown, although theories are vigorously debated. A timescale of 4 billion years is short to be altering galactic forms, and this shows that we really do not know how galaxies formed in the first place. The HST data show that spiral galaxies, at least, have been disturbed, disrupted and destroyed. 'We see so many distorted galaxies,' says Dressler. 'There are so many little shreds of galaxies, it almost looks like galactic debris flying around in these clusters.'

It is possible that tidal encounters – brought on by complex gravitational interactions between galaxies in clusters – might be what is destroying the fragile spiral shapes of galaxies. And, as mentioned above, galactic mergers and collisions might play a role in changing the familiar spiral shapes. Regardless of what is causing the disruptions, the HST images are compelling evidence that the universe is changing as it ages.

As HST directs its gaze toward more of these distant galaxy clusters, what it finds offers tantalizing clues to the evolution of galaxies. Figure 5.12 shows snapshots of galaxies ranging in age from the current time back to an epoch when the universe was a small fraction of its current age, and makes an excellent illustrative summary of galactic evolution:

(1) the ordinary changes responsible for the formation of galaxies by gravitational attraction and flattening due to rotational processes;
(2) the explosions responsible for activity in galaxies;
(3) the normal formation and evolution of stars. Massive stars will evolve quickly and recycle part of their masses into the interstellar medium in the form of gas and dust, and new stars will form in these regions. Even if the mass stays in the same general area, the appearance must change;
(4) disruption and merging through collisions, which should drastically alter their structure and appearance.

Piecing the various possible processes together to produce a generally accepted, detailed picture of the evolution of galaxies is a major challenge. Any theory will be tested by comparison with galaxies at different distances which, because of light travel time, is equivalent to looking back at galaxies in earlier and earlier times.

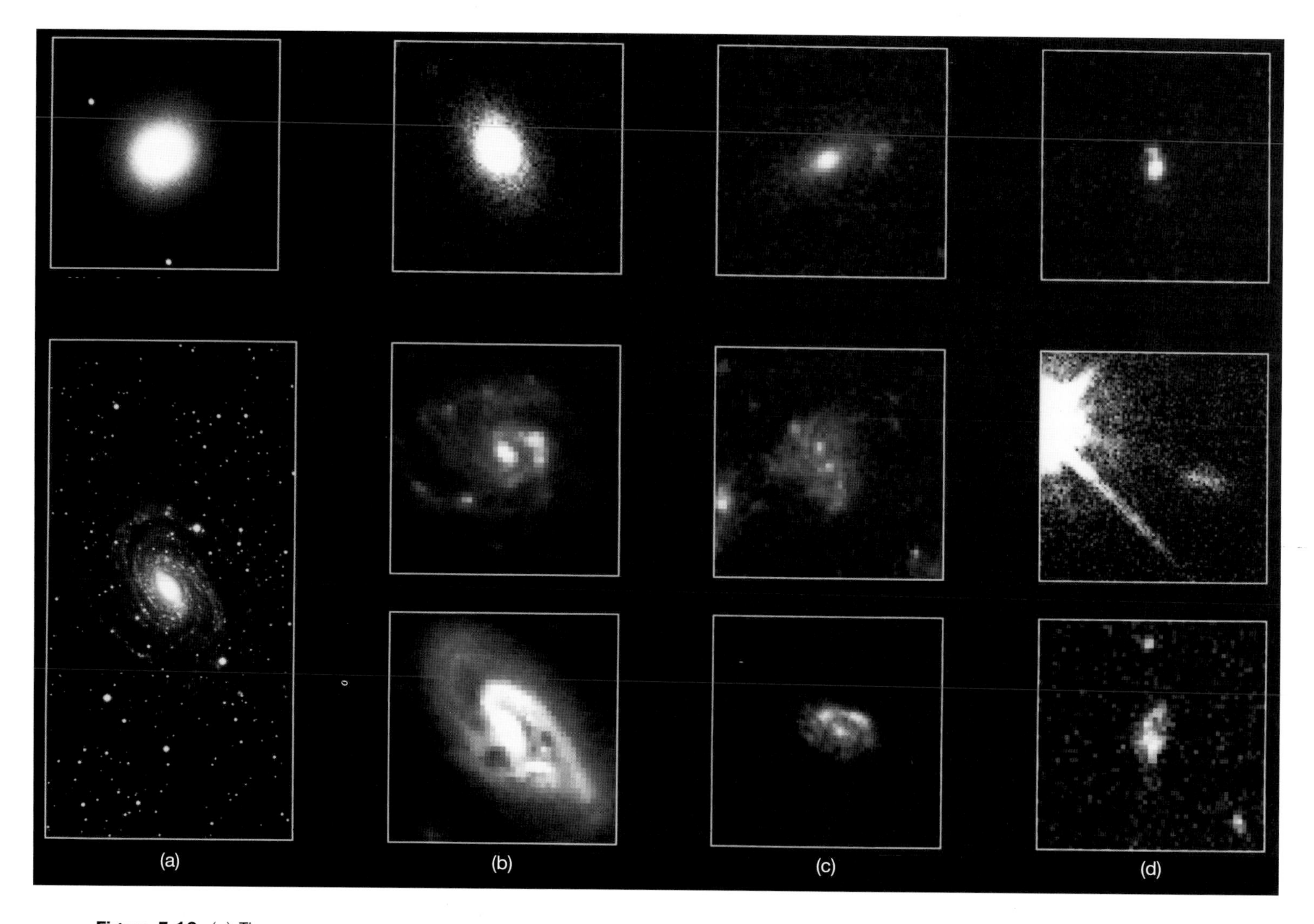


**Figure 5.12.** (a) These are traditional spiral and elliptical galaxies that make up the two basic classes that we see in our current epoch, about 14 billion years after the Big Bang. Elliptical galaxies contain older stars, while spirals have vigorous on-going star formation in their disks. Our Milky Way is a typical spiral. Both galaxies in this column are a few tens of millions of light years away. (b) These galaxies existed in

## Individual galaxies

It is hard to believe when we look at a galaxy in the sky, or an image in a book, that some galaxies are host to some of the most frenetic activity in the universe. For a long time, galaxies were thought to be quiescent. That changed in 1943, when astronomer Carl Seyfert drew attention to a handful of galaxies with unusual properties. Now called Seyfert a rich cluster when the universe was approximately two-thirds its present age. Ellipticals (top) appear fully evolved because they resemble today's descendants. By contrast, some spirals have a 'frothier' appearance, with loosely shaped arms of young star formation. The spiral population appears more disrupted due to a variety of possible dynamical effects from dwelling in a dense cluster. (c) Distinctive spiral structure appears more vague and disrupted in galaxies that existed when the universe was one-third its present age. These objects do not have the symmetry of current-day spirals and contain clumps of starburst activity. However, even this far back in time, the elliptical galaxy (top) is still clearly recognizable, but the distinction between ellipticals and spirals grows less certain with increasing distance. (d) These extremely remote objects existed when the universe was one-tenth its present age. The distinction between spiral and elliptical galaxies may well disappear at this early epoch. However, the object in the top frame has the light profile of a mature elliptical galaxy. This implies that ellipticals formed early in the universe, whereas spirals took much longer to form. (A. Dressler, Carnegie Institution of Washington; M. Dickinson, STScI; D. Macchetto, ESA and STScI; M. Giavalisco, STScI; NASA)

galaxies, they have strong emission lines from their central regions and a star-like appearance. Ordinary galaxies have only absorption lines and inconspicuous, sometimes 'blobby', nuclei that more or less blend in with the rest of the central region. It was a good bet that something energetic and unusual was going on in Seyfert galaxies.

Astronomers now recognize huge variations in the activity levels of galaxies. There is the level we observe in our own galaxy: strong radio signals emanating from the nuclear region of the Milky Way. A middle level shows clear evidence of explosive activity from nuclear regions of other galaxies; the Seyfert galaxies are the most active example of this type of galaxy. The highest level of activity includes the superluminous quasi-stellar objects – quasars, for short. These are probably the most exciting candidates for galactic studies with HST.

## Galaxies with central black holes

As little as 25 years ago, black holes were an interesting theoretical construct, despite the tendencies of Hollywood sci-fi movie makers and science-fiction writers to use them as integral parts of stories involving space travel and intergalactic intrigue.

Speculation about an object so dense that light could not escape from it goes back at least two centuries, but a formal understanding required the Theory of General Relativity as developed by Albert Einstein at the beginning of the 20th century. Today, astronomers and physicists are convinced that black holes exist, and they use them to explain many rare and exotic happenings in quasars and galactic nuclei. Black holes seem to be an important constituent of the universe, and a short review of their fascinating properties may be useful before we look at what HST has seen in its search for these exotic stellar beasts.

Although the matter in a black hole has collapsed to a point, the effective 'size' or boundary of the object is defined by the event horizon. Because light cannot escape from a black hole, happenings within the event horizon cannot be known to an observer on the outside, who can only speculate on the basis of what is occurring outside the black hole. The size of the event horizon itself is calculated by taking the mass of the black hole (in solar masses) and multiplying it by 3 kilometers.

According to theory, black holes have only three properties: their mass, their spin (a measure of rotation), and their electrical charge. The electrical charge is believed to be zero, and so the properties are even simpler. For our purposes, the property of interest is the mass, and we find that black holes come in three varieties: mini, stellar and supermassive.

*Mini black holes* could have formed in the immense pressure, temperature and turbulence shortly after the Big Bang. They would have masses comparable to mountains and, after billions of years, would vanish in a flash of energy. The properties of these mini black holes are governed by general relativity and by quantum mechanics. This accounts for their atypical behavior, which is to emit light, something the larger black holes do not do. While mini black holes may be quite common in the universe, there is currently little direct evidence for their existence.

*Stellar black holes* are formed by massive stars at the end of their lives. When a star with mass equal to $3M_s$ or greater exhausts its internal energy source, no known force can stop its collapse, and it forms a black hole. There is considerable evidence for the existence of stellar black holes.

*Supermassive black holes* are probably an essential component of the cores of many galaxies. Conditions at the centers of galaxies seem ideal for the formation of massive stellar black holes, which have merged or gathered up material to form larger and larger black holes. Supermassive black holes can grow to millions or billions of solar masses, and there is good evidence that they exist. In addition, supermassive black holes are the most likely viable source of energy that can explain the superluminosity of quasars.

You may be asking yourself the following question: if no light escapes from black holes (except for the minis), how can we see them? The answer is that the gravitational field near the event horizon is very strong, and material falling into the black hole gains energy and is compressed. Thus, we 'see' black holes by observing the hot, dense material around them; this material is a strong emitter at ultraviolet, X-ray and gamma-ray wavelengths. Evidence for the existence of a specific black hole is usually circumstantial, and not every emitter of high-energy radiation is necessarily a black hole. Generally, astronomers are comfortable when they see numerous effects that would require the existence of black holes to explain them.

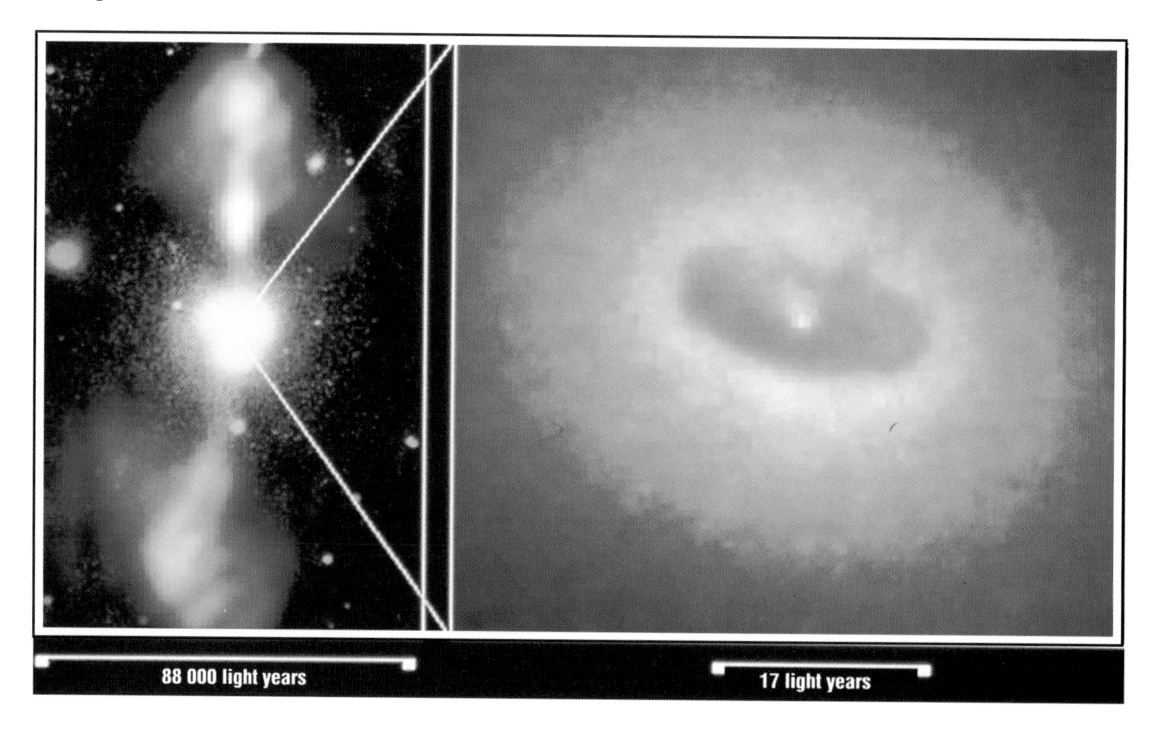


**Figure 5.13.** NGC 4261, as shown in a combined optical and radio image (left) and a WF/PC-1 image (right), clearly showing the central disk. (Walter Jaffe, Leiden Observatory; Holland Ford, Johns Hopkins University and STScI; NASA)

Many galaxies have some form of enhanced emission – radio, infrared, ultraviolet or X-ray – and often jet-like features are observed. The possibility that the 'central engine' in the form of a massive black hole powers the activity is a good one, and imaging results on the central regions of several galaxies support this. Based on a growing body of HST data, it turns out that central massive black holes may be rather common in galaxies.

The matter around black holes seems to form two common features – disks and jets. When material contracts over astronomical distances, it forms rapidly rotating disks; we expect to see these around black holes. The disks can be quite thick and dense in the plane of the disk, but the central part, perpendicular to the disk, can be relatively transparent. Thus, energetic phenomena produced by the material falling into the black hole cannot escape through the disk, but they can escape perpendicular to the disk, thus permitting the formation of 'jets'. This is the situation that HST found in the center of a galaxy called NGC 4261.

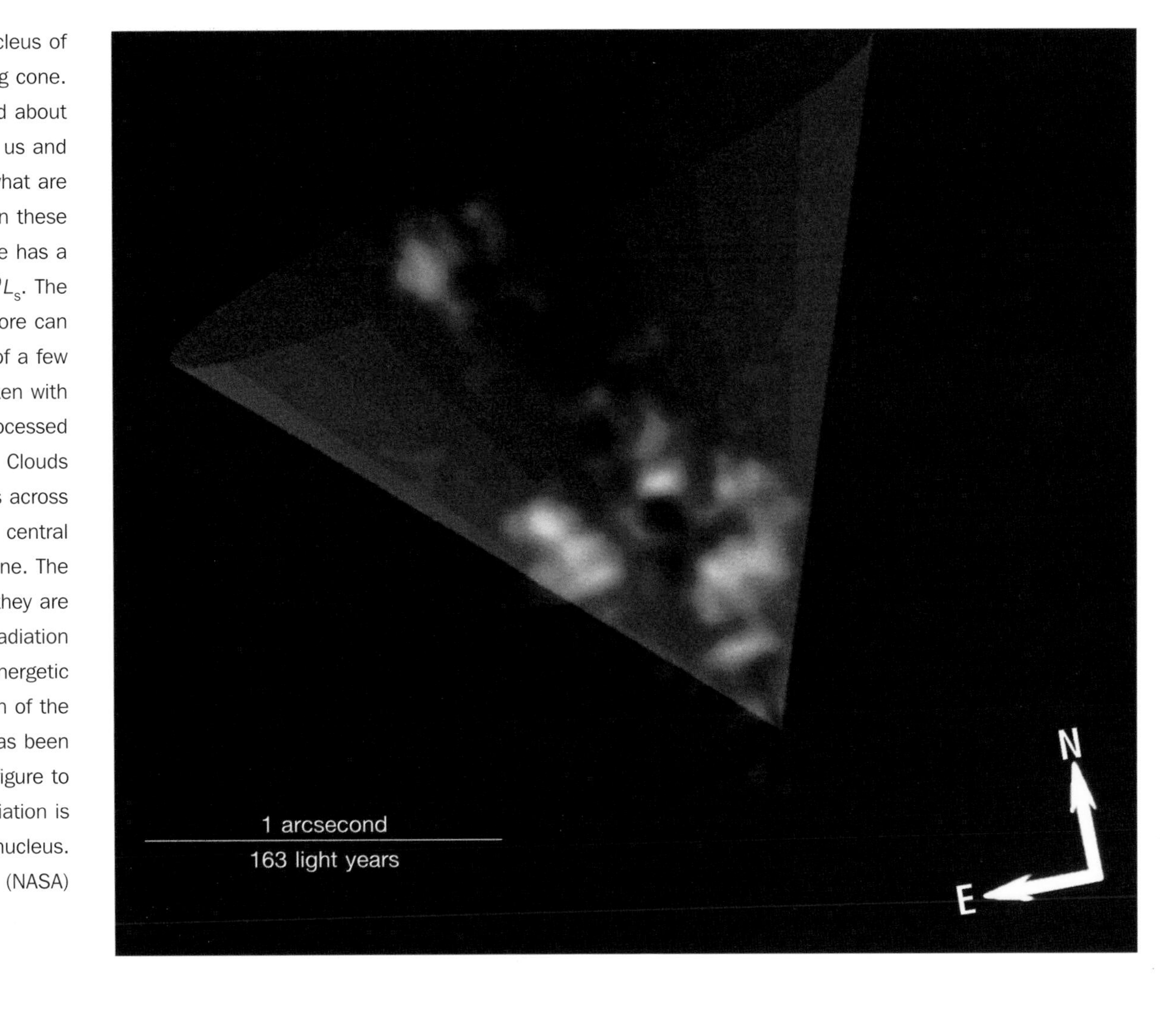


**Figure 5.14.** The nucleus of NGC 1068 and its ionizing cone. This active galaxy is located about 60 million light years from us and is the prototype of what are called Seyfert galaxies. In these active galaxies, the core has a luminosity of $10^{9} L_{s}$. The brightness of the core can fluctuate over a period of a few days. The image was taken with WF/PC-1 and computer processed to show additional detail. Clouds as small as 10 light years across are clearly resolved in the central 150 light years of the cone. The clouds glow because they are caught in a beam of radiation from the galaxy's energetic nucleus. A representation of the cone of radiation has been artificially added to the figure to illustrate how the radiation is beamed from the hidden nucleus. (NASA)

*NGC 4261* Figure 5.13 shows views of the elliptical galaxy NGC 4261, one of the 12 brightest galaxies in the Virgo Cluster. The left-hand panel shows a combined optical and radio image. The central blob is composed of stars (optical image), and the pair of opposed jets of energetic particles show up in the radio image. The WF/PC-1 image on the right shows a giant disk of gas and dust, presumably feeding a central black hole. The infalling gas is heated and compressed. Some of the hot gas squirts out from the immediate vicinity of the black hole to form the jets seen in radio emission. Particularly when the energy requirements are also considered, the evidence for a central black hole, while indirect, is compelling.

*NGC 1068* The WF/PC-1 image of NGC 1068, Figure 5.14, may also provide evidence for a central galactic black hole. Space Telescope Science Institute astronomer Holland Ford headed up a team that examined the HST data for that evidence. 'Because NGC 1068 is very bright and relatively nearby, we can determine the exact location of the black hole and see the effects of jets, winds, and ionizing radiation from the black hole,' Ford said.

This activity would affect the central area of the galaxy – a region about 300 light years across. Ford continued: 'The WF/PC images resolve the center of the galaxy. They reveal far more detail than has been seen in any pictures taken from the ground. Several small gaseous clouds, about 10 light years across, are clearly resolved in the central region, about 150 light years across.'

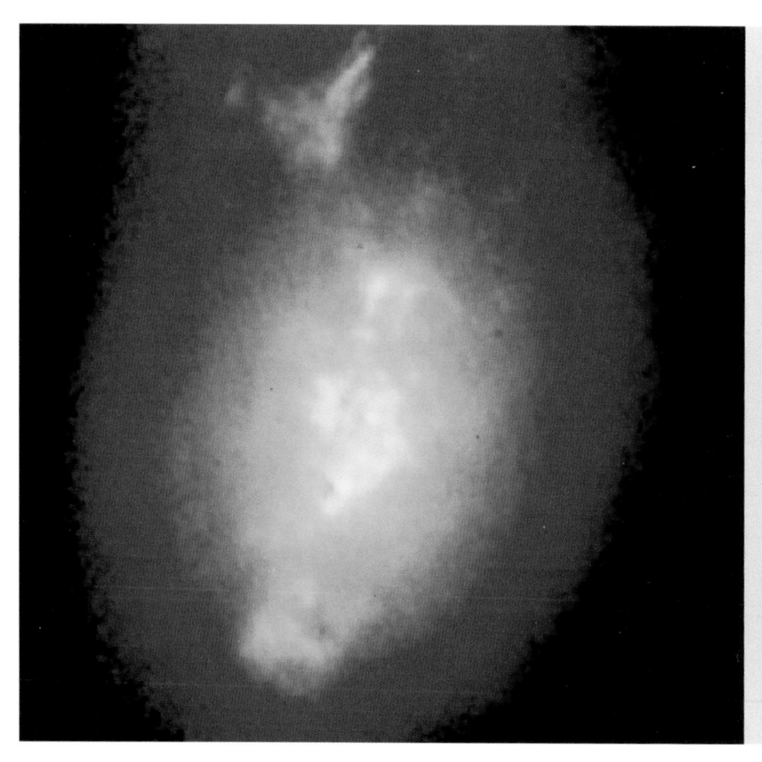

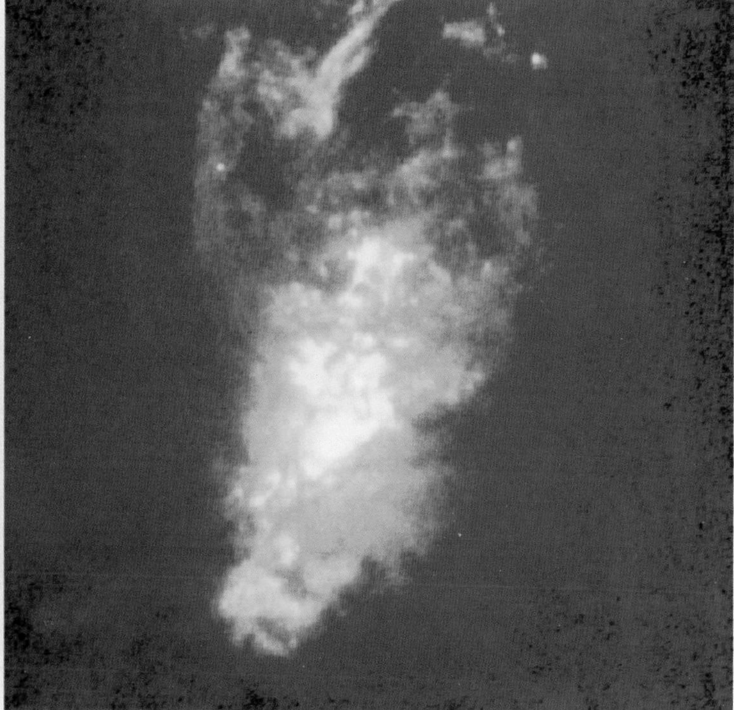

**Figure 5.15.** HST imaged NGC 1068 both before (left) and after (right) the 1993 servicing mission. Previous observations revealed the hot gas clouds in the region surrounding the core. Using the FOC, astronomers obtained images that show a high degree of filamentary detail in the clouds of gas around the core. Knots and streamers of emission belie the intricate geometry of the nucleus of the galaxy. (Duccio Macchetto, STScI and ESA; William Sparks and Alessandro Capetti, STScI)

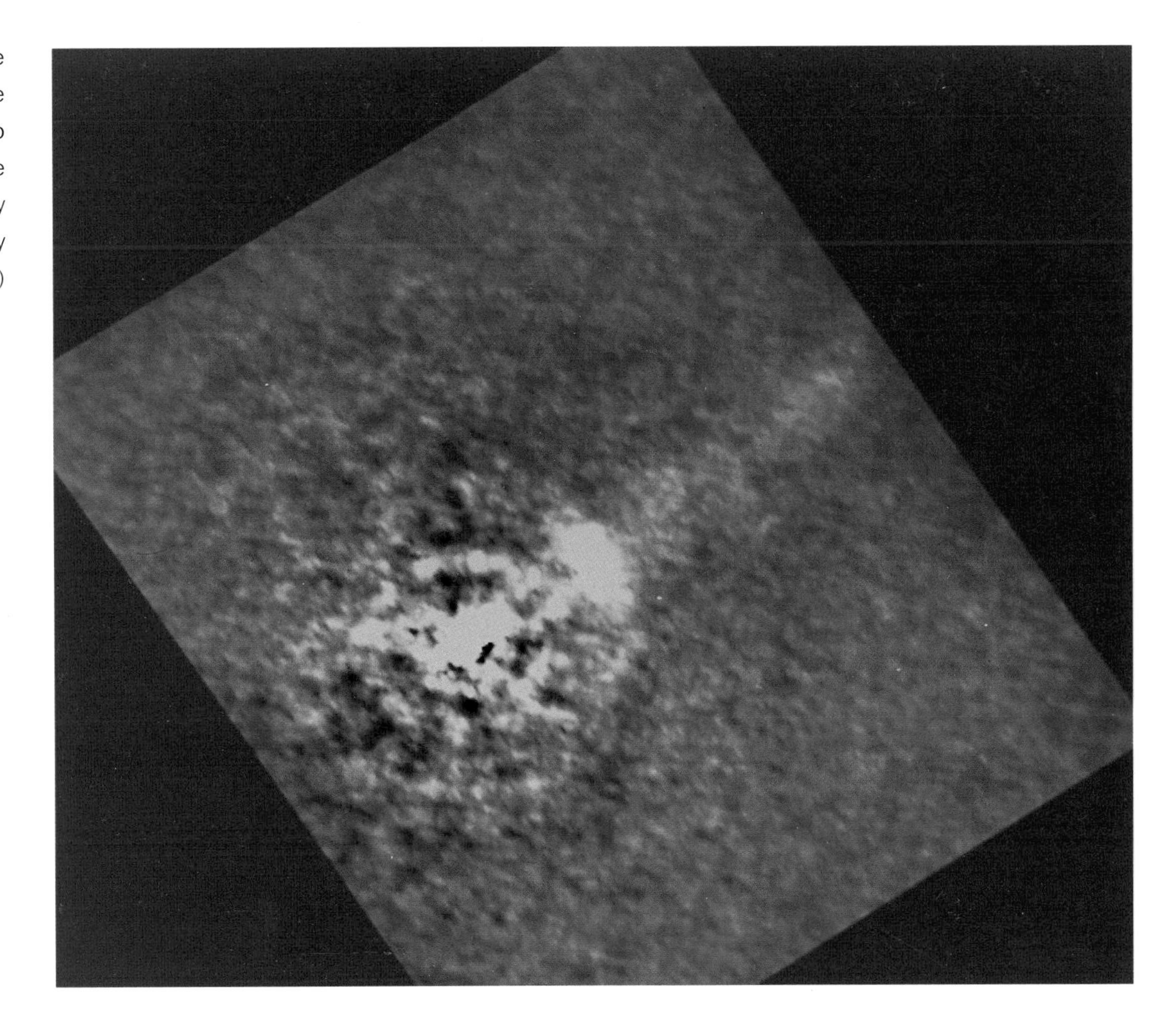

**Figure 5.16.** FOC image of the active galaxy PKS 0521-36. The galaxy has been subtracted to reveal the jet structure. These 'fireworks' are produced by electrons accelerated to nearly the speed of light. (ESA)

These clouds are ionized gas, glowing because they are being bombarded by a beam of radiation from the galaxy's energetic nucleus. A cone has been added to the image to illustrate the probable path of the radiation from the unseen nucleus. Because such beaming is expected from the nuclei of active galaxies, it is indirect evidence for a massive black hole. See Figure 5.15 for an FOC (post-COSTAR) image of NGC 1068's central region.

*PKS 0521-36* Another candidate for a supermassive black hole lies in the center of the active radio galaxy PKS 0521-36, shown in Figure 5.16. In this FOC image, the disk of the host galaxy has been subtracted using computer processing to emphasize the jet streaming out from its core. The 'engine' of the jet lies in the core of the galaxy and accelerates electrons to almost the speed of light, thus producing visible light and radio wave emission. The jet is about 30 000 light years long. To put that into perspective, the distance from Sun to its nearest stellar neighbor – Proxima Centauri – is 4.3 light years.

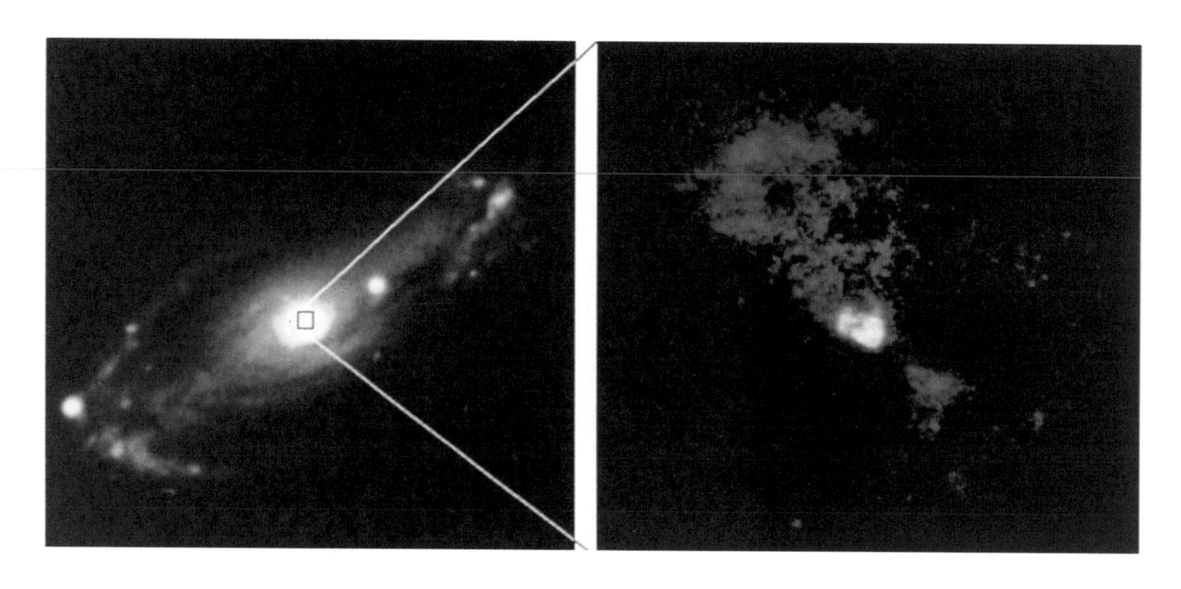

**Figure 5.17.** WF/PC-1 (in PC mode) image (right) and a ground-based image (left) of the active galaxy NGC 5728. The HST image shows the effects of the radiation escaping from the open ends of the dense, gas ring 'doughnut' and illuminating material in a cone. (Andrew S. Wilson, STScI; NASA; Allan Sandage, Carnegie Observatories)

*NGC 5728* Figure 5.17 shows both ground-based and WF/PC-1 images of NGC 5728, a barred spiral Seyfert galaxy. A dense ring prevents us from seeing the central, glowing region. Radiation escapes from the open ends of the ring to form the luminous cones. Although the central region is hidden, the activity is probably powered by a supermassive black hole.

*M31* One of the most interesting galaxies studied by HST to date is right in our neighborhood – the Andromeda Galaxy (M31). According to astronomer Sandra Faber, Andromeda is an intriguing place to study, and no one is quite sure what is happening there.

40 000 light years　　2000 light years　　40 light years

**Figure 5.18.** Multiple views of the Andromeda Galaxy, M31. The images at left and center are ground-based views of the galaxy and its core, respectively. The WF/PC-1 (in PC mode) image (at right) clearly shows a double nucleus. See the text for discussion. (T.R. Lauer, National Optical Astronomy Observatories; NASA)

Faber and her colleagues found a double nucleus at the heart of Andromeda. Figure 5.18 shows M31 with scales from 40 000 light years for the ground-based view of the entire galaxy to the 40 light years for the HST view of the nuclear region.

At this relatively early stage, Faber and her team have at least two working explanations of the situation. One is that the brighter component, apparently composed of densely packed bright stars, is a remnant of a collision between M31 and another galaxy, which it cannibalized. A black hole is believed to lie at the center of M31, and it would devour the other core in just a few hundred thousand years, according to Kitt Peak National Observatory astronomer Tod Lauer: 'This is very short in cosmic time. We would have to be looking at the galaxy at a very special time to see it now.'

The other explanation for the peculiar appearance of the M31 core is that there is a band

**Figure 5.19.** WF/PC-1 (in PC mode) image of the core of M32. The increase in brightness towards the center suggests the existence of a black hole. (T.R. Lauer, National Optical Astronomy Observatories; NASA)

of absorption by dust, which creates the illusion of two peaks of light. Whatever is really happening there, the central region of M31 is much more complex than previously thought. As Faber put it, 'I love it. This is typical astronomy. You get these little clues about something and you try to put it all together. It's wonderful.'

*M32* Andromeda has a companion galaxy called M32, and HST has imaged its central region also. Figure 5.19 shows that the stars are strongly concentrated toward the center; this density of stars is more than 100 million times the density of stars in the neighborhood of the Sun. The region shown in the figure is approximately 175 light years across. The grainy appearance near the edges of the image is caused by individual stars being resolved. Theoretical models suggest that the structure of M32 is consistent with a central black hole of mass 3 x $10^6 M_s$.

*Markarian 315* HST has also made observations of the active Seyfert galaxy designated Markarian 315. The core of this galaxy is shown in Figure 5.20, and the HST image

**Figure 5.20.** WF/PC-1 (in PC mode) image of the Seyfert galaxy Markarian 315 showing two nuclei about 6000 light years apart. (J. KacKenty, STScI; NASA)

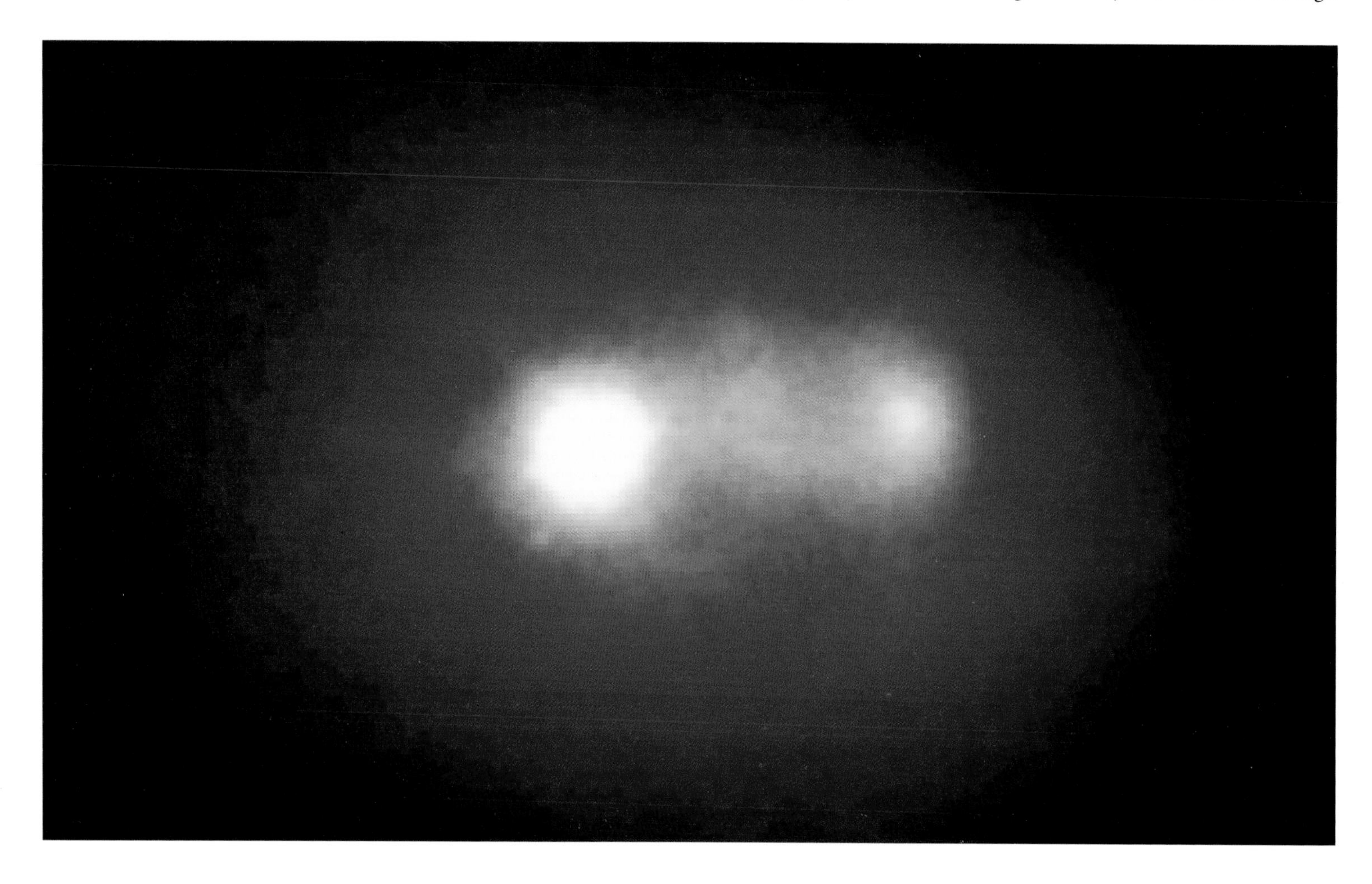

reveals two nuclei located some 6000 light years apart. The brighter one is believed to be the true center of the galaxy containing the massive black hole. Normally, the core is powered by the massive black hole accreting the available material in the vicinity. The fainter nucleus is believed to be the core of another galaxy merging with Markarian 315. This galaxy also has a jet-like feature. According to John McKenty of the Space Telescope Science Institute, 'The Hubble images provide support for the theory that the jet-like feature may be a "tail" of gas stretched out by tidal forces between the two interacting galaxies.'

Thus, galaxy mergers may be a mechanism for driving material deep into the core of a galaxy. This provides more fuel for the central engine, the massive black hole, and, hence, while the merger is underway, the activity level should be greatly enhanced.

*M87* Within the Virgo Cluster of galaxies lies one of the first known 'peculiar' galaxies – a giant elliptical galaxy identified as M87. It has been known for years to have a bright, optical jet in its central region. In 1991, the Faint Object Camera took an image that reveals unprecedented details. The jet was found to be some 40 000 light years long, and it has features as small as 10 light years across. At the time, astronomers suspected that the jet

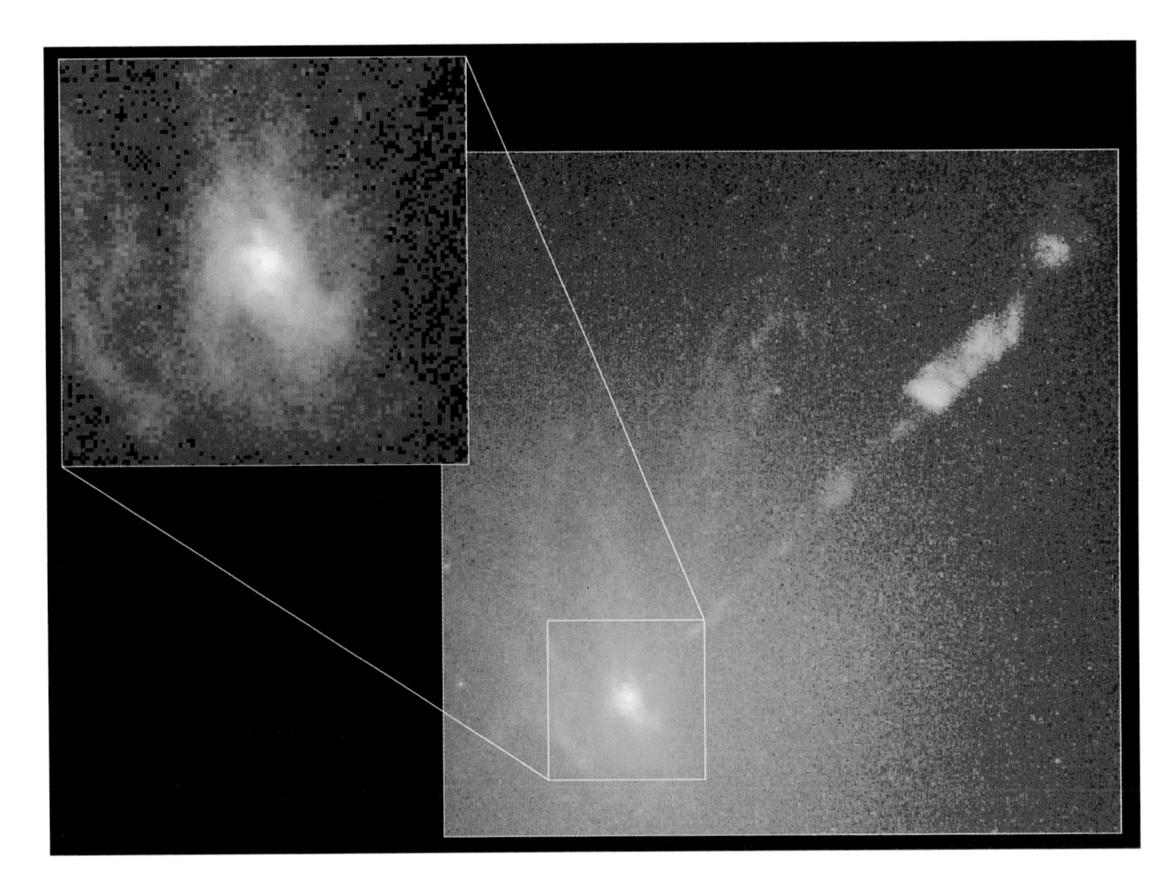

**Figure 5.21.** The gas disk in the nucleus and jet of M87 as imaged by WF/PC-2. (Holland Ford, STScI and Johns Hopkins University; Richard Harms, Applied Research Corp.; Zlatan Tsvetanov, Arthur Davidsen and Gerard Driss, Johns Hopkins University; Ralph Bohlin and George Hartig, STScI; Linda Dressel and Ajay K Kochhar, Applied Research Corp.; Bruce Margon, University of Washington-Seattle; NASA)

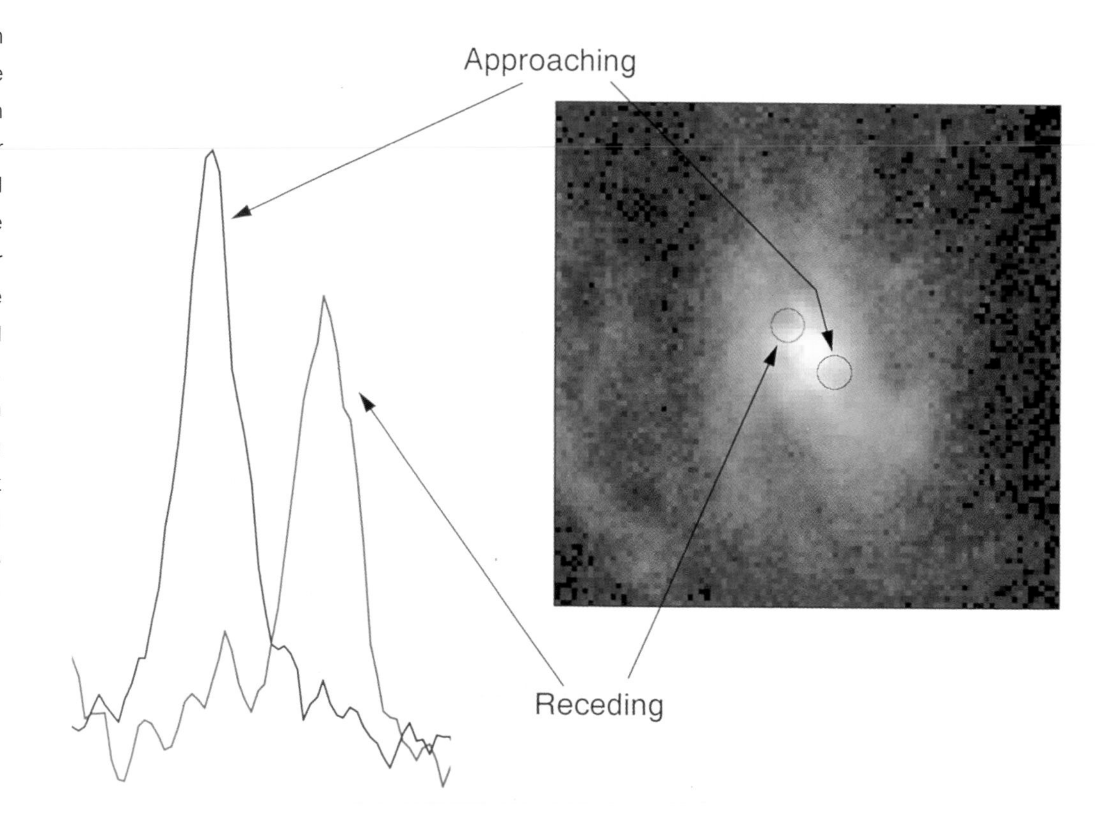


**Figure 5.22.** Schematic diagram of velocity measurements in the gas disk of M87. The speeds on opposite sides of the disk center show that the gas is receding (red spectrum) and approaching (blue spectrum) at 550 kilometers per second. These speeds are the gravitational signature of a central mass of at least 3 billion Suns. There are nowhere near enough stars to produce the attraction needed, and thus M87 must contain a black hole. (Holland Ford, STScI and Johns Hopkins University; Richard Harms, Applied Research Corp.; Zlatan Tsvetanov, Arthur Davidsen and Gerard Kriss, Johns Hopkins University; Ralph Bohlin and George Hartig, STScI; Linda Dressel and Ajay K. Kochhar, Applied Research Corp.; Bruce Margon, University of Washington-Seattle; NASA)

was emanating from a massive black hole. The WF/PC-2 view of the jet and core of M87 is shown in Figure 5.21. The faint star-like sources around the central region are globular clusters, each containing between 100 000 and 1 000 000 stars.

In 1994, HST provided what many astronomers consider conclusive evidence for a black hole at the heart of M87. The evidence is based on velocity measurements of a disk of hot gas whirling around the black hole. The presence of the disk in the images allowed for precise measurement of the black hole's mass, which was discovered to be as much as 3 billion Suns ($3 \times 10^9 M_s$), but is concentrated into a space the size of the Solar System. The speed of the mostly hydrogen gas in orbit around the object is *tremendous* – at least 550 kilometers per second! This indicates that something very massive is in the center, and, since there are few stars there, it must be a black hole. The Faint Object Spectrograph studied the spectral signature of the orbiting gas, and that observation was the clinching evidence for the black hole.

Holland Ford, of the Space Telescope Science Institute and Johns Hopkins University, and Richard Harms (who was at Applied Research Corporation at the time) were co-investigators on the observation. Harms actually thinks that the center of M87 could harbor something much stranger than a black hole, but it is hard to imagine what that might be. 'A massive black hole is actually the conservative explanation for what we see in M87,' he said. 'If it's not a black hole, then it must be something even harder to understand with our present theories of astrophysics.'

Only 25 years ago, black holes were considered to be something so strange and outside the boundaries of space and time that no one took their existence seriously. Yet, they spark the imagination. The idea of something that is so powerful that it sucks light into a gravitational well has passed into the daily usage as a reference to anything that sucks time, money and energy. It is exciting to see that an object that was once the realm of science fiction has been proven to exist. HST's role in the search for these objects will almost surely prove the existence of others embedded in the hearts of distant galaxies.

## Quasars

In Chapter 6, we describe quasars as immensely bright beacons useful for probing the universe. Here we take a look at the engine 'under the hood' of quasars. Quasars were originally detected as radio sources, and, because these objects looked starlike, they were first dubbed 'quasi-stellar radio sources' – quasars.

Observationally, quasars exhibit large brightness variations, and some show jets; see the image of 3C 273 shown in Figure 5.23. They also appear to be immensely luminous, but here a bit of deduction is required. With an observed quasar brightness, we can determine the luminosity only if we can fix the distance. For years this was a major problem. Astronomer Maarten Schmidt of the California Institute of Technology solved the problem by noting that the spectra of quasars made sense if certain emission features had very large redshifts. The only accepted way to produce large redshifts is as part of the Hubble expansion of the universe. If the distances are assigned in this way, quasars, being very distant, therefore must be immensely bright.

The fact that the brightness or luminosity of a quasar can fluctuate on a timescale of a day or less implies that quasars are actually quite small. Simply put, large objects have no way to vary synchronously (i.e. all at the same time), and the shortest time by which an object can vary is the time it takes light to travel across the entire object. The observations confine the sizes of quasars to something quite small – perhaps no more than a few light days across (not much larger than the Solar System) for the light-emitting region.

The requirement of extraordinarily large energy production in a small volume together with the existence of jets point to the conclusion that supermassive black holes exist as the central engines for quasars. A black hole with a mass of 10 billion Suns ($10^{10}M_s$) would have an event horizon 200 astronomical units across. The luminosity could be produced by only 1 solar mass of material falling into the black hole per year.

Now that we *think* we have a power source for quasars – a central, massive black hole

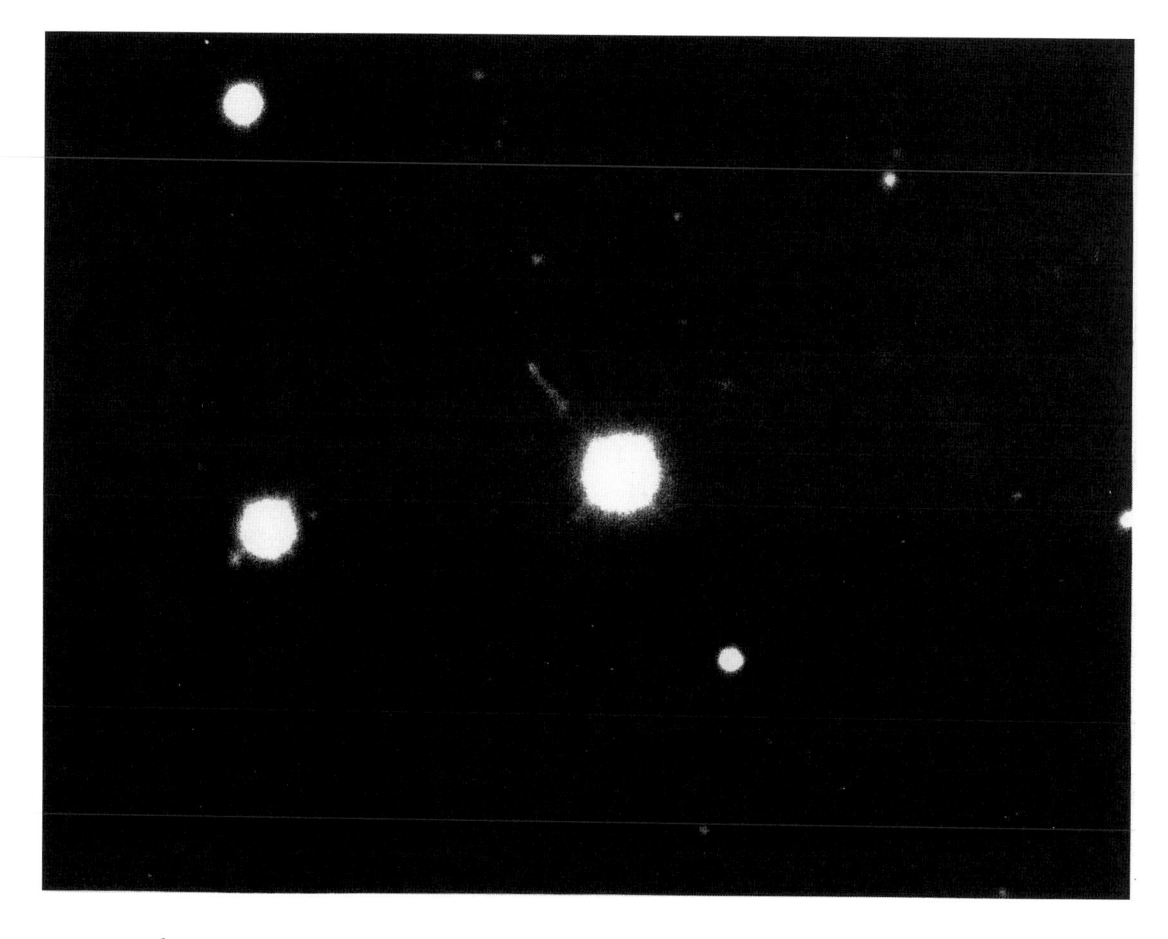

**Figure 5.23.** A 4-meter Mayall Telescope (Kitt Peak National Observatory) photograph of the quasar 3C 273. This galaxy is 100 times more luminous than the brightest normal galaxy, and the jet (shown) measures some 150 000 light years in length. (National Optical Astronomy Observatories)

– we need to press on and determine exactly how quasars work. How does the energy of matter falling into the black hole convert into the distinctive quasar spectrum we see from X-rays, to ultraviolet light, to visible light, to infrared radiation, to radio waves? Can we show that we really have a massive black hole in the core of a quasar by determining its mass? Recall from our discussion of M87 that we can only expect to establish the existence of massive black holes by gravitational effects on their surroundings. Therefore, if we had independent estimates of the mass of the black hole and the accretion mass rate, we could check the luminosity against the observations.

A comparison of nearby quasars with those existing 10 billion years ago implies that a typical quasar was then 100 times more luminous than those observed currently. This result implies that quasars – like galaxies – evolve over time. If we hypothesize higher mass accretion rates to make higher luminosities, then the masses of the central black holes in quasars must increase with time.

A fundamental understanding of quasar spectra would be a part of the process of answering questions about quasar evolution. Figure 5.24 depicts two combined FOS and ground-based quasar spectra, showing broad emission lines. In one interpretation, these are produced by clouds of gas in orbit around the black hole. If this is correct, the speeds of

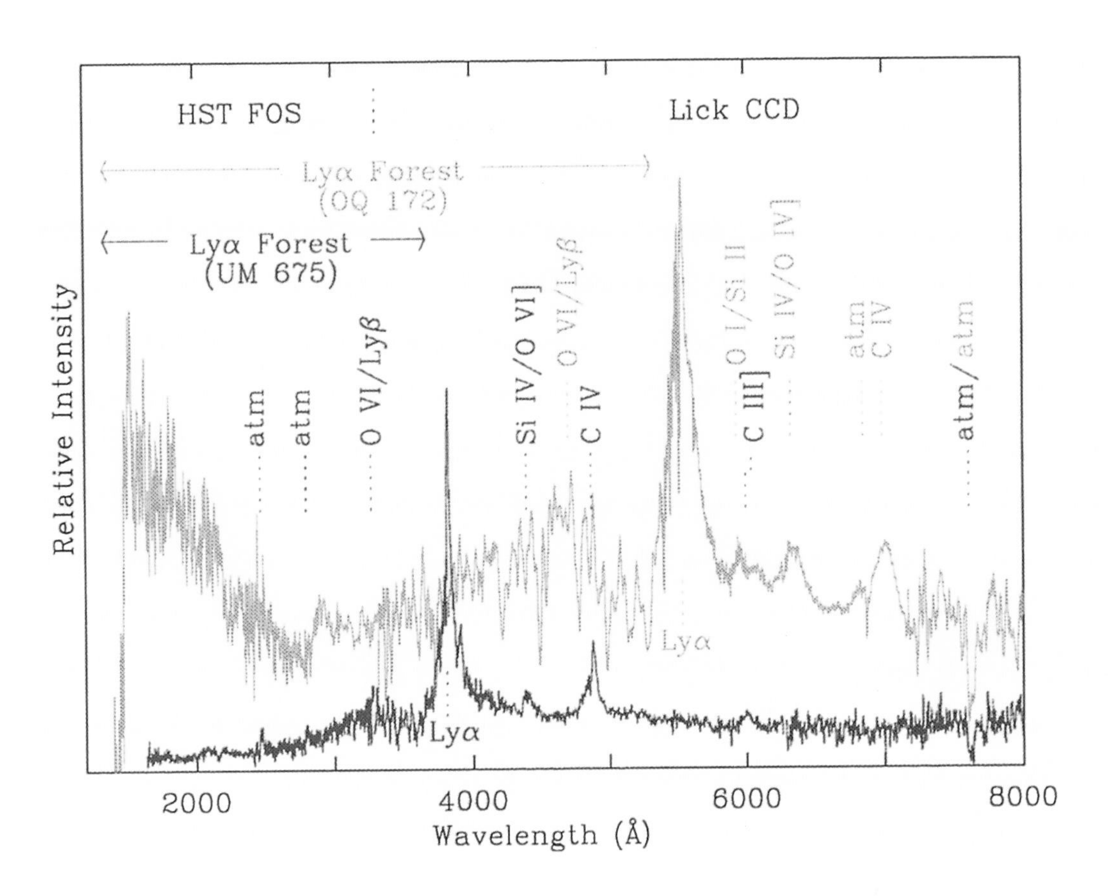


**Figure 5.24.** Combined FOS and Lick Observatory spectra of two quasars. UM 675 (green), a quasar with $z$ approximately equal to 2; and a high redshift quasar, OQ 172 (gold), with $z$ approximately equal to 3.5. The spectra have been labeled, and 'atm' indicates atmospheric features. The major feature in both spectra is the principal line of hydrogen, labeled Lyα (Lyman-alpha). (Ron Lyons, Ed Beaver, Margaret Burbidge, University of California, San Diego, and the FOS Team)

the clouds could be used to determine the mass of the black hole. The FOS spectra have observed wavelengths as low as about 1500 angstroms, and provide access to the rich spectrum of the near ultraviolet, which includes lines at lower wavelengths redshifted into this region. When combined with ground-based spectra taken at Lick Observatory, the result is a series of spectra covering thousands of angstroms.

Despite all this information, we do not yet have a breakthrough on understanding the fundamental nature of quasars. Intensive work on quasars continues, and as Robert Williams, Director of the Space Telescope Science Institute, has commented on the slow but steady progress astronomers have made in developing 'quasar engine' theory, 'Sometimes progress comes slow, sometimes serendipitous. Despite all the best efforts to really attack this in an analytical way, it may be some fluke that ultimately causes us to understand this.'

## Collisions between galaxies

*Arp 220* In the preceding discussion of both M31 and Markarian 315, we mentioned the possibility of galactic mergers as one way to explain some interesting structures seen

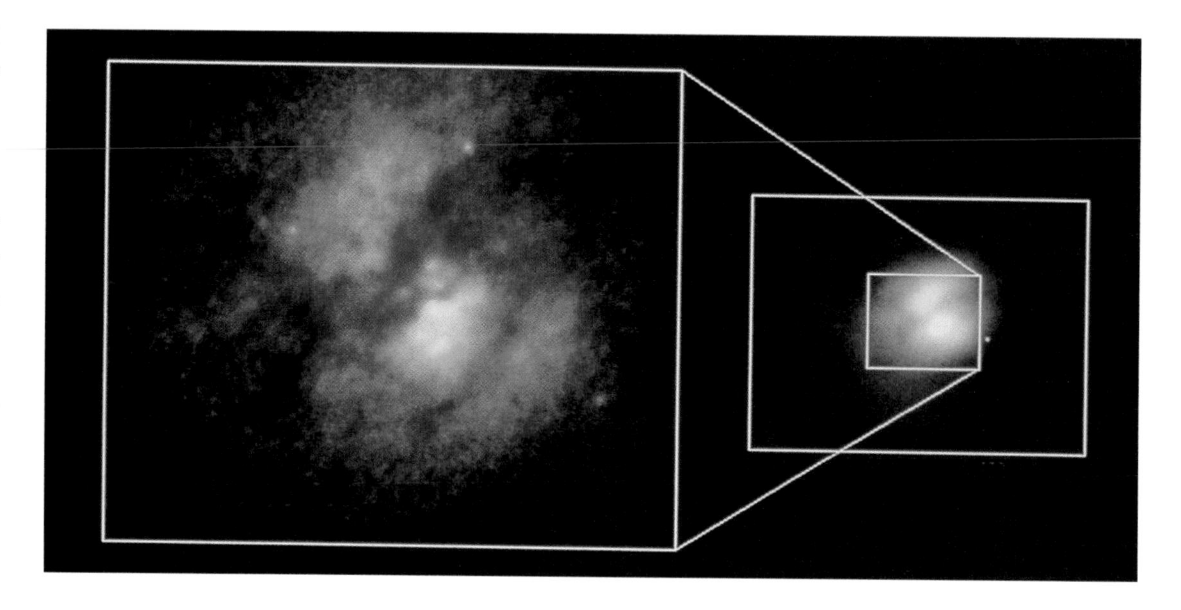

**Figure 5.25.** WF/PC-1 (left) and ground-based (right) images of the core of the galaxy Arp 220. The HST image reveals gigantic, young star clusters some ten times larger than previously observed. See Figure 5.26 for details of the image. The clusters were probably produced by the collision of two galaxies which caused a furious rate of star formation. The ground-based image was taken by K. Borne, H. Levison and R. Lucas at the USNO Flagstaff Station, Arizona. (E. Shaya and D. Dowling, University of Maryland; the WF/PC Team; NASA)

in the central parts of galaxies. In the peculiar galaxy Arp 220, shown in Figure 5.25, the HST image is compared to a ground-based photograph. We can see complex structure within about 1 arcsecond of the nucleus. Most of this structure is newly discovered, but some had been seen in maps made in radio wavelengths.

A partial interpretation of this image is shown schematically in Figure 5.26. The nucleus is marked, as are star clusters and possible supernovae. These star clusters are ten times

**Figure 5.26.** Diagram showing the location of the star clusters and other features in Arp 220. (E. Shaya and D Dowling, University of Maryland; the WF/PC Team; NASA)

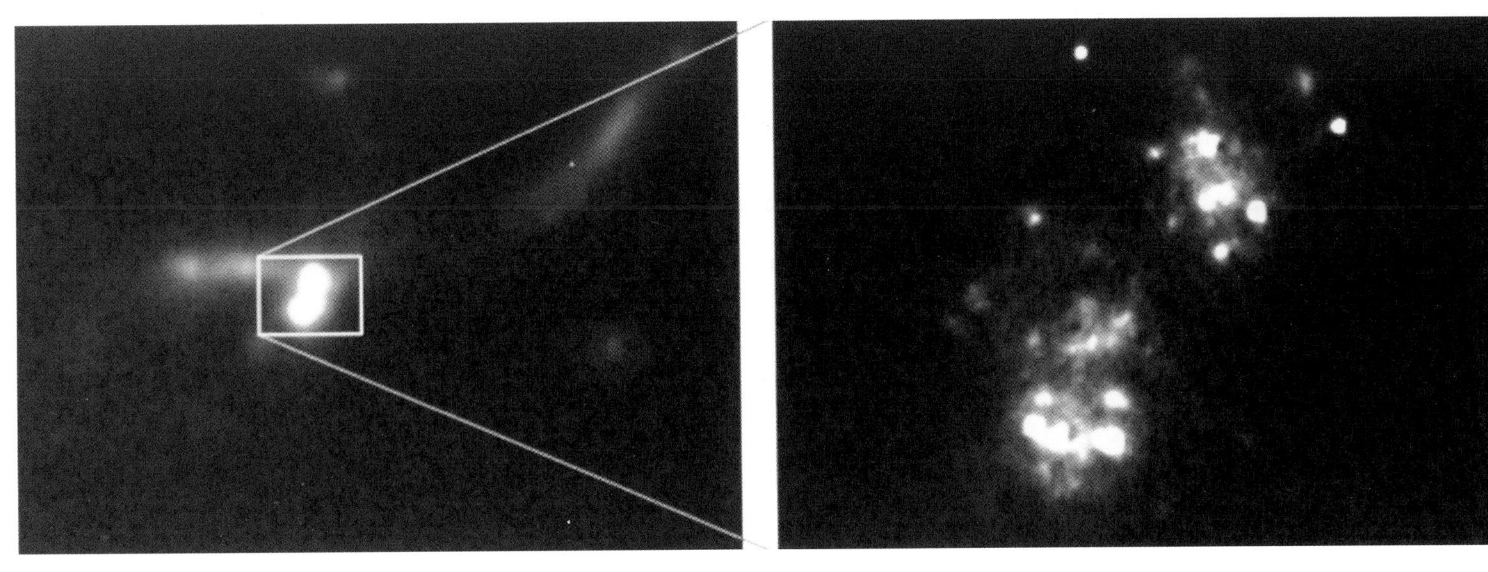

**Figure 5.27.** Starburst activity in colliding galaxies as seen in ground-based (left) and FOC (right) images of NGC 1741. The FOC ultraviolet image shows a pair of starburst centers that are separated by 3000 light years; the individual regions are 150 to 300 light years across. These are several times more luminous than the prototypical starburst cloud, 30 Doradus (see Figure 5.28) in the Large Magellanic Cloud. (Todd Small, CalTech; Peter Conti, University of Colorado; NASA; ESA)

larger than any previously observed clusters, and they are very young. Galaxy collisions could have produced regions of very dense dust and gas in which star formation – so-called 'starburst activity' – would proceed at a furious rate. The formation of very massive stars (see the discussion of the star Melnick 42 in Chapter 4) would be expected.

*NGC 1741* Colorado astronomer Peter Conti is another HST user who has been scrutinizing starburst galaxies that have undergone mergers in the recent past. A good example of the starburst activity as studied by Conti is seen in the FOC image of NGC 1741, a pair of colliding galaxies. The two centers of starburst activity are separated by 4 arcseconds, or 3000 light years, at a distance of 50 megaparsecs for the entire galaxy. The compact starburst regions are 150 to 300 light years across and contain between several hundred and several thousand very young, massive stars. A prototypical, nearby starburst cloud is the 30 Doradus complex in the Large Magellanic Cloud shown in Figure 5.28. The starburst clouds in NGC 1741 are several times more luminous, containing many more hot, young stars than 30 Doradus. The double-lobed structure seen in both the ground-based and HST images in Figure 5.27 suggest colliding galaxies, and Conti has been studying several other areas of the sky to confirm his theory. The collision would have produced very dense regions of dust and gas, conditions ideal for the formation of massive stars. Conti and his team of observers hope to use the spectrographs to obtain even higher-resolution data on just what is happening within the knots of starburst activity in the galaxy.

*NGC 1275* At least one colliding galaxy is changing our viewpoint about galaxies and their associated globular clusters. Near the elliptical galaxy NGC 1275 – also called Perseus A – lies a collection of *young* globular clusters. Astronomer Sandra Faber and a team of scientists used WF/PC-1 to look at the core of the giant galaxy. 'It's the brightest galaxy

**Figure 5.28.** The Tarantula Nebula, or 30 Doradus, as photographed with the 4-meter telescope of the Cerro Tololo Inter-American Observatory. The cluster (in the Large Magellanic Cloud) containing many young, bright stars is 30 times larger than the Orion Nebula in our galaxy. (National Optical Astronomy Observatories)

within 200 million light years,' she said. 'It's at the center of a big cluster of galaxies and in general this type of galaxy is thought to be a product of a galactic merger. They're usually kind of bland-looking ellipticals and they look like they've been there for a long time. Perseus A is not like that. It's a big radio source; it's lumpy, full of dust and there's clearly a lot of gas around it. The spectrum is full of A-type starlight, which indicates there's been recent star formation.'

According to Faber, HST made two significant discoveries about this galaxy. 'The first was that we detected tidal arms on the image and that indicates that this galaxy did undergo a merger with something really massive, let's say within the last 300 million years,' she explained. 'So a lot of this activity could be traced to that. There's growing evidence that

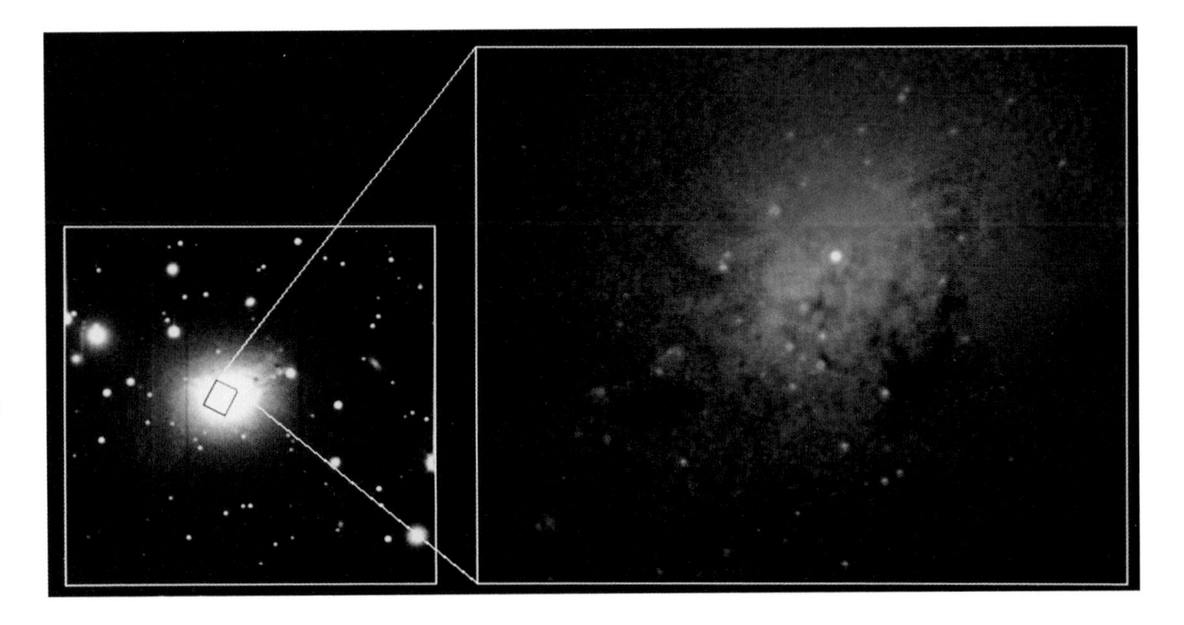

**Figure 5.29.** Ground-based view (4-meter Mayall Telescope at Kitt Peak National Observatory) and WF/PC-1 image of the core of NGC 1275. The HST image reveals individual star clusters that appear as bright blue dots. These globular clusters contain young stars rather than old stars, which is the usual circumstance. (J. Holtzman, University of California-Santa Cruz; NASA)

merging of gas-rich galaxies creates these intense starbursts, and the A-star light here. A-type stars live about 200 million years so it's perfectly consistent with a burst of star formation a few hundred million years ago triggered by a merger. That's point number one. Point number two is that we discovered a bunch of globular clusters. To be more precise, they are blue point-like sources very bright, a hundred times brighter than the brightest known globular cluster. We believe we've detected "proto-globular clusters".'

The significance of this find is important to the understanding of the origin and evolution of globular clusters. Because these clusters usually contain very old stars, Faber and her colleagues expected to see reddish stars. Instead, the HST image of NGC 1275's central region shows individual star clusters that look distinctly blue, and therefore young.

*The Cartwheel Galaxy* HST also captured the sight of a spectacular head-on collision between two galaxies located 500 million light years away in the southern hemisphere constellation of Sculptor (Figure 5.30). As with NGC 1741, the telescope resolved knots of star creation in the wake of the collision, and provides a new window on the creation of massive stars in large gas clouds. The large galaxy in the center of the ring was a normal spiral before its encounter with a smaller intruder galaxy. The ring is a direct result of another galaxy plunging through the core of the host galaxy. The collision sent a ripple of energy out from the site, plowing gas and dust in front of it. A 'firestorm' of star creation has resulted, and in other areas supernovae have been set off by the dynamics of the collision. The ring contains several billion new stars, evidence of starbirth on a grand scale and over an incredibly short time.

Like archaeologists at a dig, astronomers must now figure out the sequence of events that occurred here 500 million years ago, and determine which galaxy plowed through the

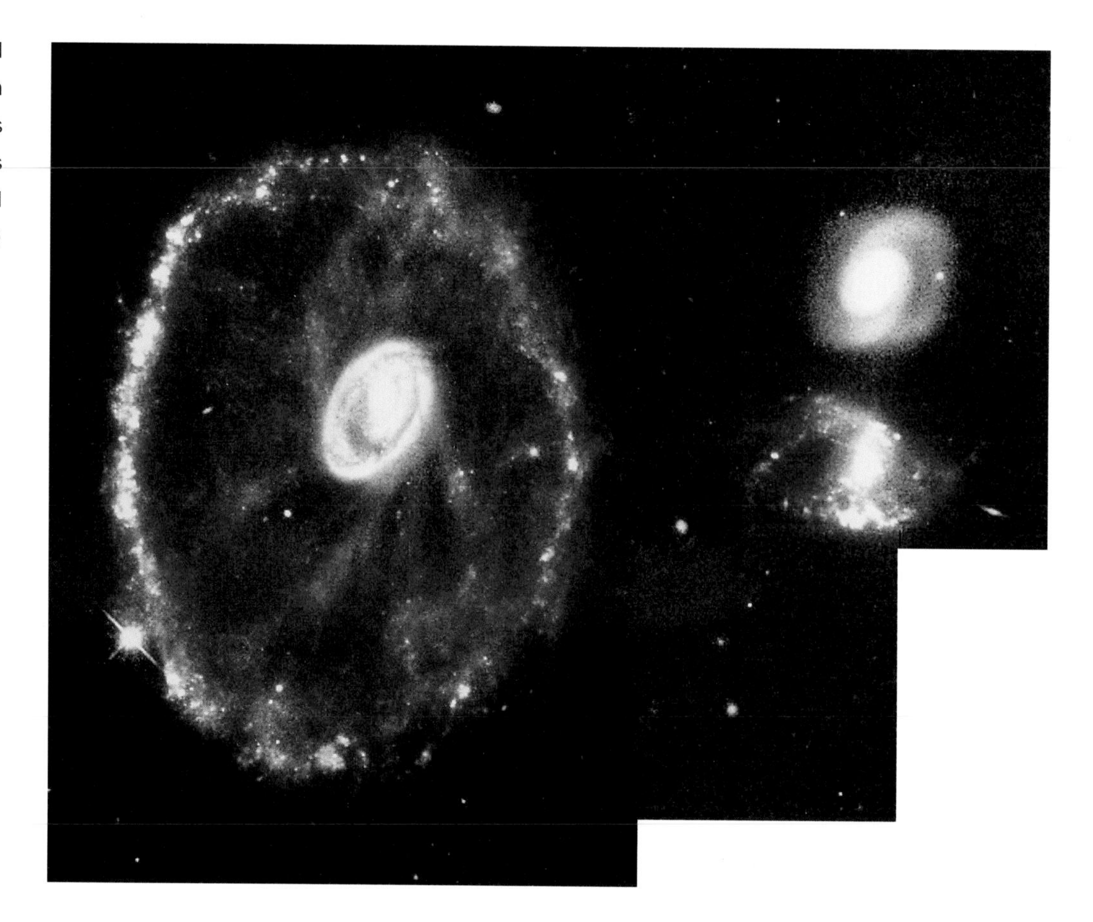

**Figure 5.30.** The Cartwheel Galaxy, taken with WF/PC-2 on October 16, 1994. This image is a combination of two images taken in blue and near-infrared light. (Kirk Borne, STScI; NASA)

heart of the Cartwheel. Two nearby galaxies could be the interlopers. The blue galaxy is disrupted and shows knots of starburst activity, which would happen if it had just completed a close encounter with its larger neighbor. On the other hand, the smoother-looking yellow galaxy shows very little evidence of gas – which means that it could well have lost all of its gas during a collision.

The high resolution and light-gathering power of the HST have provided scientists with powerful ways to study galaxies – particularly in the ultraviolet. As HST continues in its mission, new cameras and spectrographs will expand its ability to capture unique characteristics of galaxies and will give astronomers a clearer view of these giant stellar cities. Studies of the long-scale evolution of galaxies have implications for the structure of the universe. Ultimately, we can use what we learn from the evolution of clusters of galaxies to understand the origin and evolution of the universe – and it is to this that we turn our attention next.

# 6 Cosmology

Near the Sun is the center of the Universe.

*Nicolaus Copernicus*

Now it is quite clear to me that there are no solid spheres in the heavens, and those that have been devised by the authors to save the appearances, exist only in the imagination, for the purpose of permitting the mind to conceive the motion which the heavenly bodies trace in their course.

*Tycho Brahe*

We had the sky up there, all speckled with stars, and we used to lay on our backs and look up at them, and discuss about whether they were made, or only just happened.

*Mark Twain*

At last we come to cosmology – the sometimes mind-boggling, always captivating study of the entire universe – its origin, evolution, structure and ultimate fate. There are some profound ideas in cosmology, notions of causality and eternity that inevitably drift across one's mind while viewing the stars on a dark, clear night. The scientific questions at the heart of cosmology seem to be grounded in queries of a more philosophical and sometimes even a religious nature. How far away is it all? Where exactly is the 'edge' of the universe? Can we see it? And when we do, what will we find? To answer these questions, we need to answer another one – how old is the universe? which begs yet another question – will the universe just keep getting bigger, making more stars and galaxies? Or does something else happen to it?

Historically, the human view of the universe has been one of ever-expanding horizons. For millennia, people cast about for explanations of the origin and evolution of the Earth, and the genesis of anything beyond the Earth was simply a mystery. Until relatively recently, our view of the universe placed the Earth at the center, with everything else –

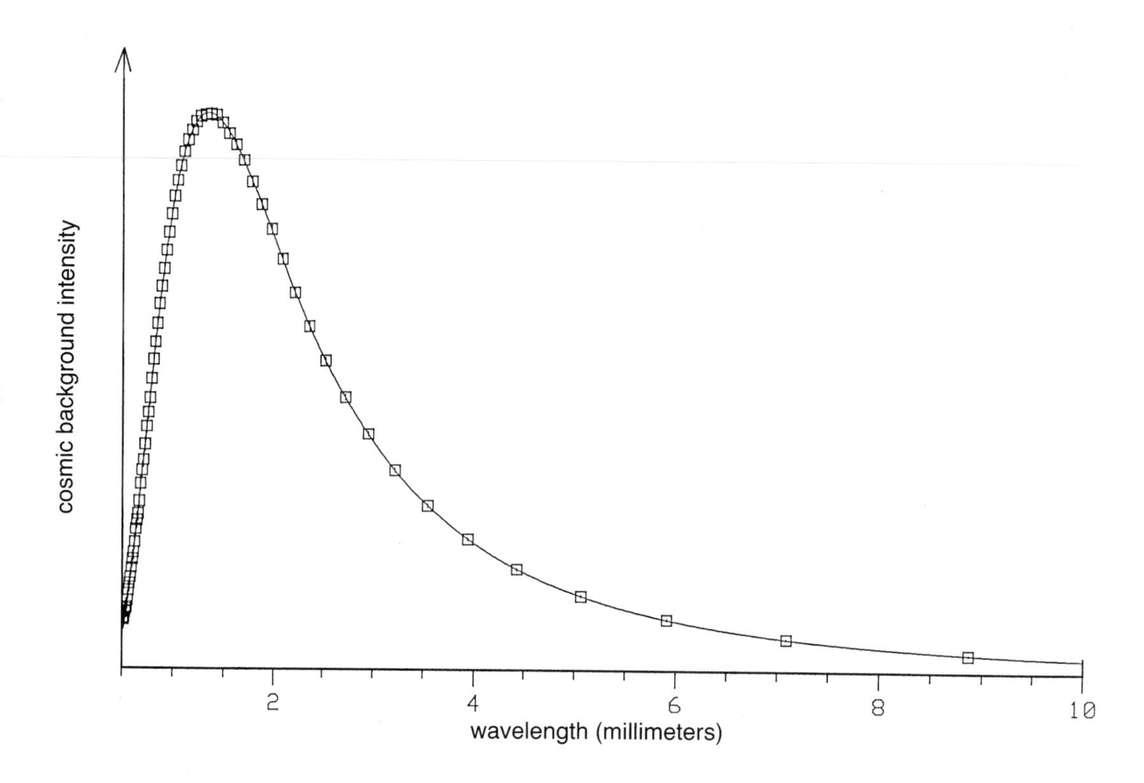


**Figure 6.1.** Cosmic Background Explorer (COBE) measurements at the north galactic pole. The cosmic background intensities are plotted against wavelength in millimeters; the observations are plotted as squares and the curve is a black body (an ideal radiator) at 2.73 kelvin. This radiation is a convincing remnant of the Big Bang. (NASA–Goddard Space Flight Center)

the Sun, Moon, planets and stars – revolving around it. Today our 'world-view' of the universe is changing again. It is as if we are standing at the foot of a cosmic ladder anchored on the Earth, looking up to a very distant top rung. We have no way of knowing exactly how far away it is yet, but we do have the means to climb the ladder. We know that our place in the universe is not central, but we are still trying to figure out just where we are in the grand scheme of things, and how far away we are from everything else. Some of this 'desire to know' is driven by simple curiosity: if we know the physical limits of our universe, we can approach an understanding of its origin and evolution.

Telescopes, cameras, spectrographs, photometers – all of our astronomy hardware – gives us a leg up on the distance ladder by making it possible to see ever more distant galaxies and quasars. Orbiting sensors such as the Cosmic Background Explorer have magnified for us the last whispers of the Big Bang echoing across the cosmos. The International Ultraviolet Explorer opened a window to a universe of frenetic activity. We have taken the step up to the first rung of the ladder, but there is still a long way to go.

Why this interest in cosmic distances? Simply put, if astronomers can determine accurate distances out to the more distant objects in the universe, they will be able to start determining the age, origin and evolution of the universe. Using HST, astronomers have been able to find some of the most distant galaxies in the universe, which are some of the youngest structures in the cosmos. Now the fun begins: determining *exactly* how far away these objects lie from us.

The Hubble Space Telescope's contribution to determining distances to galaxies is best described as a work in progress. During its first four years, Hubble's work was a veritable laundry list of cosmologically interesting observations:

- an accurate determination of the deuterium-to-hydrogen ratio in the interstellar medium;
- the detection of more than a dozen intergalactic hydrogen clouds within 1 billion light years of the Milky Way Galaxy;
- a survey of more than 300 quasars, which provided new information on the frequency of gravitational lensing by quasars;
- the setting of a new value for the Hubble Constant, from highly accurate measurements of the distance to galaxy M100 in the Virgo Cluster, using a set of 20 Cepheid variable stars;
- the discovery of a gravitational lens with the largest separation between elements ever observed.

Before we look at HST's work in detail, it is important to examine some of the basic concepts and assumptions of cosmology.

First, much of HST's work is involved with determining accurate distances to faraway objects that are difficult or impossible to see with ground-based instruments. As we have discussed elsewhere, its position above the atmosphere is ideal for ultraviolet spectroscopy, which is an important tool in cosmological research. Yet, the work is not as straightforward as targeting dim objects and making long exposures, or taking a spectrum of a distant object, measuring its redshift and then declaring that it lies billions of light years away. Many discussions of cosmology do include a statement that an object is probably at a certain distance. While the statement may be true, it hides the struggle involved in pinning down distances to anything but the nearest objects.

The other side of the cosmological coin involves determining ages of distant objects, in the hopes of discovering just how old the universe might be. Certainly, distances are one way of zeroing in on an age for the cosmos, but we can also make other types of observations that bear directly on the process of dating the universe. Using the technique of radioactive dating, geologists have determined that the Earth, the Moon and meteorites are around 4.5 billion years old. The Sun's age is taken to be approximately 5 billion years.

Analysis of starlight and our knowledge of stellar evolution gives us the age of other stars. Once we characterize other stars according to their size and luminosity, we find that their ages range from a few thousand years – for the very youngest newborns – to at least 10 billion years for the oldest red giants. Still older are stars in globular clusters near the core of the Milky Way that *appear* to be 13–15 billion years old. Since the globulars are thought to have formed at approximately the same time as the galaxy, we know that the Milky Way Galaxy is at least the same age. The Milky Way is about the same age as all the other galaxies. Paradoxically, other galaxies *appear* to be much younger than the Milky

Way because they are so distant. Some of the most distant galaxies we have seen emitted their light when the universe was only half the age it is now. Farther in space and further back in time lie other galaxies that must have formed when the universe was only one-fifth of its present age. Theoretically, the further back we look, the younger are the galaxies we see. Because the universe is expanding, finding the limit or 'edge' of the universe may well prove to be impossible. But, we can continue to look across the cosmos, finding ever more distant and older structures.

## Cosmological distances

The astronomical distance scale also starts right here on Earth. We look out and see the Sun, Moon and planets and wonder how far away they are. Celestial mechanics and simple observations determine the relative spacing of the planets, and radar observations determine the absolute scale very accurately. So, for example, the distance from the Earth to the Sun is 149 million kilometers; light takes 8 minutes to traverse that distance, making the Sun 8 light minutes away. The moon is 380 000 kilometers from Earth; light takes 1.27 seconds to make the trip, you could say that the moon is 1.27 light seconds away. Mars is around 228 million kilometers from the Sun, or 13 light minutes away. The distant outer planet Pluto is 5900 million kilometers from the Sun, or 5.5 light hours away.

These distances are relatively easy to manage, but when we start to look at nearby stars and galaxies, such as Proxima Centauri, which lies $4.1 \times 10^{13}$ kilometers away, or the Large Magellanic Cloud, which lies $1.61 \times 10^{18}$ kilometers away, you can see that the distance numbers start getting out of hand. Astronomers have adopted several shorthand 'standard' references for large distances. One method is to use the astronomical unit (or AU) – the average distance between the Earth and the Sun, which is $1.5 \times 10^{8}$ kilometers. Thus, Mars lies at 1.5 astronomical units and Pluto is about 40 astronomical units away. However, that system also breaks down when we move to stellar and galactic distance scales. Proxima Centauri is at more than 273 000 astronomical units, and the Magellanic Clouds are $10^{10}$ astronomical units away. We obviously need larger units for more distant objects. It is often easier to use a unit that gives us another way of understanding distance – the distance light travels in a year ($9.5 \times 10^{12}$ kilometers), more commonly known as a light year. Thus, Proxima Centauri is 4.3 light years away, and the Large Magellanic Cloud is just over 170 000 light years distant.

We can also use the parsec, a unit based on angular measure which equals 206 265 astronomical units, or 3.26 light years. Thus, the distance to Proxima Centauri becomes 1.3 parsecs. The Large Magellanic Cloud is 52 000 parsecs away, or 52 kiloparsecs. For even larger distances, the system once again becomes unwieldy, so we speak in terms of millions of parsecs, or megaparsecs. These are huge numbers describing incredible distances. See Table 6.1 for some distance comparisons.

Table 6.1 *A comparison of distances*

| Distance unit | Kilometers | Light-travel time |
|---|---|---|
| Light second | $3x10^5$ | 1 second |
| Light minute | $1.8x10^7$ | 1 minute |
| Astronomical unit (AU) | $1.5x10^8$ | 8 minutes |
| Light year | $9.5x10^{12}$ | 1 year |
| Parsec (pc) | $3.1x10^{13}$ | 3.3 years |
| Kiloparsec (kpc) | $3.1x10^{16}$ | 3.3 thousand years |
| Megaparsec (Mpc) | $3.1x10^{19}$ | 3.3 million years |

One might wonder just how astronomers *know* these distances. The answer is that they use a variety of methods, both direct and indirect, to measure distances to objects in the universe.

For objects relatively close to us, another technique of distance determination entails using a combination of parallaxes (the apparent shift in the position of the star when viewed from Earth over the course of a year) and a more esoteric technique (measurements of moving clusters) to measure distances to clusters of stars. If stars in other galaxies and globular clusters can be classified according to the Hertzsprung–Russell diagram, the main sequence itself can become a standard candle.

In 1924, Edwin Hubble determined the distance to a nearby, naked-eye object he knew as the Andromeda 'Nebula'. His tool was a type of star called a Cepheid variable – named after the fourth-brightest star in the constellation Cepheus, itself a variable. Cepheids have a very interesting and useful characteristic: their intrinsic brightness changes over periods of time ranging from 1 to 50 days. These brightness changes obey a period–luminosity law developed by astronomer Henrietta Leavitt in the early part of the 20th century: brighter stars vary over longer periods of time. Thus, if we measure the period of a Cepheid variable (that is, the length of time it takes to go from maximum brightness to minimum brightness, and then back again), we can determine its intrinsic brightness. Because the intensity of light fades as a function of distance from us, comparing the intrinsic brightness of the Cepheid with its apparent brightness gives us a distance. Edwin Hubble applied this method to Cepheids in Andromeda and thus estimated the distance this galaxy is from us.

Cepheids are members of a group of objects in the universe called *standard candles* that astronomers use to determine distances. Think of standard candles in the following way: imagine that you are standing at one end of a large room looking at a few light sources against the wall at the opposite end of the room. If all of the sources are 100-watt light bulbs, and they are all at the same distance, they should all look equally bright. If some bulbs are closer than others, they will look brighter, but they will not actually *be* intrinsically brighter than the more distant ones. They will always be 100-watt light bulbs no matter how far away they are. What astronomers look for are objects that have the same *intrinsic brightness*. That is, they would all be cosmic 100-watt light bulbs. Some might be a few light years away and look bright, and others might be a few megaparsecs away and look dim. Because they have the same intrinsic brightness, however, their function as distance indicators is important. Cepheid variables are one example of this type of source. And, if we can find Cepheids in galaxies, or other variable stars such as the RR Lyrae stars in globular clusters, then we can determine accurate distances to them.

Armed with these standard candles, astronomers can accurately determine the distances to nearby galaxies such as the Large and Small Magellanic Clouds, the Andromeda Galaxy (M31) and its companion, M32. Extension of the distance scale to a nearby cluster of galaxies can also be done by measuring the magnitude of the brightest stars and the size of HII regions. An entire galaxy can be a standard candle, as can the brightest galaxy in a cluster, or even a specific type of supernova, which can rival an entire galaxy in brightness.

The Hubble Constant – often called $H_0$ or 'H-naught' – is another factor in distance determinations. $H_0$ is a shorthand way of indicating how the rate at which galaxies are receding from each other scales with separation, and it is tied to the spectral redshift of these galaxies (which we will be discussing shortly). From the cosmological perspective, accurate distances are the key to the Hubble Constant.

There are other factors to be considered when determining the age of the universe that, together with the Hubble Constant, play important roles in a theoretical construct of the universe called the 'Standard Model'. This is the cosmological equivalent of a blueprint for the universe, and has inspired much of HST's cosmology work.

## The expanding universe and the Standard Model

In the early part of the 20th century, as techniques to estimate distances to faraway stars and outlying galaxies were developed, Edwin Hubble built on another fundamental discovery. The spectra of galaxies generally showed redshifts; i.e. the spectral lines appeared to be shifted to the red or longer wavelengths. If this is interpreted as a Doppler effect, this means that the galaxies are all moving away from us. Moreover, a comparison of distances and redshifts showed that the farthest galaxies appear to be moving away from us the fastest. In other words, the farther away the galaxy, the larger the redshift and thus, the higher its speed of recession.

From this observation, the idea of the 'expanding universe' was established. However, it is important to remember that, even though all the galaxies are receding from us, that does not mean that the Earth, or the Sun, is at the center of the universe. To understand this apparent paradox of thought, consider the most popular depiction of an expanding cosmos: a balloon on which we have painted dots to represent galaxies. If we blow up the balloon, its surface expands and *all* the dots recede from each other. Thus, there is no

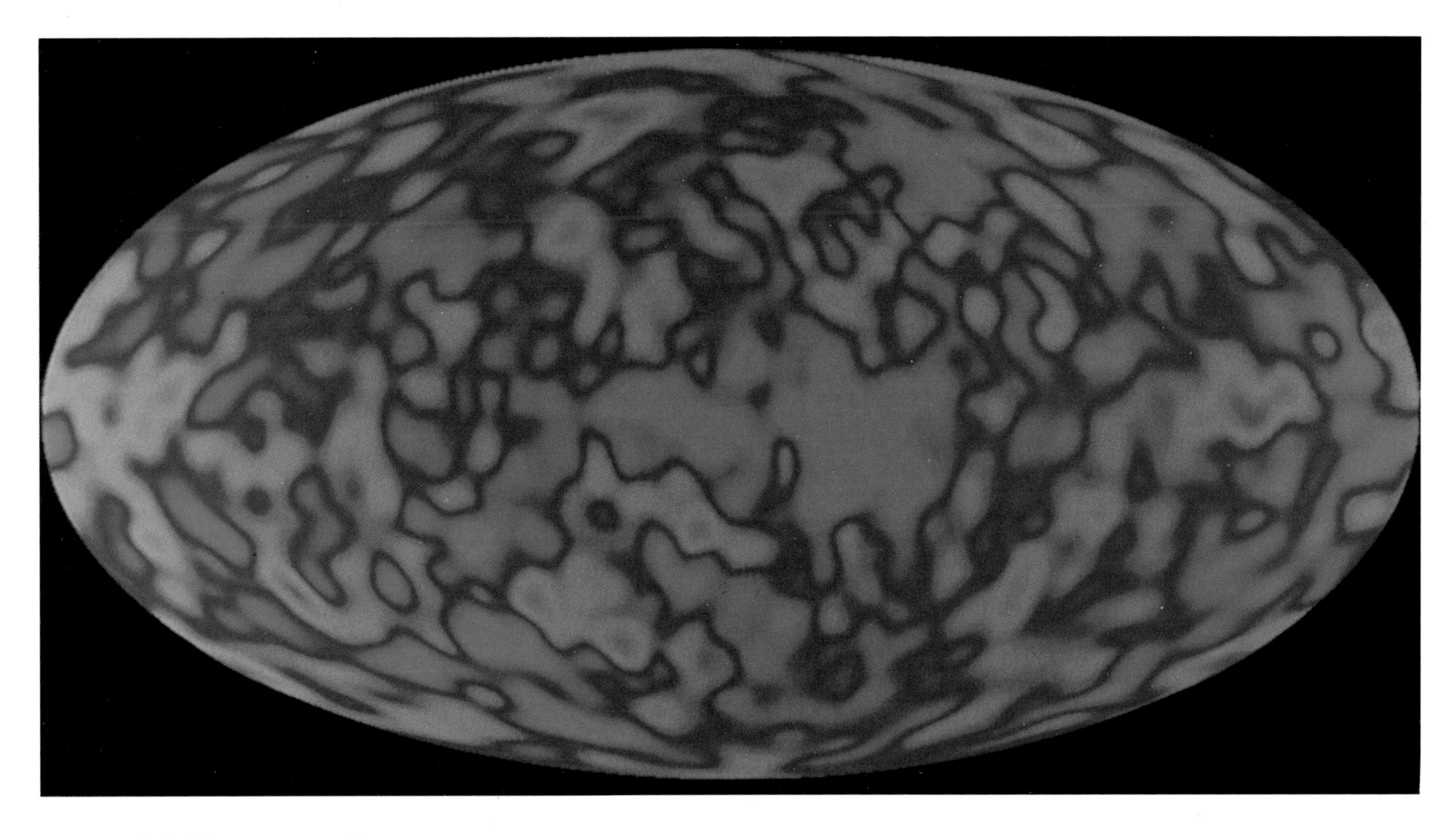

**Figure 6.2.** Microwave map of the whole sky based on 1 year of COBE data. The plane of the Milky Way Galaxy runs horizontally across the middle of the map, with the galactic center at the map center. Red indicates radiation 0.01% warmer and blue 0.01% cooler than the average of 2.73 kelvin. Computer analysis is necessary to separate instrument noise (most of the patches) from the faint cosmic signals. The patterns of cosmic fluctuations are consistent with some models of galaxy formation. (NASA–Goddard Space Flight Center)

central dot. In addition, each dot 'sees' every other dot receding, and those dots farthest away recede fastest. The correct view is that there is no center or preferred location of the universe. Specifically, the location of the Earth or the Sun, or anything else, is not unique in any way.

With the idea of an expanding universe and some reliable ways to measure distances, astronomers finally come face to face with the big question in cosmology: *how old is the universe?* This is a great question to debate at cocktail parties. Of course cosmologists (the astronomers who study the origin and evolution of the universe) do not really ask the question in that way. They are more likely to frame the questions in terms of measurable quantities, such as the density of the universe, which influences the expansion rate of the universe. So they ask their own types of cosmological questions. How long ago did the universe begin? Is the universe expanding? Is the expansion rate the same everywhere in the universe? How does the density of the universe influence its expansion rate? What is the critical density of the universe? How dense does it have to be if it is to expand forever? What density do we need if we want the universe to stop expanding forever? And so on.

As you might expect, cosmologists, being scientists, assign numbers and letters to the key ideas in these questions. If we know the distances between galaxies and the speed at which they are moving apart, then theoretically we can calculate the time at which all the galaxies were together. The crucial number we need to do this calculation is $H_0$, which gives the current expansion rate of the universe in units of kilometers per second per megaparsec.

Next, we need to estimate the age of the universe – a variable we can call 'Time Zero' or label as $T_0$. This number is directly related to the Hubble Constant for reasons that we will see shortly. For now, consider some possible numbers to 'plug into' $T_0$. For the sake of argument, if we use numbers somewhere in the range of 10 to 20 billion years, we would get Hubble Constants of between 100 kilometers per second per megaparsec and 50 kilometers per second per megaparsec, respectively.

The higher the value of $H_0$, the lower the age of the universe; i.e. the faster the rate of expansion of the universe, the less time it will have taken to reach its current size. We should realize that estimating the universe's age this way leads to an upper limit (or overestimate) because the effects of gravity could have slowed the rate of expansion. The uncertainty in $T_0$, affected principally by uncertainties in distances and density estimates, illustrates that there is still a lot of work that needs to be done when it comes to determining the true age of the universe. If the density is not accurately determined, then we will have no way of fixing $T_0$ correctly.

There are other properties of the expanding universe: $Q_0$, which is a deceleration parameter (the rate at which things slow down in the universe), which can be used to test the possibility that the exact Hubble relationship may differ for galaxies at different distances; and $\rho_0$ – the critical density of the universe, which would be exactly the average density of matter needed to eventually stop the expansion of the universe. The density is usually discussed in terms of the actual density divided by $\rho_0$ This factor is called $\Omega_0$. Thus, if $\Omega_0$ exceeds 1, the universe will stop expanding and will collapse back on itself in something called the 'Big Crunch'. If $\Omega_0$ is less than 1, the universe will expand forever. The question of the universe's ultimate fate – indefinite expansion versus collapse and possible renewal – is one of the most fiercely debated in all of cosmology.

Progress in this area of cosmology has refined the standard model of the beginning of the universe – a model which says, simply, in the beginning, things happened very quickly. The birth of our universe is often called the Big Bang, followed immediately by an extremely dense, hot universe called the *Primordial Fireball.* Theoretical studies by astrophysicists like George Gamow and his collaborators in 1948 led to the conclusion that a remnant of the Primordial Fireball existed, probably in the form of a microwave background radiation permeating all of the universe. This radiation was discovered in 1965 by Arno A. Penzias and Robert W. Wilson.

The universe as a Primordial Fireball initially (i.e. about $10^{-43}$ seconds after the Big Bang) consisted of energetic elementary particles and energetic photons. This is the domain where cosmology meets elementary particle physics. At about $10^{-34}$ to $10^{-30}$ seconds, the

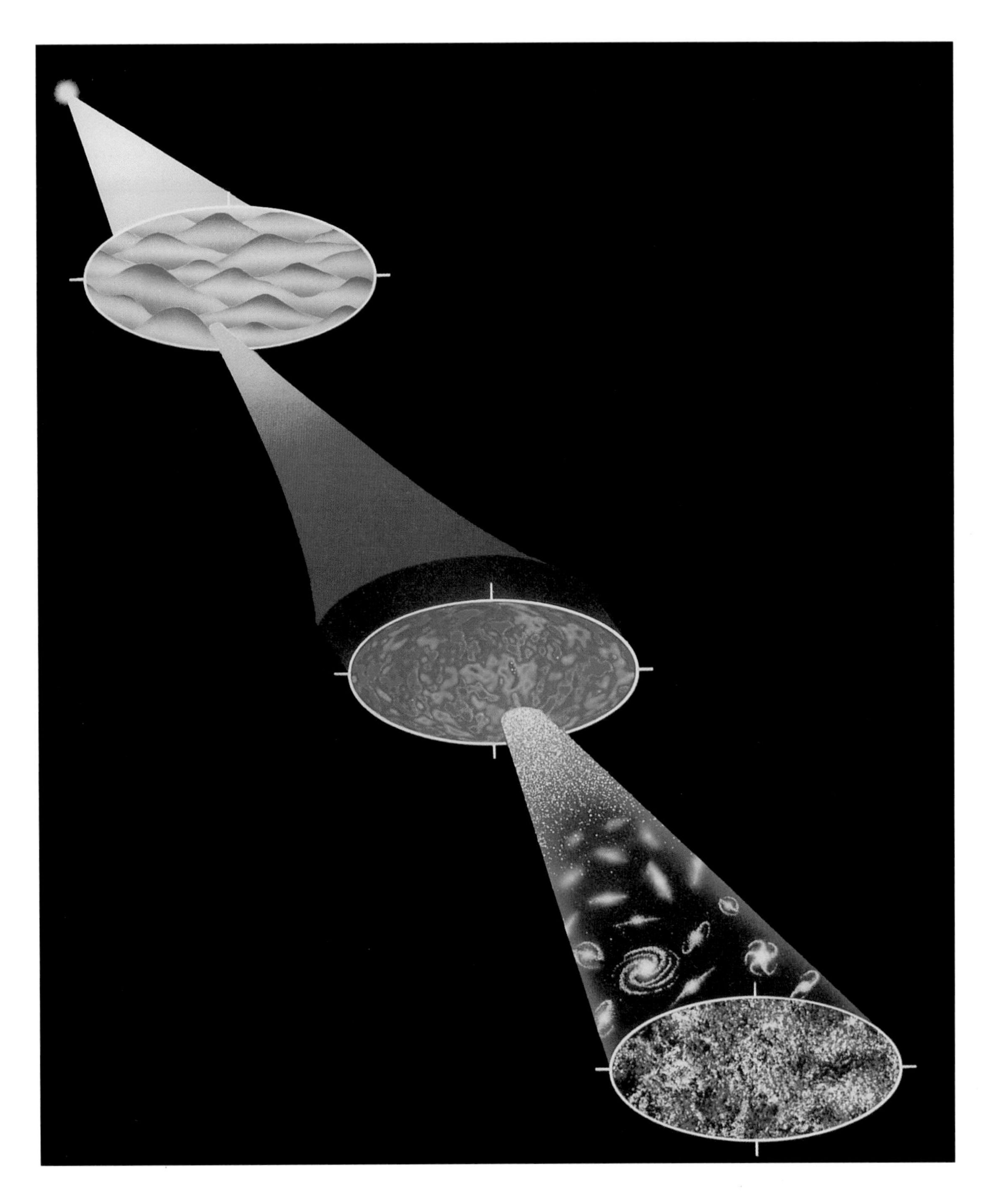

**Figure 6.3.** A summary diagram of the origin and evolution of the universe consistent with the COBE observations. Just after the Big Bang, small fluctuations existed in the universe which, after rapid expansion, became the fluctuations seen by COBE. These were indicative of favorable conditions for galaxy formation. Finally, we have the current universe some 15 billion years after the Big Bang. (NASA–Goddard Space Flight Center)

universe is thought by some to have passed through a stage of very rapid expansion known as 'inflation'. Around $10^{-6}$ seconds, the elementary particles (called quarks) combined to form neutrons and protons. By about 1000 seconds after the Big Bang, the universe had cooled, and the formation of major elements in the Primordial Fireball was complete. These elements were hydrogen, deuterium, helium and lithium, and they formed in proportions determined by the value of $\Omega_0$.

At some point, the matter and the photons (light) no longer interacted and they evolved separately. The matter simply cooled as the universe expanded, as did the electromagnetic radiation, from $10^4$ kelvin to a current, almost-uniform background radiation at 3 kelvin. This has been observed in detail by the Cosmic Background Explorer (COBE) satellite, and fine structure has been detected (see Figure 6.2). A summary of the Big Bang picture is shown in Figure 6.3.

A skeptical reader may wonder how *anyone* can talk with any certainty about events that occurred over 10 billion years ago, $10^{-43}$ seconds after the Big Bang. The answer is that events of long, long ago leave unmistakable traces even today. The 3 kelvin echo of the Big Bang is one such trace, and simple observations of the abundance of helium relative to hydrogen in the interstellar medium, and on the surfaces of many stars, provide another indirect, but valuable, way of checking on conditions in the early universe. This, in turn, allows us to start fixing some upper and lower limits on the amount of matter in the universe. When those limits are known, we can then put out some useful values for $\Omega_0$.

Knowledge of fine structure is particularly important in understanding the evolution of matter itself. Without going into too much detail, fine structure implies the existence of denser regions in the universe, which makes it easier to form matter into galaxies.

The organization of galaxies into larger units occurs in two stages: clusters and superclusters. Examples are the Virgo Cluster and the Hercules Cluster (Figure 6.4). Galaxy clusters are important both in cosmology and in the study of the evolution of galaxies (see Chapter 5). The biggest known structures are immense filaments and sheets of galactic clusters and superclusters. Some have picturesque names such as the 'Great Wall'. Undoubtedly, these large-scale structures are the remnants of the fine structure seen in the COBE results, which mirror fluctuations in the very early universe that were preserved by its inflationary expansion. Once protogalaxies have formed, condensation of gas into stars should occur, followed by stellar evolution with its implications for planetary formation and the development of life.

Not all astronomers are in agreement with the Standard Model of the universe we have just discussed. Some scientists offer competing theories for the origin and evolution of the universe: for example, the idea of the steady-state universe proposed by Herman Bondi, Thomas Gold and Fred Hoyle.

The steady-state situation is thought to be maintained by the continual production of new matter (at a very low rate) which forms galaxies that fill in the gaps produced by the expansion of the universe. This theory has produced an interesting and humorous side

**Figure 6.4.** The Hercules Cluster of galaxies as photographed by the 4-meter Mayall Telescope of the Kitt Peak National Observatory. The core of the cluster contains less than 100 galaxies. (National Optical Astronomy Observatory)

issue. The phrase Big Bang, now so inherent in our vocabulary, was actually coined by Sir Fred Hoyle to describe the 'Standard Model' possibly in an unflattering way. Hoyle still does not agree with the Big Bang picture, and a recent attempt to rename the Big Bang only served to illustrate that some questions are profound and require a long time to answer.

## Dark matter

Regardless of the state of debate over which theory best describes the origin and evolution of the universe, there are still some cosmic scores to settle. Basically, all the parameters mentioned above, $H_0$, $T_0$, $Q_0$ and $\Omega_0$, need to be determined accurately. Astronomers are making good progress with $H_0$, but the density parameter may prove a tougher nut to crack. This is because there are many unknowns about how much matter exists in the universe. The question of unseen 'dark matter' affects any meaningful estimate for the density of the universe. However, once some meaningful numbers are assigned to the Hubble Constant and $\Omega_0$, the other two unknowns in the cosmological equations will be easier to determine.

For decades, astronomers have known that much of the material in the universe is in the form of dark matter. However, what is not known is how much. The existence of dark matter is deduced as follows: gravitational effects (on the motion of stars or galaxies, for example) require a certain mass of material, but searches utilizing light-emitting sources fail to find enough to contribute significantly to the density of the universe. The remainder is blamed on dark matter.

Since we cannot directly observe dark matter, we have come up with many possibilities for what it might be, and some of the names are picturesque. For example, MACHOs, or massive compact halo objects, very faint stars or Jupiter-like objects, could be arranged in an essentially spherical cloud around galaxies. Black holes could be much more widespread than we suspect. Exotic particles could make up dark matter, and neutrinos could fill the bill if they have a small mass. Some evidence for this possibility exists, but, of course, remains to be confirmed. WIMPs, weakly interacting massive particles, might be good candidates for dark matter, and are predicted by theory. Axions are particles produced by breaking a fundamental symmetry, and are also predicted by theory. These particles, if they exist, are remnants of the very dense, hot early universe that existed shortly after the Big Bang. The searches are underway, and have helped create the relatively new field of particle astrophysics.

## HST and cosmology

HST's contribution to cosmology takes several approaches. There are the efforts to determine the density and makeup of the universe. These are observations that use quasars as candles to illuminate the intergalactic medium. In the same way, observations of relatively bright nearby stars have been used to determine the abundances of chemical elements in the interstellar medium. Studies of the amounts of deuterium and hydrogen relative to each other in the interstellar medium are critical to determining the density of the universe. Stars near the Sun come in for their share of HST's cosmological studies because their distances can be measured astrometrically. Cepheid variable stars in the galaxy M100 have been studied to determine a preliminary value for the Hubble Constant using HST. Turning its gaze toward the most distant reaches of the universe, HST is also peering farther into space, further back in time, to some of the most distant galaxies known. Finally, HST observers are using the phenomenon of gravitational lensing to test Einstein's Theory of General Relativity.

### Lyman-alpha forest in nearby quasars

Quasars, as we discussed in the preceding chapter, are interesting animals in the distant reaches of the universe. Aside from their own intrinsically interesting characteristics, it turns out that they make excellent 'candles' to help us figure out what lies between them and us. Because quasars are the brightest objects known, the lines of sight can cover a significant fraction of the universe. The spectra of quasars show numerous absorption lines

(Figure 6.5) that are now known as the 'Lyman-alpha forest'. What is happening is that a small, but important, portion of the ultraviolet light from quasars is absorbed by something lying in between us and them in the intergalactic medium – clouds of predominantly hydrogen gas. These intervening clouds produce a forest of absorption lines.

The absorption lines are redshifted due to the expansion of the universe, and the amount of shift depends on the distance between us and the object we are studying. To understand this concept, imagine listening to a group of train whistles here on Earth. Close-by, slowly moving whistles will sound different than far-away, rapidly moving whistles, and this gives you an idea of how fast the trains are traveling and how far away they are.

The line producing the absorption from the hydrogen atom is called the Lyman-alpha line. Its wavelength in the laboratory is 1216 angstroms, a value about one-quarter of the wavelength of visible light. When light is redshifted, astronomers use a quantity $z$ to describe the amount of redshift. The value of $z$ gives the shift in units of the original wavelength. For example, $z = 3$ means a shift of three times the original value or a new, total wavelength of 4864 angstroms; $z = 2$ produces a new wavelength of 3658 angstroms; $z = 1$ corresponds to a new wavelength of 2432 angstroms; and finally $z = 0$ means no shift at all or a wavelength of 1216 angstroms.

Because wavelengths less than about 3200 angstroms are absorbed by the Earth's atmosphere, earlier ground-based studies could only deal with the larger values of $z$, essentially $z = 2$ and larger. For these values, the redshift moved the wavelength into the visible part of the spectrum, and thus it could be seen from the ground. To study the low values of $z$ which are visible in ultraviolet spectra, astronomers needed an observatory above the Earth's atmosphere. Until the launch of HST, essentially nothing was known about the nearby or low-$z$ Lyman-alpha clouds. Because the number of clouds in the line of sight decreases markedly from $z = 3$ to $z = 2$, the expectation was that there would be very few lines for $z$ close to zero.

A group of researchers, led by Ray Weymann of the Observatories of the Carnegie Institution of Washington, has been using the quasar 3C 273 to study the line of sight between us and the quasar. 'For several years now, people have known that there were these clouds of hydrogen between us and distant quasars,' he said in a 1993 interview. The result of that work is that there turned out to be more clouds between us and the quasar than anyone had anticipated.

The GHRS observations of this nearby quasar ($z = 0.16$) are shown in Figure 6.5, in which the Lyman-alpha forest lines are marked; the other absorption lines are produced by the interstellar gas in our own galaxy. The result is that there is no significant difference in the number of lines from $z = 2.1$ to $z = 0.16$. Thus, the population of Lyman-alpha absorbing clouds is apparently unchanged over this interval.

What this means is that the clouds simply exist – and there may be neither creation nor destruction of clouds. Or, if some of these processes are active, the net effect is still no change. This is a beautiful example of the importance of making observations, even when

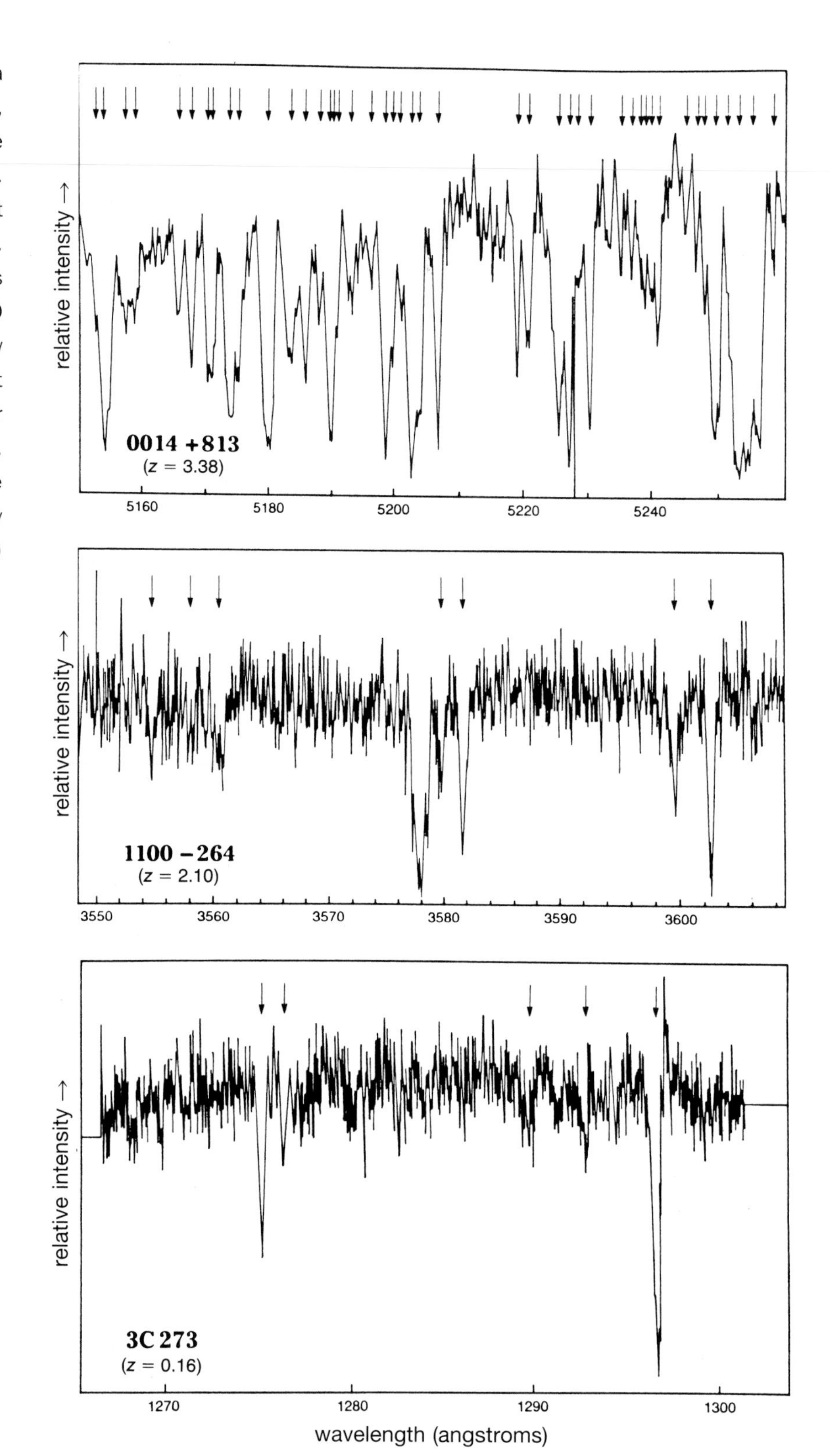


**Figure 6.5.** The Lyman-alpha forest for galaxies at $z$ = 3.38, 2.10 and 0.16. The latter is the GHRS observation of 3C 273. Individual Lyman-alpha forest lines are depicted by the arrows. The large decline in numbers between $z$ = 3.38 and $z$ = 2.10 led to the expectation of very few lines at $z$ = 0.16. This was not the case; see the text for discussion. (Simon L. Morris, Observatories of the Carnegie Institution of Washington; Sky Publishing, Inc.)

we think we know the answer. These results have also been confirmed by FOS spectra of 3C 273 and other nearby quasars.

The origin of these clouds is interesting to contemplate. Are they clouds associated with galaxies, perhaps a residue of the process of galaxy formation? Or are they unrelated to galaxies, being held together by a very hot intergalactic gas? The preliminary feeling is that the clouds could well be associated with galaxies. As Weymann (who is a GHRS Co-Investigator) wrote in a 1994 paper describing his work:

> While the luxuriant growth of the very high redshift Lyman-alpha forest has given way to a rather arid savannah at low redshifts, their continued study is proving of importance for addressing the fundamental question of the origin and evolution of galaxies and clusters of galaxies.

This is not the last we will see of these hydrogen clouds. As University of California at San Diego astronomer Margaret Burbidge, who has been trying to understand the physics of quasars and their environments for years, has said, 'There's been a lot of interest in trying to see if there are other chemical elements in these hydrogen clouds that are producing absorptions. Are they primordial clouds of gas that form galaxies? Are they forming galaxies? Are they something blown off other objects? There's a lot of fascination out there, and we do not understand them very well. Apart from the interest in what we get from the quasar itself, we found that there were some absorption features in Lyman-alpha at different redshifts, lower redshifts than the quasar, and this has been one of the major discoveries of the Space Telescope.'

## The deuterium-to-hydrogen ratio in the local interstellar medium

The abundances of elements has been a common thread in the studies carried out by HST. As we mentioned earlier, the lighter elements, hydrogen, deuterium, helium and lithium, were created in the Primordial Fireball, with their proportions depending on the density in the fireball. The density in the fireball also determines the value of $\Omega_0$. The amount of deuterium created (measured by the ratio of deuterium to hydrogen, D/H) is small and quite sensitive to this density, but the amount of the other elements is relatively insensitive.

The D/H value can be determined from observations of stars near the Sun because the light from a nearby star serves as a probe of the local interstellar medium. However, deuterium has its fundamental absorption line (Lyman-alpha again) displaced some 0.3 angstroms from the hydrogen line. Because there is so much more hydrogen than deuterium in the universe, the hydrogen Lyman-alpha line would completely swamp the deuterium Lyman-alpha line unless a nearby star was used. The GHRS measurements used the star Capella, which lies some 12.5 parsecs away from the Earth.

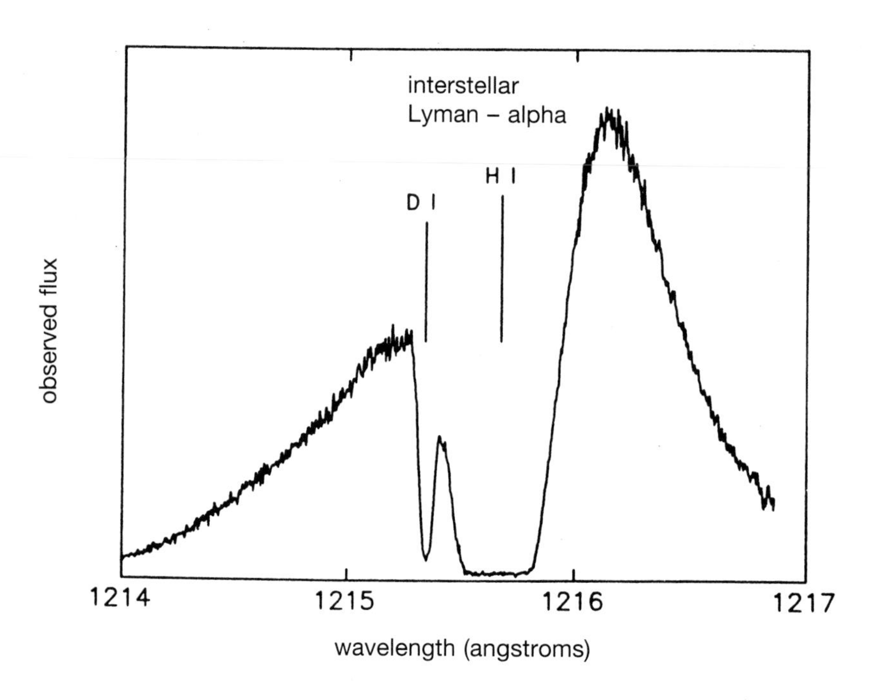


**Figure 6.6.** Spectrum of Capella taken by the GHRS in highest resolution mode. The features of interest are the absorption lines caused by hydrogen atoms (marked H) and deuterium atoms (marked D) in the interstellar medium between Earth and Capella. Analysis of the spectrum determines the D to H ratio. (J. Linsky, Joint Institute for Laboratory Astrophysics and University of Colorado)

The spectrum for these measurements is shown in Figure 6.6, and the deuterium Lyman-alpha absorption line is clearly visible. A detailed analysis has been carried out by a team of scientists led by Jeffrey Linsky of the University of Colorado and National Institute of Standards and Technology's Joint Institute for Laboratory Astrophysics.

Linsky found the value for the abundance ratio as $D/H = 1.65\ (+0.07, -0.18) \times 10^{-5}$, an amount Linsky characterized as 'the Amazing Trace'. There had been attempts to measure this ratio previously, and this new measurement is close to the average of the previous values. However, the uncertainty in the GHRS value is much less, and it places a tighter constraint on the density of the universe.

To the best of our knowledge, all the deuterium in the universe was created in the Primordial Fireball between 100 and 1000 seconds after the Big Bang. There are no other creation processes in the universe, but there are certainly destruction processes. The nuclear reactions in stars destroy deuterium, and much of the interstellar gas in our galaxy has been through several stars (see the discussion on stellar evolution in Chapter 4). Thus, the measured value needs to be increased to account for the deuterium destroyed in stars, and the estimate for the primordial value is 1.5 to 3.0 times higher. What this means is that a higher value of D/H corresponds to a lower density of the universe.

The information in Figure 6.7 gives the relative abundances of several elements (see the caption for details) as a function of the current density of baryons – a type of particle that absorbs and emits light (like protons and neutrons) – in the universe. This is now a standard result for Big Bang cosmologies. As can be easily seen from these results, the density as estimated from the D/H value is some 10 to 30 times too small to stop the expansion

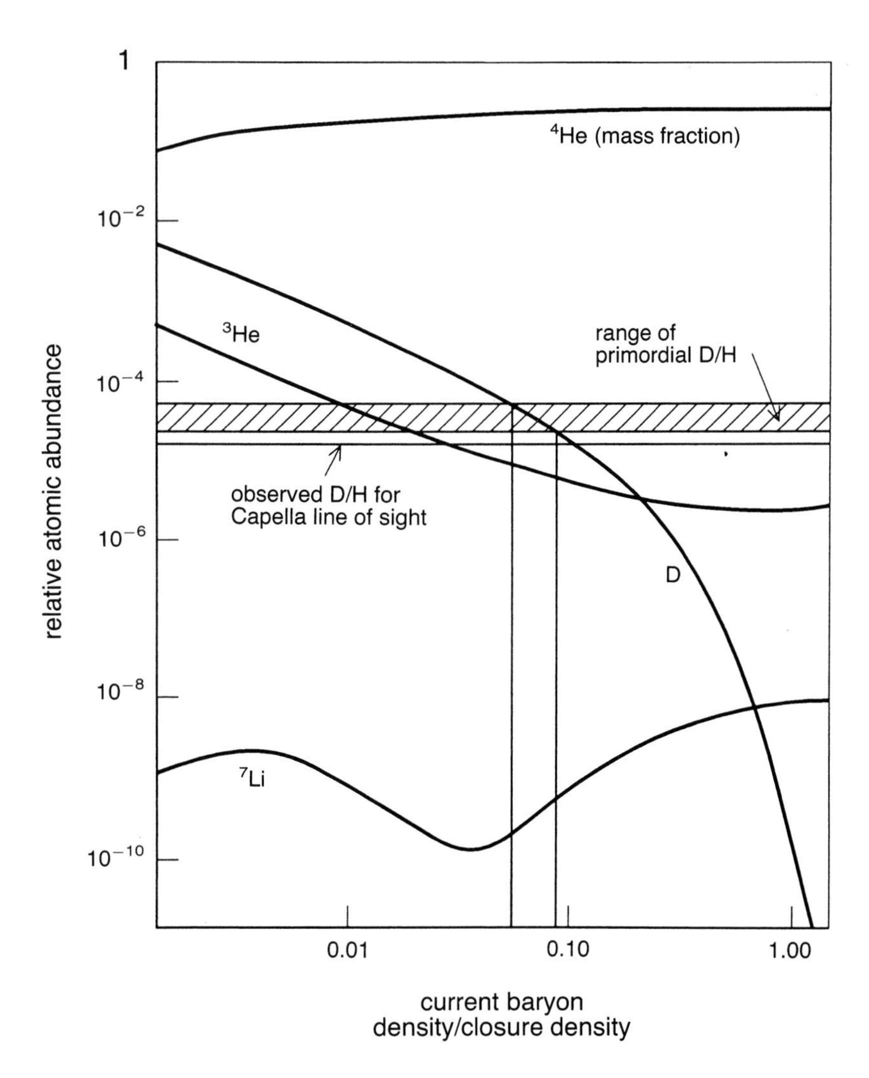


**Figure 6.7.** Abundances of elements – He (helium), D (deuterium) and Li (lithium) – as calculated from a standard Big Bang model. The abundances are plotted against the density of baryons in the universe in units of the closure density ($\Omega_0$). The observed value of D/H is plotted, as is the estimated range for primordial D/H. The highest baryon density possible is about one-tenth of the closure density. (J. Linsky, Joint Institute for Laboratory Astrophysics and University of Colorado)

of the universe. (The detailed value depends somewhat on the value of $H_0$, but the conclusions remain the same.)

This startling result leaves us with two broad possibilities. The first is the straightforward interpretation that the universe had a beginning with the Big Bang, but will simply keep expanding forever. Thus, it would have no ending, no 'Big Crunch'. This result is scientifically acceptable, but many people are uncomfortable with it on philosophical grounds.

The second possibility is that the universe will stop expanding – but if that is true, then things become bizarre. The density estimate cited here refers only to ordinary matter, composed of baryons; this so-called ordinary matter interacts with light. But is there matter that does not interact with light?

Astronomers know from the motions of galaxies in clusters, and from mass estimates of galaxies based on rotation curves, that there is a lot of dark matter in the universe. So, could the rest of the mass of the universe be in the form of exotic, non-baryonic matter that does not emit radiation at any wavelength? Such matter has not yet been detected by nuclear physics experiments. As we discussed earlier, candidates for such matter include heavy neutrinos, axions and WIMPs. The particle physics community is actively searching for such forms of matter.

If the universe is closed, then it is found that the ordinary matter that we are familiar with comprises only 2–10% of the universal total, 90–98% being in an exotic form that is presently unknown. If this scenario is the correct one, it could complete the liberation of the human species from its ancient fascination with the geocentric hypothesis. Not only are we not the center of the universe, but we may not even be made of the dominant form of material in the universe.

If the D/H measurement and the inferred value of $\Omega_0$ stand the test of time, they will have important cosmological consequences. In addition, the D/H measurement is a prime example of a very local measurement with cosmological implications.

## Determination of distances

HST's contribution to solving distance determinations in cosmology comes about in two ways: by measuring stellar parallaxes of stars relatively near the Sun, and by using standard candles to measure very great distances. In these two ways, astronomers take steps toward establishing a more accurate extragalactic distance scale. The Astrometry Team (using the Fine Guidance Sensors) is making very accurate parallax measurements, but, due to the inherent length of time required for such accuracy, they are concentrating on specific parallaxes, as opposed to the Hipparcos satellite, which is doing an automated survey. HST parallaxes with accuracies of 0.002 arcsecond should be obtainable from HST within a few years, whereas ground-based measurements of parallaxes take decades, and are less accurate.

The importance of these parallaxes for stars relatively near the Sun to the extragalactic distance scale is as follows. The scale is built up like a ladder one rung at a time. Distances to nearby stars are used to determine the measuring techniques for stars farther away. Each time distances are established, they are used to calibrate the technique for objects farther away. However, each step rests on the one below it, and an error in the distances to stars near the Sun could throw off the whole progression of distance determinations. Thus, while the astrometric work being carried out with the FGSs may not seem spectacular, it is very important.

The study of distant galaxies is one of several so-called 'Key Projects' designated as top priority on HST. As part of the on-going observations, HST has directly observed Cepheid variables in several distant galaxies. The best-known observations were in the Virgo Cluster of galaxies. Ultimately, HST will study about 50 galaxies, looking for Cepheids to determine the galactic distances. These distances, along with a series of redshift measurements, will allow astronomers to stabilize one of the shakier steps on the cosmological distance ladder and to continue to refine the value of the Hubble Constant.

**Figure 6.8.** WF/PC-2 color composite image of M100 showing the resolution of individual stars in the spiral arms. (J. Trauger, Jet Propulsion Laboratory; NASA)

In late 1994, astronomers announced that HST had measured the distance to Cepheid variables in the Virgo Cluster galaxy M100. Based on these measurements, astronomers determined that the galaxy, which was previously thought to be around 50 million light years away, is actually 56 million light years distant. On the basis of this distance, a new value was calculated for the Hubble Constant $H_0$, and this was used to calculate a new age for the universe: the current estimate of the age of the universe is 12 billion years, based on HST's newly calculated $H_0$ of 80 kilometers per second per megaparsec. This is the predicted age of an empty, low-mass universe that has expanded constantly since the Big Bang. The observations mostly support the notion of a low-density universe; if the density

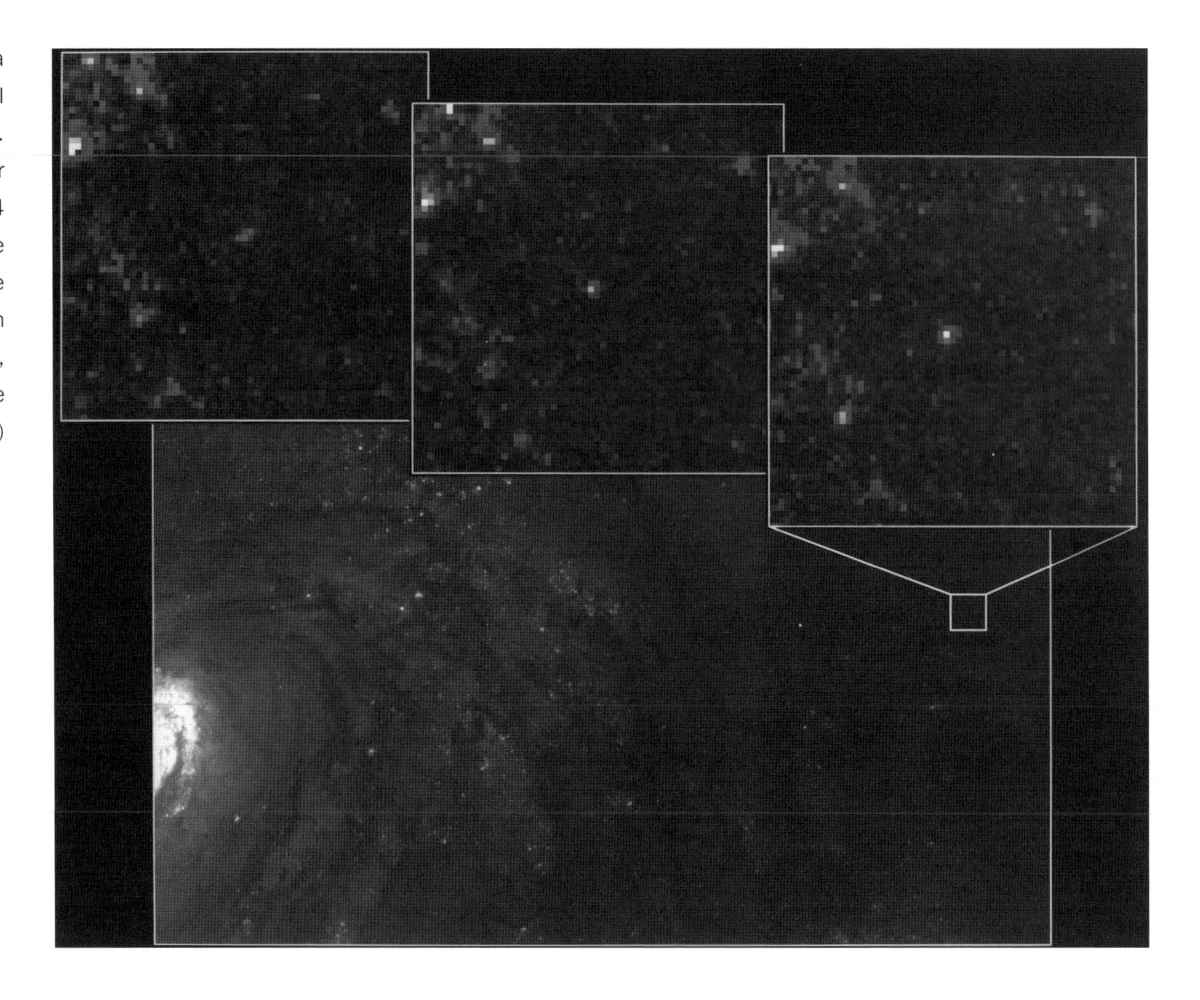

**Figure 6.9.** Identification of a Cepheid variable star in a spiral arm of M100 (bottom panel). Images of the region (upper panels) taken on May 9, May 14 and May 31, 1994, show the star's brightness variation. The images were obtained with WF/PC-2. (Wendy L. Freedman, Observatories of the Carnegie Institution of Washington; NASA)

of the universe were high, then the expansion of the universe must have slowed down. This means that the expansion rate was faster in earlier times. Cosmologists can correct for this change in the expansion rate, which for $\Omega_0=1$ gives a lower limit on the age of the universe of only 8 billion years!

Now, you might notice that 8 billion years is much less than the predicted age of the oldest stars observed in globular clusters and indeed conflicts with what we have all believed for a long time – that the universe is roughly 15 billion years old! Clearly something is not right, and, as one scientist put it, 'it's a wonderful time for cosmologists'. Either the commonly accepted ideas about the age of the universe must be discarded, or perhaps astronomers need to re-examine some of their assumptions about distances to and ages of some of the oldest stars in the universe. However, these numbers do not represent a final calculation by any means. HST will continue to scan M100 and other galaxies in an ongoing effort to pin down a more accurate figure. Finally, the role of dark matter in the universe needs to be defined before we can be sure of any final results in cosmology.

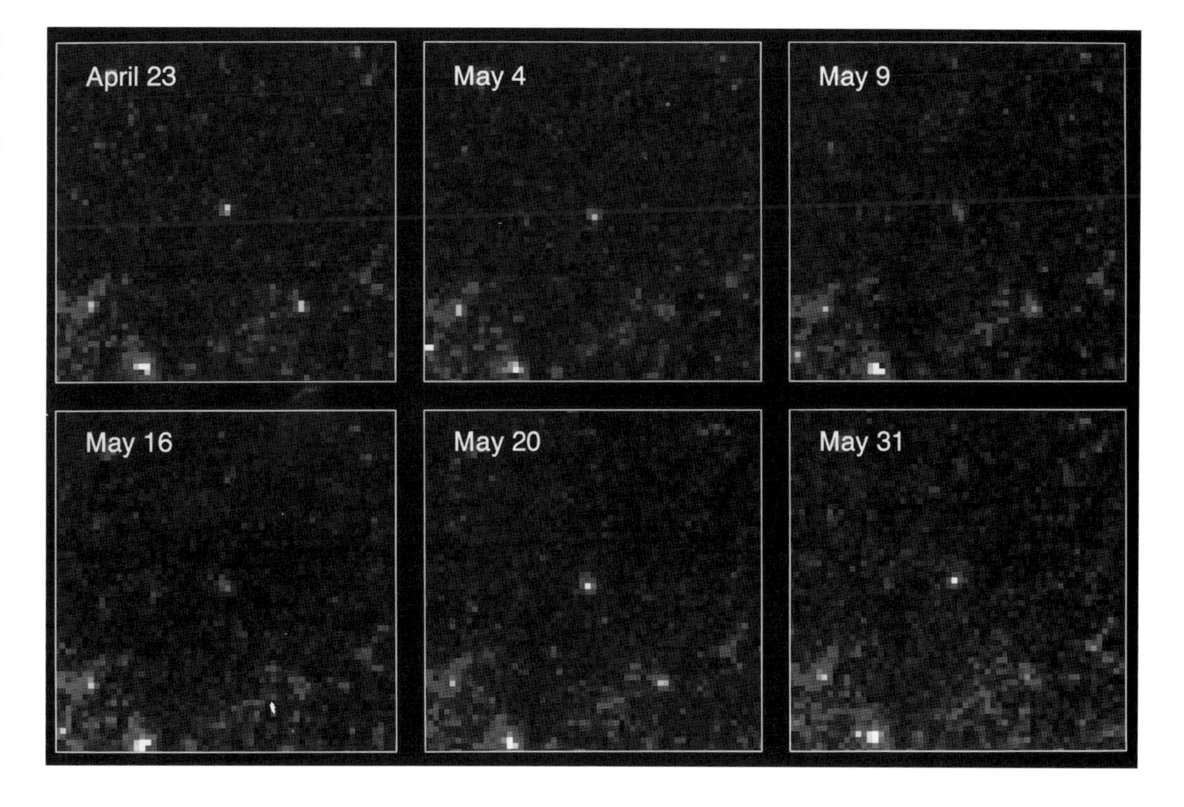


**Figure 6.10.** Detail of the light variation for the same Cepheid as shown in Figure 6.9. Such observations enable the period of variation and, hence, its intrinsic brightness to be determined. (Wendy L. Freedman, Observatories of the Carnegie Institution of Washington; NASA)

## Distant galaxies and clusters

From time to time, we have seen announcements of the discovery of the most distant galaxy known or a new observation of the most distant galaxy. Figure 6.11 is the WF/PC-1 image of 4C41.17, one of the most distant galaxies known. This galaxy appears as it did when the universe was only about one-tenth of its present age, or perhaps 1 to 2 billion years old. The image shows a clumpy structure. According to Leiden University astronomer George Miley, 'The extreme clumpiness of the visible emission suggests that the inner region of this primeval galaxy is highly disturbed.'

The HST image reveals a chain of luminous knots at the core of the galaxy. 'These knots could be giant star clusters forming,' says Miley. 'If that is so, then each would contain about 10 billion stars and be 1500 light years across.'

Miley also proposed an alternative theory stating that these knots are gas or dust clouds caught in a 'searchlight' beam of energy from a massive black hole hidden at the galaxy's core.

Figure 6.12 is a WF/PC-1 image of what may be the most distant cluster of galaxies. The larger objects are in a foreground cluster of galaxies approximately 4 billion light years away. Nearly 30 compact and fainter objects are in a cluster that lies about 7 to 10 billion light years away. Inspection of this image reveals that these galaxies do not resemble the elliptical and spiral galaxies discussed in Chapter 5.

**Figure 6.11.** WF/PC-1 image of 4C41.17, one of the most distant galaxies known. The galaxy is being observed when the universe is about one-tenth of its present age. If the light is from stars, each clump would be about 1500 light years across and contain about 10 billion stars. (George Miley, Leiden University; Kenneth Chambers, University of Hawaii; Wil van Breugel, Lawrence Livermore National Laboratories, University of California; Duccio Macchetto, STScI)

Analysis of the most distant galaxies for cosmological purposes is our best hope of directly determining the deceleration rate $Q_0$. If faint galaxies at high redshifts can be resolved, distance indicators such as a supernova can be observed, and the distance can be determined. If the density is low (corresponding to a value for $\Omega_0$ much lower than unity), the deceleration rate may also be very low. One possible complication regarding galaxies is that stellar evolution may be somewhat different in early galaxies, and it is possible that the distance indicators may be inaccurate. Future observations with HST will concentrate on finding out the differences in stellar evolution by galactic age.

## General relativity and gravitational lenses

The Theory of General Relativity is the basis for calculations of the evolution of a Big Bang universe, and confirmation of the general relativistic picture is always welcome. According to general relativity, the space (strictly speaking, the space-time) near a massive object is distorted by the object. This produces the same basic effect as a gravitational pull, and orbits can be produced. In fact, for planets, general relativity and Newtonian mechanics give essentially the same results.

**Figure 6.12.** WF/PC-1 image of a distant cluster of galaxies, possibly the most distant known. The cluster contains about 30 members, which are faint and compact. The large galaxies in the image are from a separate foreground cluster. The distant cluster is at least 7 billion light years away. (Alan Dressler, Carnegie Institution; NASA Co-Investigators: Augustus Oemler, Yale University, James E. Gunn, Princeton University, and Harvey Butcher, Netherlands Foundation for Research in Astronomy)

This distortion of space has another consequence. The light shining from an object directly behind another, more massive, object can have its direction changed so that it appears to the observer to be shifted from its real position. This is a phenomenon called a 'gravitational lens'. HST is carrying out a number of observations of these lenses, and one on-going program is being led by John Bahcall of the Institute for Advanced Study in Princeton, New Jersey. Bahcall explains, 'HST images provide unprecedented information about the frequency of gravitational lensing by galaxies.' The program is called 'Snapshot Survey' because it uses HST's Planetary Camera to take short exposures during gaps in the telescope's schedule. The resulting images are slightly trailed, but, according to Bahcall, the data are quite valuable for determining characteristics of lensing objects: 'More HST data has been studied in this survey than any other HST science program.'

HST's other gravitational lens programs are returning equally useful data. Figure 6.13 shows the situation schematically for the case of a very distant galaxy seen through a cluster of foreground galaxies which act as the lens. Figure 6.14 shows the HST view alongside the ground-based view of the galaxy cluster AC 114. The L-shaped images (upper left and lower right) are the lensed images of the background galaxy, and they appear larger and brighter than they would without the lensing. The two objects in the center of the image are physically unrelated galaxies, probably in the foreground cluster.

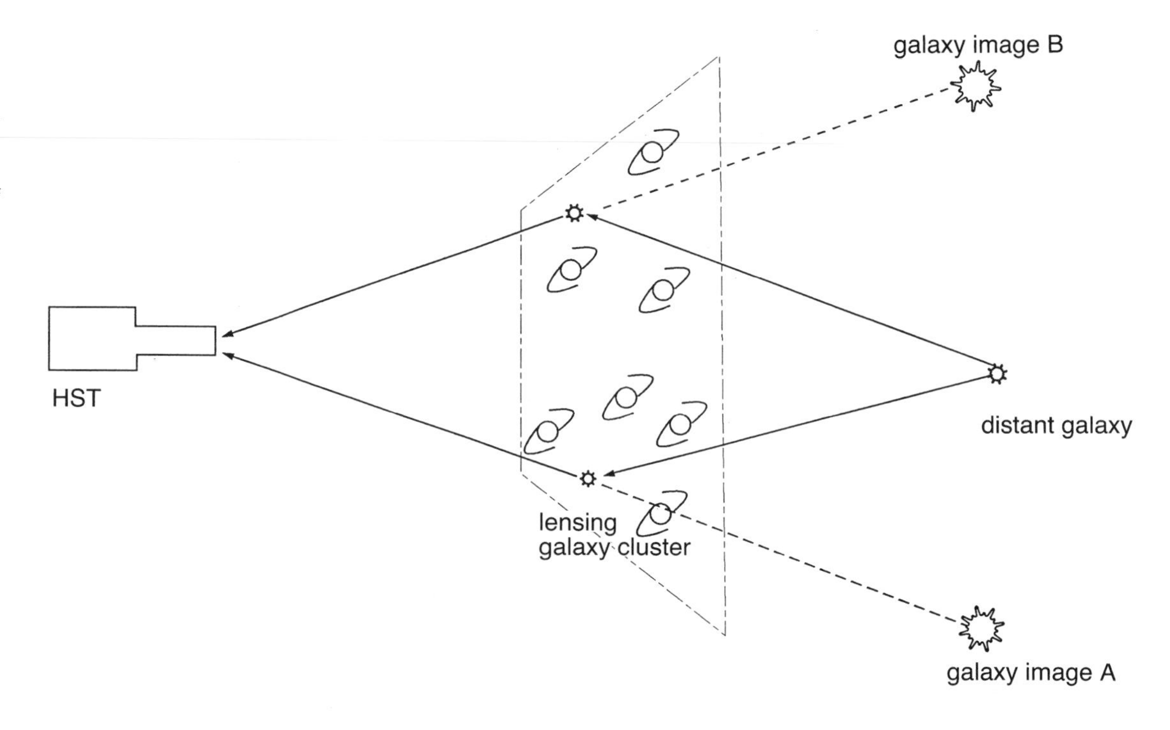


**Figure 6.13.** Schematic diagram illustrating the gravitational lens effect. Light from a distant galaxy can appear as multiple images due to the lensing effect of foreground galaxies. (Richard Ellis, Durham University; NASA)

The discovery of different gravitational lenses is exciting, and studying them offers the potential for very important cosmological results. A light variation (for example, the explosion of a supernova) in the background galaxy, or a change in quasar brightness, will be seen at Earth at different times for different gravitational-lens-produced images. If the geometrical circumstances are known, the time delay gives an independent value for the Hubble Constant, $H_o$.

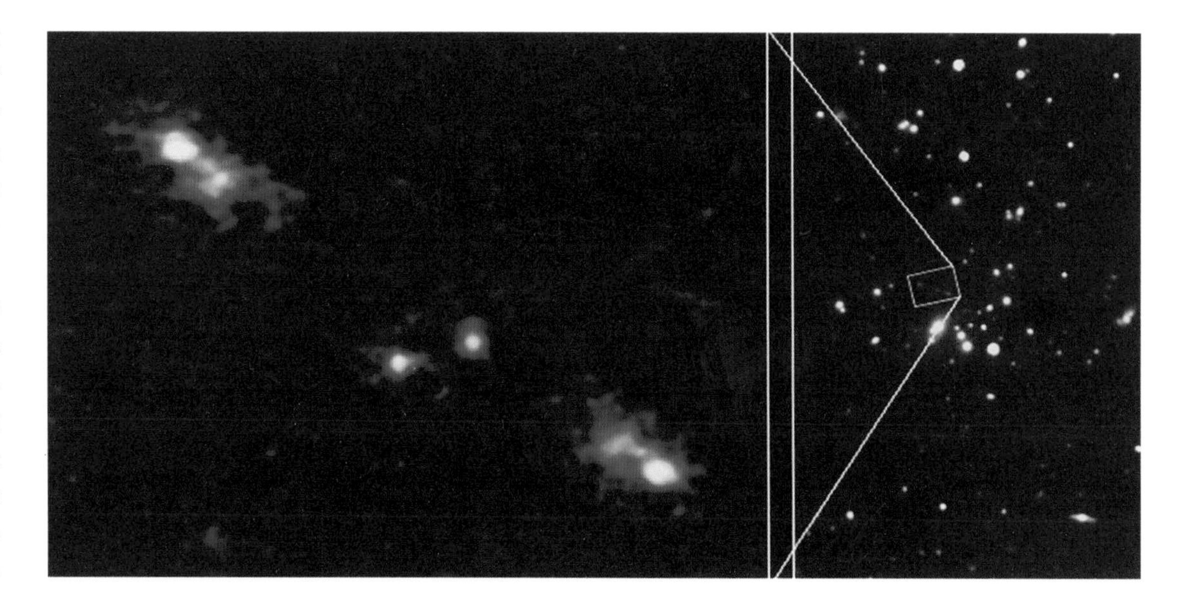

**Figure 6.14.** WF/PC-1 (left) and ground-based (right) images of gravitational lensing in galaxy cluster AC 114; see the text for a discussion. (Richard Ellis, Durham University; NASA)

**Figure 6.15.** (facing page) FOC image of gravitational lens G2237+0305, the 'Einstein Cross'. The lensing has produced four images of a very distant quasar. The central object is the intervening (lens producing) galaxy. (ESA; NASA)

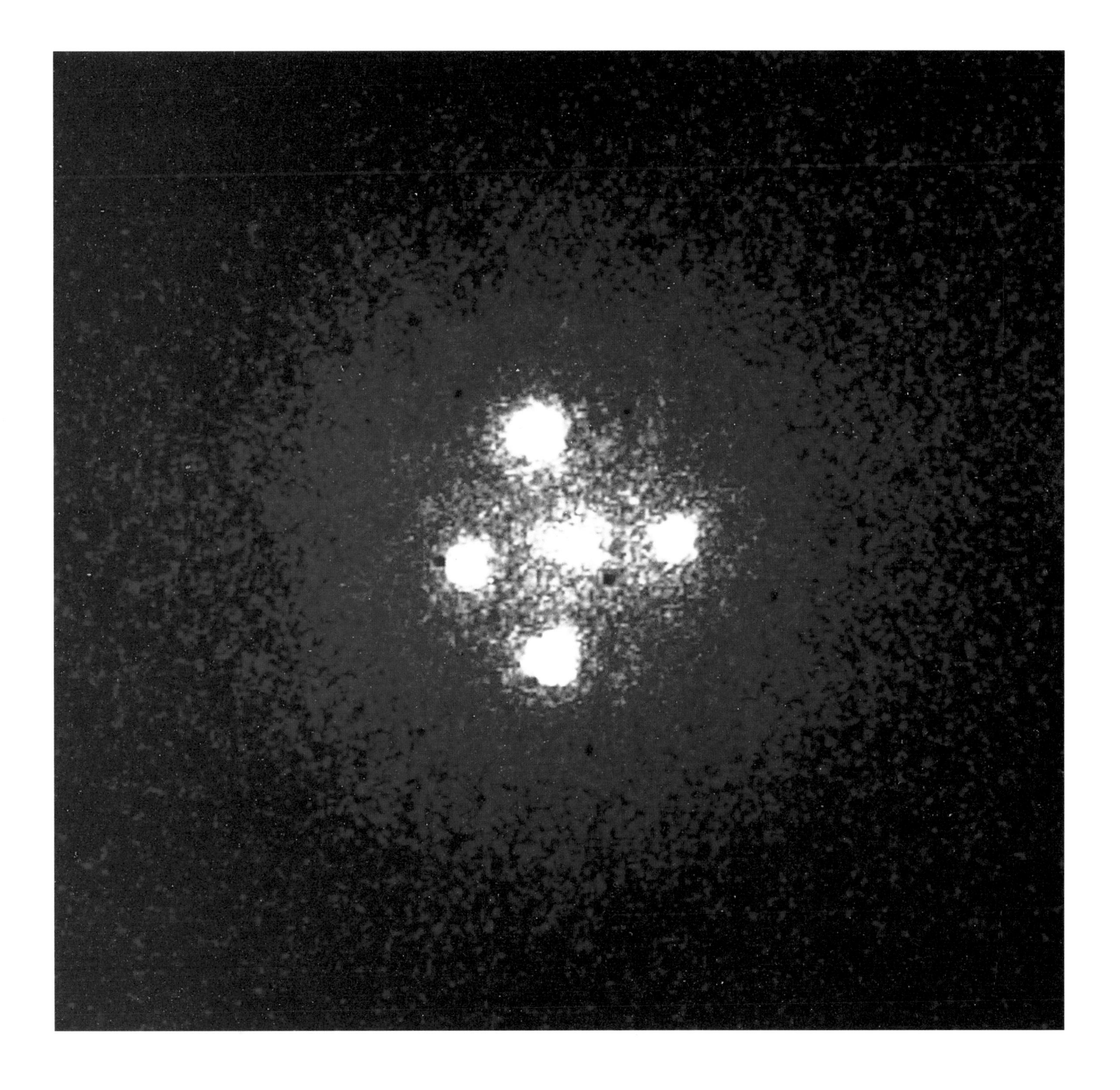

**Figure 6.16.** WF/PC-2 image of the rich galaxy cluster Abell 2218, which contains several examples of gravitational lensing. The arc-like pattern spread across the image like a spider's web is an illusion caused by the gravitational field of the cluster. (W. Couch, University of New South Wales; R. Ellis, Cambridge University; NASA)

Another example of gravitational lensing is observed in the object G2237+0305, also known as the 'Einstein Cross', shown in Figure 6.15. The angular separation between the top and bottom images is 1.6 arcseconds. The outer four images come from a distant quasar, whereas the central image comes from a galaxy about 20 times closer to us. Several images can result from a lensing situation, depending on the precise details of the alignment. The HST image is the most detailed taken of this source.

In another gravitational lens imaged by HST, the telescope's high resolution revealed a collection of ghostly looking arcs. These arcs were created by the distortion of light by the galaxy cluster Abell 2218. The cluster is so massive and compact that light rays from a very distant population of galaxies are bent to form the arcs of light shining around several of the cluster members. That distant set of galaxies, which lies directly behind the lensing cluster, existed when the universe was just one-quarter of its present age. Since we see these galaxies at a very young age, the arcs can be used to look at star forming regions in the distant galaxies. Figure 6.16 shows multiple imaging, a phenomenon that appears when the distortion is large enough to produce more than one image of the same galaxy. In this case, Abell 2218 shows seven of these multiple events. Besides studying the larger arcs, astronomers are also interested in a collection of 'arclets' that also appear in the image. Distances will be calculated for 120 of the arclets, which represent galaxies 50 times fainter than can be seen using ground-based instruments. As with the other lenses it is studying, HST's view of this very rich field of lensing features is providing astronomers with rare insights into the distribution of matter in the lensing cluster's center. Studies of the remote galaxies viewed *through* lenses like Abell 2218 promise to reveal a great deal about the evolution of normal galaxies, particularly in very earliest epochs of their existence.

## Where are we now?

We opined earlier that cosmology and general relativity are two of the most profound types of science engendered by a study of astronomy. It is important to remember that cosmology was one of the main drivers for HST back when the telescope was still a twinkle in Lyman Spitzer's eye. Scientifically, we end up back at the beginning of HST's life story. If HST did not do anything else except help solve the cosmological distance problem, it would still be counted as one of the great successes of astronomy.

As we have seen throughout this chapter, HST results have spiced up the stew that is our understanding of the universe. The refined value for the deuterium-to-hydrogen ratio, which limits the value of $\Omega_0$ for baryons in the universe to approximately 0.05, is one result with interesting implications. The value of the Hubble constant, $H_0$, of 80 kilometers per second per megaparsec, which implies an age for the universe of 8 to 12 billion years, depends on a more accurate value of $\Omega_0$ for all matter.

A reasonable, but by no means definitive or final, summary of the current situation is that we are in a low-density universe with an $\Omega_0$ value of about 0.25. The low density implies that the universe will expand forever and that the deceleration parameter ($Q_0$) is small. If true, the Hubble expansion has not been significantly slowed since the Big Bang, and hence the Hubble Constant of 80 kilometers per second per megaparsec yields an age for the universe of about 12 billion years. (This still leaves the troublesome problem of the 13- to 15-billion-year-old stars in globular clusters. However, we should not accept that figure as being etched in stone because there is some uncertainty in the ages of these stars, and indeed, in the very age of the universe itself!)

To affect the expansion of the universe, then, we need a denser universe. A large amount of exotic particles would need to exist to bring the value of $\Omega_0$ up to 1 (the critical value needed to slow the expansion of the universe), but even an $\Omega_0$ value of 0.25 requires a major contribution from exotic particles.

A major unresolved problem is how structures such as galaxies, clusters of galaxies, etc., were formed in the times available and with the density fluctuations expected. Until all the pieces of the puzzle fit together, our picture is not complete.

For now, until the HST can turn its gaze even further back in time, we are constrained to live in a universe that is expanding forever. It is a younger universe than we thought, and some of our ideas about the ages of the oldest globular cluster stars need to be re-examined. Even our studies of the Sun, a star we know better than any other in the universe, reflect a great lack of knowledge. Based on our theories of stellar evolution, we should be seeing many more neutrinos from the Sun – yet we see so few (about one-third the number expected) that we are questioning our concepts about the Sun's energy source, and our knowledge of elementary particle physics. If we can be this imprecise about the Sun, then a 2 to 3 billion year discrepancy in the ages of certain stars in the universe is understandable. What it points out is that we do not know everything.

Furthermore, an unquestioning acceptance of ideas like 13- to 15-billion-year-old stars in a 10-billion-year-old universe is a dangerous thing. Not only is the universe that HST has shown us not cooperating with the theories we have developed to explain it – it is even stranger than we ever dared imagine. If HST never accomplishes anything more, this will be its legacy – and something astronomers everywhere can look to as a source of extraordinary, exciting and dynamic discoveries.

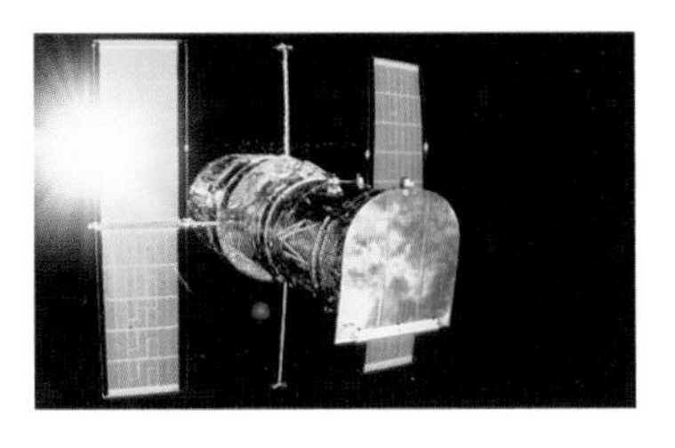

# Epilog: the future of HST

These are heady and bittersweet times for the Hubble Space Telescope. As we have shown in our scientific discussions, the machine works and it is doing cutting-edge science. The fact that HST is often doing science that could not be carried out any other way does not abate the continuing political and financial pressures on the project. As with many other Big Science projects, however, there is good news and there is bad news.

First, the bad news. The HST budget continues to come under attack. Given that it took a major expenditure of resources to carry out the First Servicing Mission, such attempted cutbacks may seem shortsighted simply because once the money was spent to fix the telescope, it seems unwise to turn around and hamstring its operation. The annual cost of HST is currently about $US235 million – and consequently draws the attention of governmental budget cutters.

Budgetary concerns certainly affected the handling of the 'Announcement of Opportunity' (AO) for future HST science instruments. The original AO was issued on March 21, 1994, and it solicited proposals for an advanced camera and for low-cost instruments (less than $US10 million). The AO explicitly noted '... the HST is severely cost constrained...'. Anyone applying to build the instruments was on notice that the 'bottom line' of their projects was a major consideration.

On March 31, 1994, the AO was 'postponed' to allow NASA to carry out a re-evaluation of the strategy for servicing over the projected 15-year lifetime of HST. The AO was reopened on September 1, 1994, but only for an advanced camera that could be built for a cost in the range $US25 to $US40 million. This level of budget is still very tight; however, the competition has been completed, and the result is the Advanced Camera (described below).

This sort of mind-changing brings up concerns that can be expressed as follows. Scientific instruments on HST can fail, and replacements need to be ready and waiting, or at least in the pipeline. If the failure rate should exceed the replacement rate, the number of productive scientific instruments on the telescope could fall to a level that would re-open an examination of the large HST annual budget. There is no magic number of instruments that is appropriate. With the replacement of HSP by COSTAR, we have already gone from five to four, and HST is still highly productive. A space telescope with three productive instruments could still be defensible, but fewer instruments than that raises some very unpleasant concerns. In a climate driven by a mania for budget cutting (in the USA), and given the current, ubiquitous disarray in NASA, the HST program would be in jeopardy.

Now for the good news. HST is the responsibility of the Astrophysics Division of NASA, and a 'Senior Review' panel recently reviewed all of the Division's operating and soon to be operating missions. HST was ranked as the number one mission by a wide margin. Specifically, the panel report states:

> We wish to emphasize that the Hubble Space Telescope was viewed, nearly unanimously, as the most valuable NASA Astrophysics mission over the next 5 years, even when normalized by the cost... HST is our flagship and its success benefits all astronomy.

Despite the gloom and doom painted in the bad news above, some of the needed replacement instruments *are* in the pipeline. HST's next servicing mission is set for February, 1997. At that time, according to the latest plan, astronaut crews will make any needed repairs, take out the Goddard High Resolution Spectrograph and the Faint Object Spectrograph and replace them with two so-called 'second-generation instruments' – the Near-Infrared Camera and Multi-Object Spectrograph (NICMOS) and the Space Telescope Imaging Spectrograph (STIS).

The two new instruments will contain optics to correct for the spherical aberration. Of course, if the Faint Object Camera should fail, the instruments removed could change. There are also concerns that the current change-out plans allow no backup for spectroscopy.

NICMOS – now being built at Ball Aerospace under the guidance of the University of Arizona's Rodger Thompson (Principal Investigator for the NICMOS Team) – will use a cryogen-cooled detector to observe the near-infrared region of the electromagnetic spectrum, i.e. those wavelengths slightly too long for humans to be able to see, out to 2.5 microns. This is another region where ground-based telescopes have had difficulty, so upgrading the HST to observe in these wavelengths will give astronomers even more detailed looks at planets, certain types of stars and galactic features, and galaxies that have so far eluded them.

STIS is also being built at Ball Aerospace, under the direction of Goddard Space Flight Center scientist Bruce Woodgate. Its wavelength range goes beyond that of both the spectrographs it replaces – and it will see ultraviolet, visible and some near-infrared wavelengths. It incorporates state-of-the-art CCDs and other detector technology, and takes advantage of advances in software and electronics. It will give a two-dimensional data format and will take simultaneous spectral measurements – of a galaxy, for example – at several different locations in the galaxy. This is an extremely demanding task, something not achievable with today's instruments. Finally, as with the instrument array in the current HST, the 1997 HST's science instruments will complement each other by sharing some wavelength ranges.

On December 21, 1994, NASA selected the proposal of a team led by Holland Ford (of Johns Hopkins University and the Space Telescope Science Institute) to build the Advanced Camera (AC). The instrument will be fabricated by Ball Aerospace and is scheduled for a November 1999 launch. The AC is expected to incorporate the very latest advances in digital technology, imaging and electronics, and is intended to replace the Faint Object Camera. The camera is planned to sample the light over a wide field at full resolution, and, when it is installed, it could extend the science return from HST until well beyond the end of the telescope's planned lifetime. The AC has several key science programs, including the study of planetary atmospheres, the evolution of galaxies and clusters of galaxies, the large-scale distribution of dark matter and other topics related to the early universe.

NASA would like to have one more replacement instrument ready for launch in 2002. To keep costs to the USA down, the instrument might come from another source, such as ESA. The scientific emphasis for this instrument is undecided at this time.

We have described the current status of HST and the prognosis for the future. The best news of all lies in the previous four chapters of this book: the telescope works, and it is turning out great science. Of course, there are discouraging aspects, and some commentators have been harshly and often unfairly critical of the project and the people involved with it. However, the problems and the criticisms pale in the face of what HST and its users have accomplished. As Sandy Faber has so eloquently noted, 'HST is just as complex as the many people who made it. Like them, it has the potential to be heroic. It is proving that now.'

In closing, it may sound hackneyed and trite, but in its first years on orbit, HST really has taken us to the 'final frontier' in astronomy. Our sincere hope is that this magnificent machine called the Hubble Space Telescope – and its users – will continue to surprise us all with major scientific discoveries as it carries out its full 15-year mission.

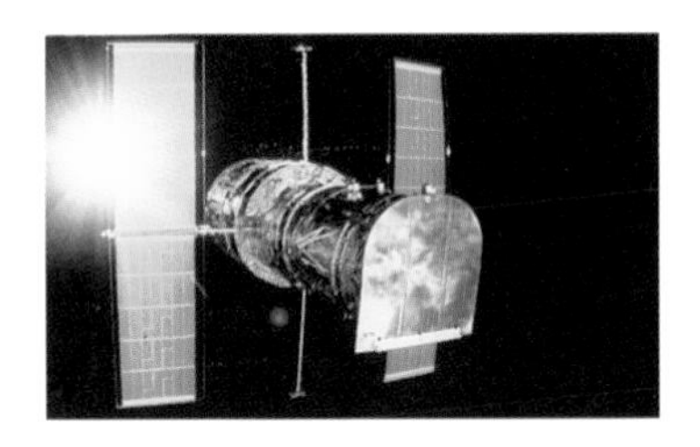

# Glossary

In addition to the usual, short entries, this section also includes extended discussions of physical concepts helpful in understanding some of the research reported. Italics denotes other entries in the glossary.

*absolute temperature scale* In the physical sciences, temperatures are related to the motions of atoms and molecules in gases. The temperature scale used is the Kelvin scale, which has the same size degrees as the centigrade scale. On the Kelvin scale, the freezing point of water is 273 kelvin, the boiling point of water is 373 kelvin, and the surface of the Sun (*effective temperature*) is 5750 kelvin. Most temperatures in this book are given in kelvin.

*absolute zero* A hypothetical concept to describe the temperature, 0 kelvin, at which the motions of atoms and molecules would stop.

*absorption line* When a continuous spectrum is viewed through a cooler, low-pressure gas, such as the interstellar gas, dark lines called absorption lines appear. The specific lines depend on the composition of the gas that is absorbing the light and causing the lines

*active galaxy* A galaxy with a bright central region and often with enhanced X-ray, ultraviolet and radio emission. Active galaxies include the Seyfert galaxies and *quasars*. The activity probably results from a massive, central *black hole*.

*angles* Angular measure is frequently used in astronomy to specify positions on the sky and to specify apparent size. The units are degrees, minutes and seconds. An entire circle contains 360 degrees; a right angle contains 90 degrees. Each degree contains 60 minutes and, in turn, each minute contains 60 seconds. Note that the minutes and seconds described here are not the same as minutes and seconds of time.

*angstrom* A unit of length equal to $10^{-10}$ meters.

*arcsecond* A commonly used, very small, angular measure in astronomy. Roughly speaking, it is one part in 200 000. To illustrate the size, a US quarter coin is about 2.5 centimeters across. If placed at a distance of 200 000 quarters or about 500 000 centimeters (5 kilometers), it subtends 1 arcsecond. If we are talking about 0.1 arcsecond, it is about the angular size of a quarter 50 kilometers distant.

*asteroids* Sometimes called 'minor planets', these are small, rocky bodies orbiting the Sun primarily between the orbits of Mars and Jupiter, the region of the 'asteroid belt'. Diameters range from the smallest detectable (at a few tenths of a kilometer) up to the largest asteroid, Ceres, at 950 kilometers.

*astrometry* The measurement of precise positions and motions, usually of stars.

*astronomical unit (AU)* The average distance between the Earth and the Sun, i.e. about 150 million kilometers.

*atomic number* The number of protons in the nucleus of an atom (and hence the number of electrons in a normal, unionized, atom).

*atomic weight* Approximately the number of protons and neutrons. For example, a helium atom, with two protons and two neutrons, has an atomic weight of 4.003.

*baryons* A class of elementary particles that includes the *proton* and the *neutron*.

*Big Bang* The postulated beginning of our universe when all the matter and radiation emerged from a point.

*Big Crunch* If the universe should stop expanding and fall back on itself, it eventually could concentrate all the matter and radiation back into a point. This event is called the Big Crunch.

*binary stars* Two stars in orbit around each other. If the separation can be resolved in the telescope, they are called visual binaries. If the double nature is revealed by motions detectable via the *Doppler effect*, they are called spectroscopic binaries. Most stars are believed to be members of binary or multiple systems.

*black body* An ideal body which absorbs all radiation incident on it. The emission from a black body depends only on its temperature.

*black hole* When massive stars collapse at the end of their lives, their mass is concentrated at a single point. The effective size of the star is called the *event horizon*, the boundary of a region with a gravitational field so strong that nothing can escape from it, even light. Hence the name 'black hole'. The lower limit for the mass of a stellar black hole is about 3 solar masses. Of course, this is the mass of the final core, not the original mass of the star on the *main sequence*, which can be much larger. Black holes with other origins are possible. So-called 'mini black holes' may have formed early in the universe, and 'massive' black holes, with masses of millions to billions of solar masses, are probably located at the centers of galaxies.

*blast wave* A traveling *shock* produced by an explosion.

*carbon cycle* Nuclear reactions that convert four hydrogen nuclei into one helium nucleus with a release of energy using carbon nuclei as a catalyst. The carbon cycle occurs at higher temperatures than the *proton–proton chain.*

*catalog numbers* Many astronomical objects in this book are designated by their catalog number. For example, M31 means the 31st object in Messier's catalog, and NGC 188 simply means the 188th object in the New General Catalog. Others found in this book are Arp numbers, CL numbers, Henize numbers, Markarian numbers, Melnick numbers, PKS numbers and 3C numbers. These are from specific catalogs or lists, often named after the astronomer who compiled them. While we give the numbers for completeness, they are not essential for understanding their role in HST science.

*Cepheid variables* Very luminous, pulsating stars that are important distance indicators. Their brightness pulsates regularly over a length of time called a period, and their *intrinsic brightness* can be determined from the period of light variation.

*cluster parallax* A distance determination method applicable to clusters with measurable motions toward a convergent point (where all the motions would intersect on the sky) and radial velocities determined by the *Doppler effect*. The distance to the Hyades *Open Cluster* has been determined using this method.

*clusters of galaxies* Distinct groups of galaxies containing from a few to thousands of galaxies.

*clusters of stars* See *globular clusters* and *open clusters.*

*comets* Solar System bodies consisting of an icy nucleus. If the comet approaches the Sun, the nucleus is heated and the cometary ices sublimate to form a cloud of gas and dust called a coma. Sometimes the gas and dust will stream out away from the comet, pushed by the solar wind (for the gas) or the Sun's *radiation pressure* (for the dust) into a long tail.

*constellation* One of the arbitrary groupings of stars in the sky, of which there are more than 80, which people imagine to look like objects, such as Orion the hunter in the winter sky (northern hemisphere). The brightest stars in each constellation are designated in order of brightness by a Greek letter in alphabetical order; for example, Alpha Orionis, Beta Orionis, etc.

*continuous spectrum* An incandescent gas under high pressure (such as the sub-surface layers of the Sun) emits a continuous spectrum, i.e. emission at all wavelengths over a wide range of the electromagnetic spectrum.

*cosmic abundances* A standard tabulation of the relative abundances of the elements in the universe compiled from solar, meteorite, stellar, nebular and earth-crust data.

*cosmic background radiation* The nearly isotropic surviving radiation from the *Primordial Fireball*. It has a *black body* temperature of 2.73 kelvin.

*cosmology* The study of the entire universe considered on a very large scale; this includes the origin, structure and evolution of the universe.

*critical density and $\Omega_0$.* The density of the universe that would just stop the expansion of the universe after a very long time is called $\rho_0$. For a *Hubble constant* of 80 kilometers per second per megaparsec, this amounts to $1.2 \times 10^{-29}$ grams per cubic centimeter, a very small number. If this were all hydrogen atoms, their density would be $0.7 \times 10^{-5}$ atoms per cubic centimeter. A more convenient quantity is $\Omega_0$, which is the ratio of the actual density to the critical density. For an actual density equal to the critical density, $\Omega_0 = 1$. For $\Omega_0 > 1$, the expansion stops and becomes a contraction, ultimately leading to the *Big Crunch*. For $\Omega_0 < 1$, the universe expands forever.

*dark matter* Unobserved, and hence 'dark', matter in the universe. This matter could be of a straightforward nature, such as very subluminous stars, and/or it could be exotic subatomic particles. The evidence for large amounts of dark matter in the universe is extensive.

*deceleration parameter, $Q_0$* A measure of the rate at which the expansion of the universe may be slowing down because of the gravitational attraction of its own mass.

*deconvolution* The sharpening of a degraded image (or spectrum), usually by computer processing. If the degradation can be accurately determined, for example by determining the degraded appearance of a star which should be nearly a point source, it can be substantially removed.

*deuterium* A heavy hydrogen atom containing one proton and one neutron in its nucleus.

*Doppler effect* A change in the wavelength of light caused by a relative motion of a source and its observer. This change in wavelength can be understood by thinking of light as a wave. If the source approaches the observer, more waves per second are seen, the frequency is therefore raised, and the wavelength is shortened; this is referred to as a blueshift – a shift towards the 'blue' end of the spectrum. When the source is moving away from the observer, the opposite occurs; this is referred to as a redshift. We hear the same effect in the sound of a siren; the pitch (frequency) is higher (corresponding to a shorter wavelength) when the vehicle approaches us, and the pitch is lower (corresponding to a longer wavelength) when the vehicle is going away from us, so, as it passes, a distinct change in the pitch of the siren is noticed.

*effective temperature* The temperature that an ideal body, called a *black body*, would have if it were to emit the same amount of energy per unit area. The term is usually applied to stars; for example, the Sun's effective temperature is about 5750 kelvin.

*electromagnetic spectrum (EMS)* The entire spectrum of all forms of electromagnetic radiation, including light, from gamma rays to radio waves.

*electron* The light, negatively charged particle in the atom.

*emission line* Incandescent gases at low pressure, such as a gaseous nebula, produce a spectrum composed of individual bright lines called emission lines. The concept of a 'line' means that the emission occurs at a specific wavelength, or over a narrow range of wavelengths, a situation in contrast to the *continuous spectrum*. The specific lines depend on the composition of the gas. See also *absorption line*.

*event horizon* The surface, or effective boundary, of a *black hole*. No material or light can escape from within the event horizon. The radius is approximately 3 kilometers times the mass of the black hole (given in solar units).

*fission* The break up of atomic nuclei into lighter nuclei.

*fusion* The merging of atomic nuclei into heavier nuclei.

*galaxy* One of the very large celestial objects consisting of stars (up to roughly a trillion, $10^{12}$) and often vast quantities of dust and gas. The Sun is in the Milky Way Galaxy. External galaxies are often described by their appearance, such as spirals, barred spirals, ellipticals and irregulars.

*general relativity* Albert Einstein's generalization of *Newtonian mechanics*, which is used in cosmological applications.

*giant* A star roughly 100 times the Sun's intrinsic brightness with a radius of roughly 100 times the Sun's radius.

*globular clusters* Tightly packed, spherical groupings of old stars in the *Milky Way Galaxy*. Globular clusters are also observed in other galaxies.

*gravitational lens* The general relativistic effect whereby the enormous mass of celestial objects changes the path of light passing by. The effect is to focus the light and thus, in essence, is a lens.

*gravity* The attraction of all bodies in the universe for all other bodies. Two bodies attract each other with a force proportional to the product of their masses and inversely proportional to the square of the distance between them. For example, the solar gravitational force acting on the Earth would be one-quarter the present value if the Earth were twice as far from the Sun as it is now.

*Helmholtz–Kelvin contraction* The stellar contraction that produces heating of the interior. The process is important in stellar evolution prior to the *main sequence* and in later stages of stellar evolution when nuclear sources of energy are exhausted.

*H–R or Hertzsprung–Russell diagram* A display of stellar properties using a plot of *effective temperature* (or surrogates such as color or spectral type) versus *luminosity* (or surrogates such as *intrinsic brightness*).

*Hubble constant, $H_0$* The constant of proportionality between the recession speeds of galaxies and their distances from each other. Current estimated values range between 50 and 100 kilometers per second per *megaparsec*.

*inflation* A hypothetical period of extraordinarily rapid expansion early in the universe, from roughly $10^{-34}$ to $10^{-30}$ seconds after the *Big Bang*. This expansion may be necessary to explain properties of the universe, such as the existence of galaxies.

*interstellar medium* The gas and dust located between the stars in the Milky Way Galaxy.

*intrinsic brightness* The brightness of an object, such as a star, that is independent of distance. This brightness can either refer to light in a specific color or to all the light, when it is the same as the *luminosity*.

*ion* An atom or molecule that has lost one or more *electrons* and thus has an electrical charge.

*isotope* Atoms of the same chemical element, but with different numbers of neutrons in the nucleus.

*kiloparsec* 1000 *parsecs* or $3.1 \times 10^{16}$ kilometers.

*Kirchhoff's laws* Experimentally determined ideas based on the absorption and emission of light. See *absorption line*, *continuous spectrum* and *emission line*.

*light-travel time* The time it takes for light, traveling at about 300 000 kilometers per second, to travel a certain distance.

*light year* The distance that light travels in one year at about 300 000 kilometers per second, i.e. $9.5 \times 10^{12}$ kilometers.

*look-back time* Because light travels through space at a constant speed (300 000 kilometers per second), it takes a finite time to travel from distant objects. Hence we 'see' distant objects at a time in their past. See *light-travel time*.

*luminosity* The total light output of a star over all wavelengths. Because this quantity is independent of distance, it is an *intrinsic brightness*.

*Lyman-alpha* The principal light-absorbing or -emitting energy transition of the hydrogen atom occurring at about 1216 *angstroms*.

*Magellanic Clouds* Two irregular galaxies, the Large Magellanic Cloud (LMC) and the Small Magellanic Cloud (SMC), easily visible to the unaided eye in the southern hemisphere. They are named after explorer Ferdinand Magellan.

*magnitudes* Stellar brightness is often described in terms of a historical system that seems confusing. The magnitude system is somewhat like standings in sports leagues where the best (and brightest) are first, the lesser (or fainter) are second, third, etc. Historically, the brightest stars (collectively) in the sky are 1st magnitude and the faintest stars visible to the naked eye are 6th magnitude. The modern definition is that stars differing by a factor of 100 in brightness differ by 5 magnitudes. This works out to a factor of 2.512 in brightness for 1 magnitude, and the fainter stars have the larger magnitude. The magnitudes as observed for stars on the sky are called apparent magnitudes. If the magnitude is referred to the standard distance of 10 parsecs (and thus is indicative of the star's *intrinsic brightness*), it is called an absolute magnitude. Finally, the magnitude can refer to all the light from a star or to the brightness of light measured through a specific filter, such as a blue filter.

*main sequence* The principal band of stars on the *H–R diagram*. Most stars appear on the main sequence after nuclear burning of hydrogen has begun, and they spend most of their lives there.

*megaparsec* One million *parsecs* or $3.1 \times 10^{19}$ kilometers.

*Milky Way Galaxy* Our own galaxy, consisting of about 100 billion stars plus gas and dust. The galaxy is disk-shaped; when we see the Milky Way in the sky, we are looking at the disk edge on.

*nebula* A cloud of gas and dust in a galaxy, such as the Orion Nebula. Before the early part of the 20th century, this term was used to describe any hazy patch in the sky.

*neutrino* A stable particle with no charge and no mass that is produced in nuclear reactions. The neutrino interacts very weakly with all other particles.

*neutron* The heavy, electrically neutral elementary particle in the nucleus of an atom.

*neutron star* A very dense star composed of *neutrons* and having a diameter of about 30 kilometers. The pressure in the star is so great that *electrons*, which usually orbit the nucleus of atoms, are pressed into the nucleus where they merge with the protons to form neutrons. The mass of neutron stars falls between 1.4 and 3.0 solar masses. Smaller stars become *white dwarfs*, while larger stars become *black holes*. Neutron stars are believed to result from *supernovae*. *Pulsars* are believed to be rapidly rotating neutron stars.

*Newtonian mechanics* Theory (or equations) for calculating the positions of celestial bodies such as planets and stars. This theory is based on Isaac Newton's laws of motion and gravity.

*open cluster* A stable grouping of young stars in the *Milky Way Galaxy*. Other open clusters present in other galaxies are good indicators of spiral arms.

*parallax* The apparent shift in position of a nearby object projected on a background when viewed from different positions. Simply holding a finger at arm's length and viewing it with only one eye and then the other illustrates the effect. In astronomy, the parallax is the shift in position of a star when viewed from positions separated by 1 *astronomical unit*. For nearby stars, an accurate parallax fixes their distances.

*parsec* The distance, $3.1 \times 10^{13}$ kilometers, at which a star has an astronomical parallax of 1 arcsecond.

*period* The length of time it takes for a body to go from a starting position or condition and return to that position or condition. Period can be used to describe such actions as the rotation of a planet on its axis or its orbit around a star, and the length of time it takes a star to vary from maximum to minimum brightness and back to maximum.

*planetary nebula* A *nebula* which resembled a planetary disk in early telescopes. Now known to be an expanding gas cloud surrounding a hot star.

*planets* The major Solar System bodies in orbit around the Sun. In order of increasing distance from the Sun, they are: Mercury, Venus, Earth, Mars, Jupiter, Saturn, Uranus, Neptune and Pluto.

*populations* A concept used to describe different kinds of stars, which are usually distinguished by age and abundance of heavy elements. Population I stars are young and have a high abundance of heavy elements, while Population II stars are old and have a low abundance of heavy elements.

*Primordial Fireball* The state of the universe that existed immediately after the *Big Bang*, consisting of energetic elementary particles and radiation.

*proton* The heavy, positively charged elementary particle in the nucleus of the atom.

*proton–proton chain* Nuclear reactions that convert four hydrogen nuclei into one helium nucleus with a release of energy. This chain occurs at lower temperatures than the *carbon cycle* and is the principal energy source for the Sun.

*protoplanets* Planets at an early stage of formation from clumps of gas and dust.

*protostars* Stars at an early stage of formation from interstellar clouds of gas and dust.

*pulsar* A rapidly rotating *neutron star* which emits characteristic pulses of radio radiation and visible light.

*quantum mechanics* The physical laws that describe the motions of electrons in atoms. These laws are radically different from *Newtonian mechanics* in that only certain electron orbits or energies are allowed. Thus, atoms emit or absorb light in discrete amounts called quanta. See *absorption line*, *continuous spectrum* and *emission line*.

*quasar* Also called quasi-stellar objects or QSOs, quasars are starlike objects which produce emission lines. Their redshifts can be large and their brightnesses vary. They are believed to be objects a little larger than the Solar System that have an intrinsic brightness some 100 times that of bright galaxies.

*radiation pressure* The tiny force exerted by photons when they bounce off small dust particles or when they are absorbed by atoms. The force is so small that it has no effect on the large objects encountered in everyday life, but it can be important in astronomy.

*resolution* See *spatial resolution*, *spectroscopy* and *temporal resolution*.

*RR Lyrae stars* Pulsating variable stars of luminosity roughly 100 times the *intrinsic brightness* of the Sun. They are useful distance indicators and are often found in *globular clusters*.

*shock* A sharp change in the properties (density, pressure, temperature) of a gas.

*signal-to-noise* Measurements of light in astronomy are always composed of two parts. The first is the signal, the part produced by the pure light from the target. The second is the noise, the part that comes from other sources, such as the telescope, the detectors, etc. The ratio of signal-to-noise indicates the quality of the measurement.

*solar nebula* The cloud of gas and dust from which the Solar System, including the Sun, formed.

*solar wind* The low-density gas consisting primarily of *protons* and *electrons* flowing away from the Sun at about 400 kilometers per second.

*South Atlantic Anomaly* A region of intense charged particle fluxes (caused by an irregularity in the Earth's magnetic field) over the south Atlantic Ocean. Spacecraft data collected in this region are often unreliable.

*spatial resolution* The measure of the ability of a telescope and camera to separate objects clearly. Roughly, the smallest detail that can be seen in an image.

*spectral lines* Because electrons in the cloud around the nucleus of an atom can have only a restricted number of specific energies, changes in energy produce photons of energy unique to that particular atom. This is the origin of spectral lines. These specific lines show that an element is present, and the strength of the line can be used to determine how abundant the element is. Shifts of the line in wavelength via the *Doppler effect* give the motion (speed) toward or away from the observer. The motion of atoms (with a speed depending on temperature) will cause a broadening of the line because some of the motions are away from and some are toward the observer. Thus, line widths can yield temperatures. Densities can be determined from the strengths of several different lines or from line widths. Finally, magnetic and electrical fields, if strong enough, can affect the the appearance of the lines. See *absorption lines*, *emission line* and *spectroscopy*.

*spectral resolution* See *spectroscopy*.

*spectroscopy* The analysis of light to determine the properties of the medium producing it and/or influencing it. In principle, spectroscopy can yield compositions, abundances, speed (via the *Doppler effect*), temperatures, densities, and, if they are strong enough, magnetic and electric fields. The rise of modern astrophysics clearly parallels the development of spectroscopy. To carry out spectroscopy, we need to disperse, or split up, the light into its component wavelengths or colors. This can be done by means of a prism, but in modern spectrographs, the dispersive optical element is a diffraction grating, which consists of a system of precisely aligned, narrowly spaced grooves. When carrying out spectroscopy, sufficient spectral resolution is needed to separate features in the spectrum. Numerically, the spectral resolution is given by the wavelength ($\lambda$) divided by the difference in wavelength to the closest wavelength that can be separated ($\Delta\lambda$). Thus, $\lambda/\Delta\lambda$ gives us a measure of the spectral resolution of instruments, including the HST spectrographs. The concept of spectral resolution is the spectral analogue of *spatial resolution* that determines the smallest detail that can be seen in ordinary images.

*spiral arm* The region in a *spiral galaxy* (such as the *Milky Way Galaxy*) that contains concentrations of gas, dust and young stars.

*stellar wind* The general, steady flow of gas away from stars resulting in loss of mass. Winds range from the gentle *solar wind* to vigorous flows some 100 million times stronger in mass loss, such as those from Melnick 42.

*supergiant* A star with maximum intrinsic brightness and low density. The radius of a supergiant can be as large as 1000 times that of the Sun.

*supernova* An immense stellar explosion which can increase a star's intrinsic brightness by as much as a billion times. The explosion blows off a major fraction of the star to form an expanding gas cloud (such as the Crab Nebula). The remaining material forms a dense object such as a *neutron star* or a *black hole*.

*synchrotron radiation* The radiation emitted when accelerated charged particles are spiraling in a magnetic field.

*temporal resolution* The measure of the ability of an optical system to clearly separate events in time. Roughly, the shortest time interval that can be determined between two different events.

*triple-alpha process* The nuclear reactions that transform three helium nuclei (often called alpha particles) into a carbon nucleus, with a release of energy. The process occurs at high temperatures and is important in red *giant* stars.

*white dwarf* An approximately Earth-sized star that does not have a source of nuclear energy in its interior. The star is supported by means of a form of pressure that arises when the densities at the star's interior are so high that the usual orbits of electrons in atoms around the nucleus cannot exist and the electrons are pushed much closer to the nucleus. A white dwarf can be supported in this way as long as its mass does not exceed about 1.4 solar masses. For stars with greater masses, *neutron stars* or *black hole*s are formed.

# References and further reading

## Books and proceedings


*Astronomy!*, by James B. Kaler (Harper Collins, 1994).

*Burnham's Celestial Handbook*, by Robert Burnham, Jr (Dover, 1978).

*Choosing and Using a CCD Camera*, by Richard Berry (Willmann-Bell, Inc., 1992).

*Cosmic Landscape: Voyages Back Along the Photon's Track*, by Michael Rowan-Robinson (Oxford University Press, 1979).

*Cosmos,* by Carl Sagan (Random House, 1980).

*Electronic and Computer-Aided Astronomy*, by Ian S. McLean (Wiley and Sons, 1989).

*First Light*, by Richard Preston (Urania, Inc., 1987).

*Light From the Depths of Time*, by Rudolf Kippenhahn (Springer-Verlag, 1987).

*Man Discovers the Galaxies*, by Richard Berendzen, Richard Hart and Daniel Seeley (Watson Academic Publishers, 1976).

*Man Into Space*, by Hermann Oberth (Harper and Brothers, 1957).

*Mars*, edited by H. H. Kieffer, B. M. Jakosky, C. W. Snyder and M. S. Mathews (University of Arizona Press, 1992).

*New Horizons in Astronomy* (2nd edition), by J. C. Brandt and S. P. Maran (W. H. Freeman, 1979).

*Science With the Hubble Space Telescope*, edited by P. Benvenuti and E. Schreier. ESO Conference and Workshop Proceedings No. 44, 1992.

*Stars and Planets*, by Jay M. Pasachoff and Donald H. Menzel (Houghton Mifflin, 1992).

*The Astronomy and Astrophysics Encyclopedia,* edited by Stephen P. Maran (van Nostrand Reinhold, 1992).

*The Decade of Discovery in Astronomy and Astrophysics* (National Academy Press, 1991).

*The Discovery of Our Galaxy*, by Charles A. Whitney (Alfred Knopf, 1971).

*The First Year of HST Observations*, edited by A.L. Kinney and J.C. Blades (NASA Publications, 1991).

*The Hubble Wars*, by Eric Chaisson (Harper Collins, 1994).

*The Milky Way* (5th edition), by Bart J. Bok and Priscilla F. Bok (Harvard University Press, 1981).

*The New Solar System,* (3rd edition), edited by J. Kelly Beatty and Andrew Chaikin (Sky Publishing and Cambridge University Press, 1990).

*The Space Telescope,* by Robert W. Smith (Cambridge University Press, 1993).

## Selected papers

Bahcall, John N. and Spitzer, Lyman. The Space Telescope, *Scientific American*, **247**, no. 1, July 1982.

Brandt, J.C. *et al.* The Goddard High Resolution Spectrograph: Instrument, Goals and Science Results, *Publications of the Astronomical Society of the Pacific*, **106**, 702, 1994.

Elliott, J.L. *et al.* An Occultation by Saturn's Rings on 1991 October 2-3 Observed with the Hubble Space Telescope, *Astronomical Journal*, **106**, no. 6, Dec. 1993.

Hensen, B.W. ESA's First In-Orbit-Replaceable Solar Array, *ESA Bulletin No. 61*, February 1990.

Linsky, J.L. Ultraviolet Observations of Stellar Coronae: Early Results from HST, *Memoire della Societa Astronomica Italia*, **63**, no. 3, 1992.

Powell, C.S. Super Loops, Scientific American, **271**, no. 3, September 1994.

Spitzer, Lyman. Astronomical Advantages of an Extra-Terrestrial Observatory, *The Astronomical Quarterly*, 7, 131–42, 1990. (Reprint of Appendix V of Spitzer's 1946 RAND Report.)

## Other articles and publications of interest

*Exploring the Universe with the Hubble Space Telescope*, NASA NP-126.

Fixing Hubble: NASA Sends in the Repair Crew, *Astronomy*, **22**, no. 1, Jan. 1994.

The Great Crash of 1994: A First Report, *Sky & Telescope*, **88**, no. 4, October 1994.

How We'll Fix the Hubble Space Telecope, *Sky & Telescope*, **86**, no. 5, Nov. 1993.

Hubble Space Telescope Update: 18 Months in Orbit, NASA, 1991.

Hubble's Road to Recovery, *Sky & Telescope*, **86**, no. 5. Nov. 1993.

Putting Hubble Right, *Astronomy*, **22**, no. 3. March 1994.

Inside the Crab Nebula, *Astronomy*, **22**, no. 12. Dec. 1994.

Scientific Uses of the Large Space Telescope, National Academy of Sciences, Washington, D.C., 1969 (Committee Report, Lyman Spitzer, Chairman.)

Time, Money and Millionths of an Inch, Hartford Courant, March 31–April 3, 1991.

Interested readers should also consult *The Space Telecope Science Institute Newsletter*, published quarterly by the Space Telescope Science Institute.

Selected issues of the *Astrophysical Journal Letters* devoted to HST results are

vol. **369**, no. 1, March 10, 1991;

vol. **377**, no. 1, August 10, 1991;

vol. **345**, no. 1, November 1, 1994.

The quotes at the beginnings of the chapters come from the following sources: Chapter 1, Hart Crane's poem *The Bridge* and Oscar Wilde's play *Lady Windermere's Fan*; Chapter 2 Michael Rowan-Robinson's book, *Cosmic Landscape: Voyages back along the photo's track*; Chapter 4, Galileo Galilei's work *Siderius Nuncius*, Heinz Pagels, quoted in *Peter's Quotations*, and Isaac Asimov, from *Isaac Asimov's Book of Science and Nature Quotations*; Chapter 5, Isaac Asimov, from *Isaac Asimov's Book of Science and Nature Quotations*, and Carolyn Collins Petersen, from Petersen's planetarium 1985 show *Gateway to Infinity*, written for the St Louis Science Center, St Louis, Missouri; Chapter 6, Nicolaus Copernicus, from *De Revolutionibus*, Tycho Brahe, from *Isaac Asimov's Book of Science and Nature Quotations*, and Mark Twain, from *The Adventures of Huckleberry Finn*.

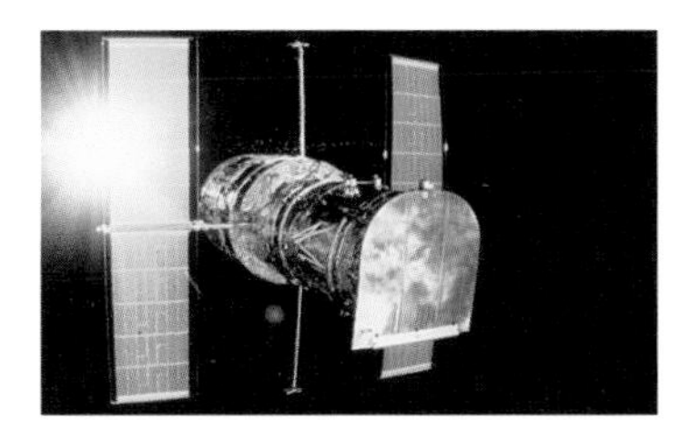

# Index

Page numbers of Glossary entries are given in italics.